Leo Laporte's 2006 Technology Almanac

Leo Laporte and Michael Miller

800 East 96th Street, Indianapolis, Indiana 46240 USA

LEO LAPORTE'S 2006 TECHNOLOGY ALMANAC

Copyright © 2006 by Que Publishing

All rights reserved. No part of this book shall be reproduced, stored in a retrieval system, or transmitted by any means, electronic, mechanical, photocopying, recording, or otherwise, without written permission from the publisher. No patent liability is assumed with respect to the use of the information contained herein. Although every precaution has been taken in the preparation of this book, the publisher and author assume no responsibility for errors or omissions. Nor is any liability assumed for damages resulting from the use of the information contained herein.

International Standard Book Number: 0-7897-3397-8

Library of Congress Catalog Card Number: 2005925003

Printed in the United States of America

First Printing: October 2005

08 07 06 05   4 3 2 1

Trademarks

All terms mentioned in this book that are known to be trademarks or service marks have been appropriately capitalized. Que Publishing cannot attest to the accuracy of this information. Use of a term in this book should not be regarded as affecting the validity of any trademark or service mark.

Warning and Disclaimer

Every effort has been made to make this book as complete and as accurate as possible, but no warranty or fitness is implied. The information provided is on an "as is" basis. The authors and the publisher shall have neither liability nor responsibility to any person or entity with respect to any loss or damages arising from the information contained in this book.

Bulk Sales

Que Publishing offers excellent discounts on this book when ordered in quantity for bulk purchases or special sales. For more information, please contact

> **U.S. Corporate and Government Sales**
> 1-800-382-3419
> corpsales@pearsontechgroup.com

For sales outside of the U.S., please contact

> **International Sales**
> international@pearsoned.com

ASSOCIATE PUBLISHER
Greg Wiegand

EXECUTIVE EDITOR
Rick Kughen

DEVELOPMENT EDITOR
Rick Kughen

MANAGING EDITOR
Charlotte Clapp

PROJECT EDITOR
Tonya Simpson

INDEXER
Chris Barrick

PROOFREADER
Suzanne Thomas

PUBLISHING COORDINATOR
Sharry Lee Gregory

INTERIOR DESIGNER
Anne Jones

COVER DESIGNER
Anne Jones

PAGE LAYOUT
Kelly Maish

CONTENTS AT A GLANCE

Introduction xxxi

January 2006 1

February 2006 31

March 2006 57

April 2006 87

May 2006 115

June 2006 145

July 2006 173

August 2006 201

September 2006 231

October 2006 259

November 2006 289

December 2006 317

Index 345

TABLE OF CONTENTS

Introduction xxxi

January 2006 1

Testing Your Connection Speed 3
On This Day: Millennium Bug Doesn't Bite (2000) 3
Blog of the Week: Slashdot 3
Broadband Internet 4
On This Day: Micro-Soft Named (1975) 4
Fact of the Week 4
Internet Explorer Keyboard Shortcuts 5
On This Day: TIME Names Computer "Man of the Year" (1983) 5
Gadget of the Week: Citizen TASER X26c Pistol 5
Mozilla Firefox: A Better Browser 6
On This Day: HP-35 Calculator Introduced (1972) 6
Hardware of the Week: Linksys WRT54GS Wireless Router 6
Firefox Tips and Tricks 7
On This Day: Amy Johnson Dies (1941) 7
Software of the Week: WS_FTP 7
The Wayback Machine: A Blast from the Past 8
On This Day: Peter Denning Born (1942) 8
Website of the Week: THOMAS 8
Anybody Remember Archie and Veronica? 9
On This Day: Hollerith Tabulating Machine Patented (1889) 9
Blog of the Week: Boing Boing 9
How Digital Sampling Works 10
On This Day: Galaxies Are Accelerating (1998) 10
Fact of the Week 10
Lossy Digital Audio Formats 11
On This Day: 45 RPM Record Introduced (1949) 11
Gadget of the Week: StikAx Music Mixer 11
Lossless Digital Audio Formats 12
On This Day: Smoking Causes Cancer (1964) 12
Hardware of the Week: Creative I-Trigue Speaker System 12
Online Music Stores 13
On This Day: First X-Ray Photo (1896) 13
Software of the Week: Album Art Fixer 13
File Trading Networks 14

On This Day: Adding Machine Patented (1874) 14
Website of the Week: Allmusic 14
Internet Radio 15
On This Day: NCSA Opens (1986) 15
Blog of the Week: Fluxblog 15
Yahoo! Search Shortcuts 16
On This Day: Dian Fossey Born (1932) 16
Fact of the Week 16
Navigating! Yahoo! 17
On This Day: Computer Discovers New Planets (1996) 17
Gadget of the Week: Survival Kit in a Sardine Can 17
Yahoo! Maps 18
On This Day: Harvard Mark I Outlined (1938) 18
Hardware of the Week: Logitech MediaPlay Cordless Mouse 18
Yahoo! Search Meta Words 19
On This Day: Apple Lisa Introduced 19
Software of the Week: Yahoo! Toolbar 19
Yahoo! My Web 20
On This Day: Buzz Aldrin Born (1930) 20
Website of the Week: My Yahoo! 20
Yahoo! Briefcase 21
On This Day: First Commercial 747 Flight (1970) 21
Blog of the Week: Yahoo! Search Blog 21
Why You Need Service Pack 2 22
On This Day: Integrated Circuit Conceived (1959) 22
Fact of the Week 22
Using the Control Panel 23
On This Day: IBM Dedicates the SSEC (1948) 23
Gadget of the Week: Global Pet Finder 23
Enabling ClearType 24
On This Day: Flouridated Drinking Water (1945) 24
Hardware of the Week: APC Uninterruptible Power Source 24
Using a Web Page as a Desktop Background 25
On This Day: Internal Combustion Auto Patented 25
Software of the Week: Theme Manager 2005 25
Adding Special Effects 26
On This Day: Jim Clark Leaves Silicon Graphics (1994) 26
Website of the Week: Microsoft 26
Moving and Hiding the Taskbar 27
On This Day: First Gaslit Street (1807) 27

Blog of the Week: LonghornBlogs.com 27
Show Album Art in My Music Folders 28
On This Day: Douglas Englebart Born (1925) 28
Fact of the Week 28
Remove the Shortcut Arrow from Desktop Icons 29
On This Day: First Primate in Space (1961) 29
Gadget of the Week: SCOTTeVEST Clothing 29

February 2006 31

Closing Stuck Programs in XP 33
On This Day: Java Development Starts (1991) 33
Hardware of the Week: USB Mini-Aquarium 33
Add a Background Graphic to the IE Toolbar 34
Use a Web Page as a Desktop Background 34
On This Day: First Use of Lie Detector (1935) 34
Software of the Week: Tweak UI 34
Activating XP Special Effects 35
On This Day: First U.S. Weather Satellite (1966) 35
Website of the Week: TweakXP.com 35
Change Owner Information 36
Adding Programs to the XP Start Menu 36
On This Day: First Removable Shirt Collar (1825) 36
Blog of the Week: TechWhack 36
What's a Blog? 37
On This Day: Golf on the Moon (1971) 37
Fact of the Week 37
Why Blog? 38
How Blogs Are Organized 38
On This Day: First Untethered Spacewalks (1984) 38
Gadget of the Week: Precision Shots Laser Slingshot 38
Creating Your Own Blog 39
Best Sites for Blogging 39
On This Day: Harvard Mark I Patent (1945) 39
Hardware of the Week: AeroCool AeroBase UFO Gaming Pad 39
Exploring the Blogosphere 40
On This Day: David Wheeler Born (1927) 40
Software of the Week: Newsgator 40
Political Blogs 41
On This Day: Wilhelm Roentgen Dies (1923) 41
Website of the Week: The Weblog Review 41

Moblogs and Photo Blogs 42
Leo's Favorite Blogs 42
On This Day: Boston Computer Society Founded (1977) 42
Blog of the Week: Photoblogs.org 42
All About Podcasts 43
On This Day: France Goes Atomic (1960) 43
Fact of the Week 43
Preparing to Podcast 44
On This Day: ENIAC Revealed (1946) 44
Gadget of the Week: Edirol R-1 Portable Music Recorder 44
Recording a Podcast 45
On This Day: Mustard Arrives (1758) 45
Hardware of the Week: Griffin radio SHARK AM/FM Radio/Recorder 45
Publishing Your Podcast 46
Learn More About Podcasting 46
On This Day: First Man-Made Diamonds (1953) 46
Software of the Week: Audacity 46
Where to Find Podcasts 47
On This Day: First Car Starter (1911) 47
Website of the Week: iPodder 47
Podcasts and iTunes 48
On This Day: Vacuum Cleaner Patented (1901) 48
Blog of the Week: Adam Curry's Weblog 48
POP vs. Web-Based Email 49
On This Day: Elevator Patented (1872) 49
Fact of the Week 49
Email Acronyms and Protocols 50
On This Day: First Flying Car (1937) 50
Gadget of the Week: VoiSec Refrigerator Voice Recorder 50
HTML Email 51
On This Day: Amerigo Vespucci Dies (1512) 51
Hardware of the Week: Seagate External Hard Drives 51
Managing Mass Mailings in Outlook Express 52
On This Day: First Cloned Sheep (1997) 52
Software of the Week: Mozilla Thunderbird 52
Troubleshooting Email Problems 53
On This Day: Steve Jobs Born 53
Website of the Week: Gmail 53
Emailing Anonymously 54
On This Day: SGI Buys Cray (1996) 54

Blog of the Week: Gawker 54
Before You Upgrade 55
On This Day: David Sarnoff Born (1891) 55
Fact of the Week 55
Upgrading System Memory 56
On This Day: Core Memory Patented (1956) 56
Gadget of the Week: Ultra Antistatic Wrist Straps 56

March 2006 57

Adding New Ports 59
On This Day: Direct-Dial Transatlantic Phone Service (1970) 59
Hardware of the Week: Sunbeam 20-in-1 Superior Panel 59
Upgrading for Digital Video Editing 60
On This Day: First Push-Button Phone (1959) 60
Software of the Week: Aloha Bob PC Relocator 60
Upgrading Your Hard Disk 61
On This Day: First Meeting of the Homebrew Computer Club 61
Website of the Week: Tom's Hardware Guide 61
USB: The Easy Way to Upgrade 62
On This Day: Stapler Patented (1868) 62
Blog of the Week: Tech Blog 62
How Big Is Big Enough? 63
On This Day: Michelangelo Virus Strikes (1992) 63
Fact of the Week 63
CRT Projection 64
On This Day: Cornflakes First Served (1897) 64
Gadget of the Week: Couch Potato Tormentor 64
DLP Rear Projection 65
On This Day: von Zeppelin Dies (1917) 65
Hardware of the Week: ATI HDTV Wonder 65
LCD Rear Projection 66
On This Day: PowerOpen Association Formed 66
Software of the Week: Cinemar MainLobby 66
Plasma Flat-Screen Displays 67
On This Day: Uranus Has Rings (1977) 67
Website of the Week: AVS Forum 67
LCD Flat-Screen Displays 68
On This Day: Luddite Riots (1811) 68
Blog of the Week: Home Theater Blog 68
Identity Theft 69

On This Day: Henry Shrapnel Dies (1842) 69
Fact of the Week 69
Preventing Identity Theft 70
On This Day: Giant Brain Completed (1955) 70
Gadget of the Week: TrimTrac GPS Security Locator 70
Dealing with Identity Theft 71
On This Day: Adobe and Aldus Merge (1994) 71
Hardware of the Week: Keyspan USB Mini Hub 71
Privacy in the Workplace 72
On This Day: First Liquid-Fuel Rocket Flight (1926) 72
Software of the Week: NSClean 72
Old Postings Can Come Back and Bite You 73
On This Day: Rubber Band Patented (1845) 73
Website of the Week: Electronic Frontier Foundation 73
Phishing Schemes 74
On This Day: First U.S. Railroad Tunnel (1834) 74
Blog of the Week: Privacy.org 74
Five Ways to Speed Up Your PC 75
On This Day: Theory of Relativity Published (1916) 75
Fact of the Week 75
Maintaining Your System Unit 76
On This Day: Teaching of Evolution Prohibited (1925) 76
Gadget of the Week: TomTom Navigator 2004 76
Maintaining Your Keyboard 77
On This Day: Pentium Chip Ships (1993) 77
Hardware of the Week: IOGEAR MiniView Micro KVM Switch 77
Maintaining Your Mouse 78
On This Day: Cold Fusion Claimed (1989) 78
Software of the Week: PC Certify 78
Maintaining Your Printer 79
On This Day: Tuberculosis Bacillus Announced (1882) 79
Website of the Week: PC Pitstop 79
More Basic Maintenance 80
On This Day: Titan Discovered (1655) 80
Blog of the Week: Doc Searl's IT Garage 80
Why You Need a PC in Your Living Room 81
On This Day: Alaska Earthquake (1965) 81
Fact of the Week 81
Buying a Media Center PC 82
On This Day: Three Mile Island Accident (1979) 82

Gadget of the Week: Philips RC9800i WiFi Remote Control 82
Building Your Own Media Center PC 83
On This Day: Coca Cola Created (1886) 83
Hardware of the Week: Denali Edition Media Center 83
Using Media Center as a DVR 84
On This Day: Meter Defined (1791) 84
Software of the Week: Tweak MCE 84
Media Center My Music 85
On This Day: Daylight Savings Time Begins (1918) 85
Website of the Week: The Green Button 85

April 2006 87

Media Center Add-Ins 89
On This Day: First Photograph of the Sun 89
Blog of the Week: Matt Goyer's Windows Media Center Blog 89
Skeletal Systems 90
On This Day: First Mobile Phone Call (1973) 90
Fact of the Week 90
The Death Clock 91
On This Day: Netscape Founded (1994) 91
Gadget of the Week: Motorola Ojo Personal Videophone 91
Dumb Auctions 92
On This Day: Oppenheimer Wins Fermi Award (1963) 92
Hardware of the Week: ActionTec Internet Phone Wizard 92
UglyDress.com 93
On This Day: Windows 3.1 Released (1992) 93
Software of the Week: Marine Aquarium Screen Saver 93
Museum of Bad Album Covers 94
On This Day: IBM Announces System/360 Mainframes (1964) 94
Website of the Week: Today's Front Pages 94
Daily Rotten 95
On This Day: Presper Eckert Born (1919) 95
Blog of the Week: Dave Barry's Blog 95
DVD Easter Eggs 96
On This Day: First Synthetic Rubber (1930) 96
Fact of the Week 96
CD Easter Eggs 97
On This Day: Apollo 13 Launched (1970) 97
Gadget of the Week: Eva Solo Magnetimer Kitchen Timer 97
Game Easter Eggs 98

On This Day: First Internet Spam (1994) 98
Hardware of the Week: USB Cafè Pad 98
Software Easter Eggs 99
On This Day: First Elephant in America (1796) 99
Software of the Week: Coffee Break Worm 99
Literary Easter Eggs 100
On This Day: China's Great Software Purge Begins (1995) 100
Website of the Week: The Easter Egg Archive 100
Sexually Explicit Easter Egg 101
On This Day: LISP Language Unveiled 101
Blog of the Week: Engadget 101
eBay News and Announcements 102
On This Day: Benjamin Franklin Dies (1790) 102
Fact of the Week 102
Using Feedback 103
On This Day: Albert Einstein Dies (1955) 103
Gadget of the Week: SportVue MC1 Heads-Up Display 103
Birddogging for Bargains 104
On This Day: First FORTRAN Program (1957) 104
Hardware of the Week: Logitech diNovo Cordless Desktop 104
eBay PowerSellers 105
On This Day: Whirlwind Computer on TV (1951) 105
Software of the Week: eBay Toolbar 105
Finding Stuff to Sell 106
On This Day: First Revolving Restaurant Opens (1962) 106
Website of the Week: Ándale 106
eBay Trading Assistants 107
On This Day: Wilhelm Schickard Born (1592) 107
Blog of the Week: Outrageous eBay Auctions 107
iPods Rule! 108
On This Day: Apple IIc Introduced 108
Fact of the Week 108
Using Your iPod with a Windows PC 109
On This Day: Integrated Circuit Patented 109
Gadget of the Week: Griffin BlueTrip Audio Hub 109
Creating Playlists for Your iPod 110
On This Day: Chernobyl Nuclear Disaster (1986) 110
Hardware of the Week: Apple AirPort Express 110
iPod Speakers 111
On This Day: Microsoft Purchase of Intuit Blocked (1995) 111

Software of the Week: iArt 111
Cool iPod Accessories 112
On This Day: Kurt Gödel Born (1906) 112
Website of the Week: iLounge 112
More Cool iPod Accessories 113
On This Day: George Stibitz Born (1904) 113
Blog of the Week: iPoditude 113

May 2006 115

How Instant Messaging Works 117
On This Day: First Computer Timesharing (1964) 117
Fact of the Week 117
Instant Messaging Clients 118
On This Day: Microsoft's Two-Button Mouse (1983) 118
Gadget of the Week: Tivoli Model One 118
Group Chats in AOL Instant Messenger 119
On This Day: First Comic Book (1934) 119
Hardware of the Week: Logitech QuickCam Orbit 119
Sending Files via Instant Messaging 120
On This Day: Commodore Sold (1995) 120
Software of the Week: mIRC 120
Tips for Chat and IM 121
On This Day: Scopes Arrested for Teaching Evolution (1925) 121
Website of the Week: Cybermoon Studios 121
Online Chat Communities 122
On This Day: Hindenburg Disaster (1937) 122
Blog of the Week: Jake Ludington's Media Blab 122
Digital Television 123
On This Day: Metric System Established (1790) 123
Fact of the Week 123
Comparing Digital Television Formats 124
On This Day: Birth Control Pill Approved (1960) 124
Gadget of the Week: Monster PowerCenter 124
Why HDTV Is Better 125
On This Day: Transcontinental Railroad Completed (1869) 125
Hardware of the Week: TViX Digital Movie Jukebox 125
How HDTV Works 126
Learn More About HDTV 126
On This Day: VisiCalc Demonstrated (1979) 126
Software of the Week: Movie Label 2006 126

Progressive Scanning 127
On This Day: Dvorak Keyboard Patented (1936) 127
Website of the Week: HDTVoice.com 127
Square Programs on a Rectangular Screen 128
On This Day: Table Knife Invented (1637) 128
Blog of the Week: High-Definition Blog 128
I Hate Spam! 129
Learn More About Spam 129
On This Day: Listerine Trademarked (1923) 129
Fact of the Week 129
How Spammers Get Your Address 130
On This Day: Root Beer Invented (1866) 130
Gadget of the Week: Oregon Scientific AWS888 Weather Forecaster 130
Protecting Your Email Address 131
On This Day: John Deere Dies (1886) 131
Hardware of the Week: Griffin PowerMate USB Controller 131
Blocking Spam in Microsoft Outlook 132
On This Day: Mount St. Helens Erupts (1980) 132
Software of the Week: Email Sentinel Pro 132
Anti-Spam Software 133
On This Day: Lawrence of Arabia Dies (1935) 133
Website of the Week: Spamhaus 133
Trace and Report 134
On This Day: Levis Patented (1873) 134
Blog of the Week: Spam Kings Blog 134
Advanced Google Searching 135
On This Day: Toothpaste Tube Invented (1892) 135
Fact of the Week 135
Google Images 136
On This Day: Bifocals Invented (1785) 136
Gadget of the Week: Wallflower 2 Multimedia Picture Frame 136
Google Maps 137
On This Day: First Telegraph Message (1844) 137
Hardware of the Week: Western Digital Media Center 137
Personalize Your Google Home Page 138
On This Day: First International WWW Conference (1994) 138
Software of the Week: Google Desktop Search 138
Gmail 139
On This Day: First Middle East Oil Strike (1908) 139
Website of the Week: Google Language Tools 139

Other Cool Google Sites 140
On This Day: Jell-O Introduced (1897) 140
Blog of the Week: The Unofficial Google Weblog 140
Choosing a Digital Camera 141
On This Day: Hilary Conquers Mount Everest (1953) 141
Fact of the Week 141
Prosumer Cameras 142
On This Day: Krypton Discovered (1898) 142
Gadget of the Week: SanDisk Photo Album 142
Digital SLRs 143
On This Day: John Kemeny Born (1926) 143
Hardware of the Week: Epson Perfection 4180 Photo Scanner 143

June 2006 145

Better Lighting = Better Pictures 147
On This Day: FM Stereo Broadcasting Begins (1961) 147
Software of the Week: Paint Shop Pro 147
How to Take Great Digital Pictures 148
On This Day: Velveeta Cheese Invented (1928) 148
Website of the Week: Digital Photography Review 148
Digital Photo Vaults 149
On This Day: Robert Noyce Dies (1990) 149
Blog of the Week: PhotographyBLOG 149
Digital Video Formats 150
On This Day: Apple II Introduced (1977) 150
Fact of the Week 150
Choosing the Right Video Camera 151
On This Day: First Household Detergent Introduced (1907) 151
Gadget of the Week: Sunpak Readylight 20 151
High-Definition Camcorders 152
On This Day: First Solar Power Plant (1980) 152
Hardware of the Week: HP dc5000 Movie Writer 152
Prosumer Camcorders 153
On This Day: Alan Turing Found Dead (1954) 153
Software of the Week: Adobe Premiere Elements 153
Recording Better Sound with Your Movies 154
On This Day: First Automat Opens (1902) 154
Website of the Week: CamcorderInfo.com 154
Stop Shaky Pictures! 155
On This Day: Speak & Spell Introduced (1978) 155

Blog of the Week: PVRblog 155
Fun with GPS 156
On This Day: 3Com and U.S. Robotics Merge (1997) 156
Fact of the Week 156
High-Tech Running 157
On This Day: First V1 Bomb Attack (1944) 157
Gadget of the Week: SoundSak SonicBoom 157
High-Tech Fishing 158
On This Day: Univac 1 Dedicated (1951) 158
Hardware of the Week: Act Labs Light Gun 158
High-Tech Boating 159
On This Day: Ben Franklin Flies a Kite (1752) 159
Software of the Week: Fish-N-Log Professional Suite 159
High-Tech Scuba Diving 160
On This Day: CTR Incorporated (1911) 160
Website of the Week: GORP 160
High-Tech Camping 161
On This Day: Amelia Earhart Crosses the Atlantic (1928) 161
Blog of the Week: Die Is Cast 161
Apple History (Part I) 162
On This Day: Blaise Pascal Born (1623) 162
Fact of the Week 162
Apple History (Part I-A) 163
On This Day: Trans-Alaska Pipeline Opens (1977) 163
Gadget of the Week: GarageBand Guitar Cable 163
Apple History (Part II) 164
On This Day: First Ferris Wheel Opens (1893) 164
Hardware of the Week: Apple 30-inch HD Cinema Display 164
Apple History (Part III) 165
On This Day: Konrad Zuse Born 165
Software of the Week: SuperDuper 165
Apple Portables 166
On This Day: Saxophone Patented (1848) 166
Website of the Week: Think Secret 166
Troubleshooting Mac OS X 167
On This Day: Barbed Wire Patented (1867) 167
Blog of the Week: The Unofficial Apple Weblog 167
I Spy...Spyware! 168
On This Day: Bar Code Debuts (1974) 168
Fact of the Week 168

What Damage Can Spyware Do? ... 169
On This Day: Chlorophyll Synthesized (1960) ... 169
Gadget of the Week: JB1 James Bond 007 Spy Camera ... 169
How to Defeat Spyware ... 170
On This Day: Mackinac Bridge Dedicated (1958) ... 170
Hardware of the Week: Panasonic BL-C30A Network Camera ... 170
Anti-Spyware Utilities ... 171
On This Day: Pygmy Mammoth Discovered (1994) ... 171
Software of the Week: Active@ Eraser ... 171
Legitimate Spyware Programs ... 172
On This Day: Leap Second Added (1972) ... 172
Website of the Week ... 172

July 2006 ... 173

Official U.S. Government Spyware ... 175
On This Day: IBM 650 Announced (1953) ... 175
Blog of the Week: Spyware Warrior ... 175
Creating Numbered Lists ... 176
On This Day: Foam Rubber Developed (1929) ... 176
Fact of the Week ... 176
Using Styles ... 177
On This Day: First Direct Keyboard Input (1956) ... 177
Gadget of the Week: VIOlight Toothbrush Sanitizer ... 177
Adding Background Colors and Graphics ... 178
On This Day: Intel Price Cuts (1994) ... 178
Hardware of the Week: Logitech Laser Cordless Mouse ... 178
Using Section Breaks ... 179
On This Day: AOL Settles Lawsuits ... 179
Software of the Week: Wordware ... 179
Using Headers and Footers ... 180
On This Day: Phillips-Head Screw Patented (1936) ... 180
Website of the Week: Woody's Office Portal ... 180
Creating a Multiple-Column Layout ... 181
On This Day: *Tron* Released (1982) ... 181
Blog of the Week: The Office Weblog ... 181
Why It's So Hard to Find What You Want Online ... 182
On This Day: Telstar Launched (1962) ... 182
Fact of the Week ... 182
Search Engines and Directories ... 183
On This Day: Surgical Zippers Announced (1985) ... 183

Gadget of the Week: Oakley THUMP 183
Top Search Sites 184
On This Day: Smoking (Officially) Causes Cancer (1957) 184
Hardware of the Week: Deck Keyboard 184
Metasearch Engines 185
On This Day: Erno Rubick Born (1944) 185
Software of the Week: WebFerret 185
The Correct Way to Search 186
On This Day: Dynamite Demonstrated (1867) 186
Website of the Week: Search Engine Watch 186
Boolean Searching 187
On This Day: ENIGMA Encodes First Message (1928) 187
Blog of the Week: Search Engine Blog 187
Using a Home Page Community 188
On This Day: Major Email Disruption (1997) 188
Fact of the Week 188
Finding a Web Host 189
On This Day: Intel Founded (1968) 189
Gadget of the Week: ProAim Golfing Goggles 189
Understanding HTML 190
On This Day: Samuel Colt Born (1814) 190
Hardware of the Week: Griffin AirClick USB Remote 190
Common HTML Tags 191
On This Day: Man Walks on Moon (1969) 191
Software of the Week: HotDog PageWiz 191
Inserting Hyperlinks 192
On This Day: Xerox Withdraws from Computer Market (1975) 192
Website of the Week: Web Pages That Suck 192
Changing Font Type and Size 193
On This Day: Typewriter Patented (1829) 193
Blog of the Week: Thomas Hawk's Digital Connection 193
Choosing a Video Game System 194
On This Day: Instant Coffee Introduced (1938) 194
Fact of the Week 194
Video Games: The Next Generation 195
On This Day: Microsoft Revenues Exceed $1 Billion (1990) 195
Gadget of the Week: Spherex RX2 Game Chair 195
Building the Perfect Gaming PC 196
On This Day: Carl Jung Born (1875) 196
Hardware of the Week: Atari Flashback 196

Choosing a Game Controller 197
On This Day: Grasshopper Plague (1931) 197
Software of the Week: RollerCoaster Tycoon 3 197
Popular Game Controllers (Part I) 198
On This Day: Tricycle Crosses English Channel (1883) 198
Website of the Week: IGN.com 198
Popular Game Controllers (Part II) 199
On This Day: First Asphalt Pavement (1870) 199
Blog of the Week: Game*Blogs 199
Hard Drive Players 200
On This Day: First Close-Up Moon Pictures (1964) 200
Fact of the Week 200

August 2006 201

MicroDrive Players 203
On This Day: Atomic Energy Commission Established (1946) 203
Gadget of the Week: AudioTronic iCool Scented MP3 Players 203
Flash Memory Players 204
On This Day: Greenwich Mean Time Adopted (1880) 204
Hardware of the Week: Razer Diamondback Mouse 204
Portable Video Players 205
On This Day: Radio Shack Announces TRS-80 (1977) 205
Software of the Week: Blaze Media Pro 205
Better Earphones = Better Sound 206
On This Day: Champagne Invented (1693) 206
Website of the Week: MP3.com 206
Celebrity Playlists 207
On This Day: First Electric Traffic Lights (1914) 207
Blog of the Week: MP3 Player Blog 207
Citywide WiFi 208
On This Day: Harvard Gets a Giant Brain (1944) 208
Fact of the Week 208
Finding a WiFi Hotspot 209
On This Day: Netscape Goes Public (1995) 209
Gadget of the Week: Canary Wireless Digital Hotspotter 209
Troubleshooting a Bad Connection 210
On This Day: Marvin Minksy Born (1927) 210
Hardware of the Week: Linksys Wireless Adapter 210
Setting Up a Wireless Home Network 211
On This Day: Leo Fender Born (1909) 211

Software of the Week: LucidLink Wireless Client 211
WiFi Security 212
On This Day: Steve Wozniak Born (1950) 212
Website of the Week: Wi-Fi Alliance 212
What's With All Those Letters? 213
On This Day: IBM PC Introduced (1981) 213
Blog of the Week: WiFi Networking News 213
Matrix Surround Sound 214
On This Day: Social Security Act Signed (1935) 214
Fact of the Week 214
Discrete Surround Sound 215
On This Day: Panama Canal Opened (1914) 215
Gadget of the Week: Sony Surround Sound Headphones 215
Comparing Surround Sound Formats 216
On This Day: Queen Telegraphs President (1858) 216
Hardware of the Week: USB Alarm Clock 216
Fantasia—The First Surround-Sound Movie 217
On This Day: Pierre de Fermat Born (1601) 217
Software of the Week: Cubase SX 217
The History of Surround Sound (Part I) 218
On This Day: Hewlett-Packard Incorporated (1947) 218
Website of the Week: The Surround Sound Discography 218
The History of Surround Sound (Part II) 219
On This Day: Condensed Milk Patented (1856) 219
Blog of the Week: wilwheaton.net 219
Troubleshooting Tips 220
On This Day: Adding Machine Patented (1888) 220
Fact of the Week 220
Dealing with a Dead Computer 221
On This Day: First Computer User Group Founded (1955) 221
Gadget of the Week: Venexx Perfume Watch 221
Troubleshooting in Safe Mode 222
On This Day: Galileo Demonstrates Telescope (1609) 222
Hardware of the Week: Marathon Computer RePorter 222
Page Fault Errors 223
On This Day: Windows 95 Ships (1995) 223
Software of the Week: PC Certify Lite 223
Creating an Emergency Startup Disk 224
On This Day: Paris Liberated (1944) 224
Website of the Week: Microsoft Knowledge Base 224

PC Survival Checklist 225
On This Day: Compaq Introduces Presario (1993) 225
Blog of the Week: RescueComp Blog 225
First Generation (1972–1977) 226
On This Day: Worcester Sauce Introduced (1837) 226
Fact of the Week 226
Second Generation (1977–1982) 227
On This Day: First Russian Atomic Bomb (1949) 227
Gadget of the Week: Jakks TV Games 227
Third Generation (1982–1984) 228
On This Day: First African-American Astronaut (1983) 228
Hardware of the Week: Classicade Upright Game System 228
Fourth Generation (1985–1989) 229
On This Day: First Car on the Moon (1971) 229
Software of the Week: Atari 80 Classic Games 229

September 2006 231

Fifth Generation (1989–1995) 233
On This Day: Virtual Library Project Starts (1994) 233
Website of the Week: Video Game Museum 233
Sixth Generation (1995–1998) to Today 234
On This Day: Ferdinand Porsche Born 234
Blog of the Week: Video Game Blog 234
Different Types of Notebooks 235
On This Day: Geronimo Surrenders (1886) 235
Fact of the Week 235
Shopping for a New Notebook 236
On This Day: *On the Road* Published (1957) 236
Gadget of the Week: Targus DEFCON MDP Motion Sensor 236
Notebooks for College 237
On This Day: Margaret Sanger Dies (1966) 237
Hardware of the Week: LapCool 2 Notebook Cooler 237
Upgrading Notebook Memory 238
On This Day: First Baby Incubator (1888) 238
Software of the Week: RMClock 238
Printing on the Road 239
On This Day: Ford Pardons Nixon (1974) 239
Website of the Week: NotebookReview.com 239
Stretching Notebook Battery Life 240
On This Day: Lincoln Highway Opens (1913) 240

Blog of the Week: Laptop Review 240
Encourage Safe Computing 241
On This Day: First Remote Computation (1940) 241
Fact of the Week 241
Safe Searching for Kids 242
On This Day: Test of First Integrated Circuit (1958) 242
Gadget of the Week: Hasbro VideoNow Color 242
Online Libraries 243
On This Day: First Auto Accident Death (1899) 243
Hardware of the Week: Logitech Football Mouse 243
Help with Homework 244
On This Day: First Lobotomy (1936) 244
Software of the Week: Microsoft Student 2006 244
Kids' Sports 245
On This Day: ACM Founded (1947) 245
Website of the Week: Kids' Space 245
Computers Make Kids Dumber? 246
On This Day: RCA Exits Computer Market (1971) 246
Blog of the Week: "Hey That Smells Great" 246
Audio/Video Receivers 247
On This Day: Jimi Hendrix Dies (1970) 247
Fact of the Week 247
Digital Video Recorders 248
On This Day: First Underground Nuclear Test (1957) 248
Gadget of the Week: Logitech Harmony 880 248
Shopping for a DVD Player 249
On This Day: First FORTRAN Program (1954) 249
Hardware of the Week: Sonos Digital Music System 249
Choosing a Speaker System 250
On This Day: Galileo Ends Mission (2003) 250
Software of the Week: Meedio Essentials 250
Setting Up Your System 251
On This Day: Peace Corps Authorized (1961) 251
Website of the Week: AudioREVIEW.com 251
Universal Remote Controls 252
On This Day: Dirigible Demonstrated (1852) 252
Blog of the Week: eHomeUpgrade 252
HDTV Programming 253
On This Day: IBM Announces MCA 253
Fact of the Week 253

xxi

Showcase DVDs (Part I) 254
On This Day: Jean Hoerni Born (1924) 254
Gadget of the Week: Laserpod Light 254
Showcase DVDs (Part II) 255
On This Day: Kevin Mitnick Indicted (1996) 255
Hardware of the Week: Logitech Z-5500 Speakers 255
Showcase DVDs (Part III) 256
On This Day: Seymour Cray Born (1925) 256
Software of the Week: ReaderWare 256
Showcase CDs (Part I) 257
On This Day: Enrico Fermi Born (1901) 257
Website of the Week: Metacritic 257
Showcase CDs (Part II) 258
On This Day: First Nuclear Sub Commissioned (1954) 258
Blog of the Week: Jazz & Blues Music Reviews 258

October 2006 259

CD Ripping and Burning Software 261
On This Day: Concorde Breaks the Sound Barrier (1969) 261
Blog of the Week: Online Music Blog 261
Configuring Windows Media Player for Ripping 262
On This Day: ENIAC Retired (1955) 262
Fact of the Week 262
Ripping CDs with WMP 263
On This Day: Transistor Patented (1950) 263
Gadget of the Week: Darth Tater 263
Burning Your Own Music CDs 264
On This Day: John Atanasoff Born (1903) 264
Hardware of the Week: Pocket CDRW 264
Burning CDs with Windows Media Player 265
On This Day: First Non-Stop Pacific Flight (1931) 265
Software of the Week: Easy Media Creator 265
Copying Entire CDs 266
On This Day: Thor Heyerdahl Born (1914) 266
Website of the Week: freedb.org 266
Writing to a Data CD 267
On This Day: Carbon Paper Patented (1806) 267
Blog of the Week: PostSecret 267
Cellular Network Technologies 268
On This Day: Calliope Patented (1855) 268

Fact of the Week 268
Dual Band and Dual Mode Phones 269
On This Day: Billiard Ball Patented (1865) 269
Gadget of the Week: FreePlay FreeCharge 269
Choosing a Mobile Phone 270
On This Day: H.J. Heinz Born (1844) 270
Hardware of the Week: Verbatim Store 'n' Go Pro 270
Smartphones 271
On This Day: NeXT Computer Introduced (1988) 271
Software of the Week: Hitchhiker's Guide to the Galaxy Game ... 271
Choosing a Headset 272
On This Day: First Aerial Photograph (1860) 272
Website of the Week: The Mobile Phone Directory 272
Favorite Cell Phones 273
On This Day: Edison Electric Light Company Founded (1878) 273
Blog of the Week: Mobile Burn 273
Are You Infected? 274
On This Day: First U.S. Birth Control Clinic (1916) 274
Fact of the Week 274
Practicing Safe Computing 275
On This Day: Mae Jemison Born (1956) 275
Gadget of the Week: ScanGauge Automotive Computer 275
Antivirus Utilities 276
Learn More About Protecting Your Computer 276
On This Day: Babbage, Binet, and Edison Die 276
Hardware of the Week: WD Passport Pocket Hard Drive 276
Repairing an Infected Computer 277
On This Day: Sir Thomas Browne Born—and Died 277
Software of the Week: NOD32 Antivirus 277
Finding New Viruses 278
On This Day: Monterey Bay Aquarium Opens (1984) 278
Website of the Week: Virus Bulletin 278
Signature Scanning 279
On This Day: First Trimline Telephone (1963) 279
Blog of the Week: Worm Blog 279
Fixing Only Part of the Picture 280
On This Day: Mr. Tornado Born (1920) 280
Fact of the Week 280
Soften Your Edges 281
Learn More About Fixing Bad Digital Pictures 281

On This Day: First Nylon Stockings (1939) 281
Gadget of the Week: Bose Noise Canceling Headphones 281
Removing Red Eye 282
On This Day: First Microwave Oven Sold (1955) 282
Hardware of the Week: Graphire Bluetooth Tablet 282
Removing Wrinkles 283
On This Day: C.W. Post Born (1854) 283
Software of the Week: Adobe Photoshop Elements 283
Removing Skin Blemishes 284
On This Day: ARPANET Crashes (1980) 284
Website of the Week: Planet Photoshop 284
Create a Glamour Glow 285
On This Day: Statue of Liberty Unveiled (1886) 285
Blog of the Week: The Unofficial Photoshop Blog 285
The War of the Worlds 286
On This Day: Ballpoint Pen Patented (1888) 286
Fact of the Week 286
Virus Hoaxes 287
On This Day: Vatican Admits Galileo Was Right (1992) 287
Gadget of the Week: Suunto Smart Watch 287

November 2006 289

How Urban Legends Become Legendary 291
On This Day: First H-Bomb (1952) 291
Hardware of the Week: LaCie Bigger Disk Extreme 291
Favorite Urban Legends 292
On This Day: George Boole Born (1815) 292
Software of the Week: Wallperizer 292
More Favorites 293
On This Day: First Frozen Food (1952) 293
Website of the Week: Snopes 293
ICE on Your Cell Phone 294
On This Day: Ancient Beer Discovered (1992) 294
Blog of the Week: UFO News Blog 294
Cycle Your Background Colors 295
On This Day: Microsoft Contracts with IBM (1980) 295
Fact of the Week 295
Add a Watermark to Your Web Page 296
On This Day: Tacoma Narrows Bridge Collapses (1940) 296
Gadget of the Week: palmOne LifeDrive 296

Make the Page Background Scroll—Automatically ... 297
On This Day: Jack Kilby Born (1923) ... 297
Hardware of the Week: Dell UltraSharp 24" LCD Display ... 297
Display a Mouseover Alert ... 298
On This Day: Carl Sagan Born (1934) ... 298
Software of the Week: JavaScript Menu Master ... 298
Open Link in New Window ... 299
On This Day: First Motorcycle (1885) ... 299
Website of the Week: JavaScript.com ... 299
Add Music to Your Web Page ... 300
On This Day: Jack Ryan Born (1926) ... 300
Blog of the Week: The JavaScript Weblog ... 300
Create a Self-Running Presentation ... 301
On This Day: Artificial Snow Produced (1946) ... 301
Fact of the Week ... 301
Create a Branching Presentation ... 302
On This Day: First BBC Broadcast (1922) ... 302
Gadget of the Week: Dell Axim X50v ... 302
Adding Music to Your Presentation ... 303
On This Day: Tobacco Discovered (1492) ... 303
Hardware of the Week: Wacom Touchscreen Display ... 303
Add Depth to a 2-D Chart ... 304
On This Day: Gene Amdahl Born (1922) ... 304
Software of the Week: Crystal Graphics PowerPlugs ... 304
Publish Your Presentation to the Web ... 305
On This Day: Mouse Patented (1970) ... 305
Website of the Week: Clipart Connection ... 305
Printing Professional Handouts ... 306
On This Day: Vitamin C Wards Off Colds (1970) ... 306
Blog of the Week: Beyond Bullets ... 306
How Satellite Radio Works ... 307
On This Day: Edwin Hubble Born (1889) ... 307
Fact of the Week ... 307
XM Satellite Radio ... 308
On This Day: Edison Invents Phonograph (1877) ... 308
Gadget of the Week: Delphi XM MyFi ... 308
SIRIUS Satellite Radio ... 309
On This Day: S.O.S. Adopted (1906) ... 309
Hardware of the Week: Soundblaster Audigy 4 ... 309
Satellite Radio in the Car ... 310

On This Day: Pencil Sharpener Patented (1897) 310
Software of the Week: Replay Radio 310
Satellite Radio at Home 311
On This Day: *Origin of Species* Published (1859) 311
Website of the Week: XMFan.com 311
HD Radio: Digital Radio Without the Satellite 312
On This Day: First Lion in America (1716) 312
Blog of the Week: Orbitcast 312
Why Online Shopping Is (Generally) Safe 313
On This Day: Electric Motor Invented (1834) 313
Fact of the Week 313
Product Reviews Online 314
On This Day: Mariner 4 Launched (1964) 314
Gadget of the Week: GoDogGo Automatic Fetch Machine 314
Comparison Shopping Online 315
On This Day: PONG Announced (1971) 315
Hardware of the Week: MSI Bluetooth Star USB Hub 315
Using Online Coupons 316
On This Day: Mason Jar Patented (1858) 316
Software of the Week: Best Price 316

December 2006 317

Reduce Your Shipping Costs 319
On This Day: First White House Telephone (1878) 319
Website of the Week: SalesHound 319
Tracking Your Shipment 320
On This Day: John Backus Born (1924) 320
Blog of the Week: Shopping Blog 320
Shopping for Your Favorite Geek 321
On This Day: Omar Khayyam Dies (1131) 321
Fact of the Week 321
Shopping for a Computer Geek 322
On This Day: Folding Chair Patented (1854) 322
Gadget of the Week: Scooba Robotic Floor Washer 322
Gifts for the Road Warrior Geek 323
On This Day: First Sound Recording (1877) 323
Hardware of the Week: Buffalo TeraStation Storage Drive 323
Gifts for the Hard-Driving Geek 324
On This Day: Jet Stream Discovered (1934) 324
Software of the Week: Picasa 324

Gifts for the Music-Lovin' Geek 325
On This Day: Eckert-Mauchly Computer Corp. Incorporated (1947) 325
Website of the Week: Craig's List 325
Gifts for the Home Theater Geek 326
On This Day: First Non-Solar Planet Discovered (1984) 326
Blog of the Week: Gizmodo 326
Navigating Amazon 327
On This Day: Last Moon Mission (1972) 327
Fact of the Week 327
Today's Deals 328
On This Day: Golf Tee Patented (1899) 328
Gadget of the Week: Robosapien V2 Toy Robot 328
Clearance Merchandise and the Friday Sale 329
On This Day: Werner von Siemens Born (1816) 329
Hardware of the Week: Creative GigaWorks PC Speaker System 329
Buy It Used—and Save 330
On This Day: Planck's Quantum Physics (1900) 330
Software of the Week: FeedDemon 330
Save Money on Shipping 331
On This Day: First Street Cleaning Machine (1854) 331
Website of the Week: Amazon Watch 331
Become an Amazon Associate 332
On This Day: End of the World (1919) 332
Blog of the Week: Blog of a Bookslut 332
Gift Ideas 333
On This Day: First Celestial Photograph (1839) 333
Fact of the Week 333
Christmas Crafts 334
On This Day: Altair 8800 Introduced (1974) 334
Gadget of the Week: Sleeptracker Watch 334
Holiday Greeting Cards 335
On This Day: Broadway Lights Up (1880) 335
Hardware of the Week: USB Christmas Tree 335
Online Greeting Cards 336
On This Day: *Snow White* Premieres (1937) 336
Software of the Week: Creating Keepsakes Scrapbook Designer 336
Santa Claus on the Web 337
On This Day: Lincoln Tunnel Opens (1937) 337
Website of the Week: Christmas.com 337
Holiday Meals Online 338

On This Day: Man Orbits the Moon (1968) 338
Blog of the Week: Make Blog 338
There's No Need to Fear.... 339
Learn More About Your New PC 339
On This Day: Centigrade Scale Devised (1741) 339
Fact of the Week 339
Taking a Look Inside 340
On This Day: Radium Discovered (1898) 340
Gadget of the Week: Egg & Muffin Toaster 340
Your PC's Microprocessor: The Main Engine 341
On This Day: Johannes Kepler Born (1571) 341
Hardware of the Week: Logitech MX3100 Cordless Desktop 341
Computer Memory: Temporary Storage 342
On This Day: First U.S. Test Tube Baby (1981) 342
Software of the Week: WinZip 342
Hard Disk Drives: Long-Term Storage 343
On This Day: First Transistorized Hearing Aid (1952) 343
Website of the Week: Tech Support Guy 343
Turning On Your PC—For the First Time 344
On This Day: Monopoly Patented (1935) 344
Blog of the Week: Biggeststars Blogs 344

Index **345**

ABOUT THE AUTHORS

Leo Laporte is the former host of two U.S. shows on TechTV: *The Screen Savers* and *Call for Help*. Leo is a weekend radio host on Los Angeles radio KFI AM 640 and co-hosts *Call for Help* on Canada's TechTV network. He also appears regularly on many other television and radio programs, including ABC's *World News Now* and *Live with Regis and Kelly* as "The Gadget Guy." He is the author of four recent bestsellers: *Leo Laporte's 2005 Gadget Guide*, *Leo Laporte's Mac Gadget Guide*, *Leo Laporte's Guide to TiVo*, and *Leo Laporte's 2005 Technology Almanac*.

In January 1991 he created and co-hosted *Dvorak on Computers*, the most listened-to high-tech talk radio show in the nation, syndicated on more than 60 stations and around the world on the Armed Forces Radio Network. Laporte also hosted *Laporte on Computers* on KSFO and KGO Radio in San Francisco.

On television, Laporte was host of *Internet!*, a weekly half-hour show airing on PBS in 215 cities nationwide. He reported on new media for *Today's First Edition* on PBS, and did daily product reviews and demos on *New Media News*, broadcast nationally on Jones Computer Network and ME/U and regionally on San Francisco's Bay TV.

Michael Miller has authored more than 70 bestselling books over the past 15 years, including *Absolute Beginner's Guide to Computer Basics*, *Absolute Beginner's Guide to eBay*, and *The Complete Idiot's Guide to Home Theater Systems*. Mr. Miller has established a reputation for clearly explaining technical topics to non-technical readers and for offering useful real-world advice about complicated topics. More information can be found at the author's website, located at www.molehillgroup.com.

WE WANT TO HEAR FROM YOU!

As the reader of this book, *you* are our most important critic and commentator. We value your opinion and want to know what we're doing right, what we could do better, what areas you'd like to see us publish in, and any other words of wisdom you're willing to pass our way.

As an associate publisher for Que Publishing, I welcome your comments. You can email or write me directly to let me know what you did or didn't like about this book—as well as what we can do to make our books better.

Please note that I cannot help you with technical problems related to the topic of this book. We do have a User Services group, however, where I will forward specific technical questions related to the book.

When you write, please be sure to include this book's title and author as well as your name, email address, and phone number. I will carefully review your comments and share them with the author and editors who worked on the book.

Email: feedback@quepublishing.com

Mail: Rick Kughen
Executive Editor
Que Publishing
800 East 96th Street
Indianapolis, IN 46240 USA

For more information about this book or another Que title, visit our website at www.quepublishing.com. Type the ISBN (excluding hyphens) or the title of a book in the Search field to find the page you're looking for.

INTRODUCTION

This is a book for people who love computers and technology but hate computer books. I'm not all that fond of them, myself, although I have to read a lot of them for my job.

Allow me to introduce myself. My name is Leo and you might have heard me on the radio or seen me on TV talking about computers and technology. I read all those computer books and magazines so you don't have to. Every week on my radio show I try to distill gallons of information into a fun, fast-paced, and informative brew. It's my goal to keep you up to date on what's happening with technology while showing you a darn good time.

This book is all those shows put on paper. Inside these pages is a year's worth of information—stuff you can really use, surrounded by stuff that's not so useful but fascinating and fun to know nevertheless. Mary Poppins had it right: Just a spoonful of sugar helps the medicine go down. There's lots of sugar mixed in with the medicine in this book.

This is the 2006 edition of the Almanac, brand new for the brand new year. There's a page for every day of the year (well, sort of; Saturday and Sunday share a page each week). But here's the neat thing: There's no need to read the pages in order, or to wait for the calendar to read that day's entry. We designed this book so you could jump around in it, read a little bit whenever you have the time or inclination, or devour it all at once, if you have a mind to. Each page stands alone, with a feature article, plus a look at this date in technology history, downloads, favorite websites, fun gadgets, and more.

Each week focuses on a single area in technology: eBay auctions, computer hardware, Windows, digital photography, satellite radio, and so on. But it's an almanac, not an encyclopedia, so you won't find an exhaustive (or exhausting) discussion of any topic inside. You will come away from each week knowing a lot more about the subject than you did before. I've included plenty of links to Web pages where you can learn more if you want to.

A note about those links: To save space (and to save you typing) I've eliminated the redundant `http://` from the web addresses in the Almanac. Type the address as printed into your browser, and unless the page has moved or disappeared, it will work. Don't type any punctuation after the address. Many URLs in the book are followed by commas or periods—they're not part of the address, they're to keep Mrs. Kandel, my sixth-grade English teacher, happy. I checked every single web address just before publication, and they were all working, but the web being the web, it's possible that some of these addresses will not be working by the time you get around to trying them. I apologize if that happens, but don't forget that you can always find a similar page by going to Google or some similar search engine. Just leave the period off the end.

Before I wrap this up, I want to thank you—not just for buying this book, but for wanting to learn more about how this stuff works. Technology is a wonderful thing, and computers are remarkable tools. They're probably the most complex machines humankind has ever invented, and yet a six-year-old can use one with seeming ease. For those of us over the age of six, it takes a little more effort, but that effort pays off handsomely. The computer is an amplifier for the mind, giving any individual the power to change the world. It's my mission in life to show people just what buttons to push and dials to twist so they can begin to use technology to make a difference in their own lives. Thanks for your willingness to try. Now, let's get going—I can't wait to see what you're going to do with the stuff you learn in here!

IS THIS BOOK FOR YOU?

If you're just picking this book up in the store, you might wonder whether it's for you. I'll save you the trouble of reading the next paragraph. Yes, this book is for you. In fact, it's for everyone you know. I suggest you buy copies for all your friends. Buy a copy for that guy standing next to you looking at that *JavaScript Programming for Nudniks* book. Buy a copy for the nice bookstore clerk. In fact, buy every copy on the shelf and hand them out to people on the street as you walk by. Spread the Almanac goodness! Hallelujah!

Well, okay, no book is for everyone, not even this fine volume. *Leo Laporte's 2006 Technology Almanac* is for people who spent a lot of money on a computer or some other fancy high-tech device and are now wondering what to do with that expensive piece of plastic and metal. Every page contains something fun or useful you can do right now.

It's not just for super geeks, but even technology experts will learn something in here. It's not just for novices, but even a beginner will be able to understand and use the tips inside. It's really for anyone who wants to bring the fun back into tech. We live in an age when the best toys are being designed for grown-ups. This book will help you rediscover how to play with them.

By the way, this book is designed for users of all mainstream computer platforms: Windows, Mac OS, and Linux. In fact, most of the computer-related tips and info are applicable cross-platform—especially the Internet-related information. And there's a ton of stuff inside that doesn't have anything to do with computers at all. Technology today is more than just computers, and I try to cover it all—from digital cameras to high-definition TVs!

HOW THIS BOOK IS ORGANIZED

Inside, you'll find a page for every day of the year (Saturdays and Sundays sharing a page, as I mentioned earlier). Each week has a primary focus: online shopping, podcasting, Photoshop, and so on. On each page, you'll find a short article related to the week's subject. There's also a short feature for each day, detailing my favorite websites, software, gadgets, and so on. It's my evil plan to have you shouting "A-ha!" and "Wow, I never knew that!" on every page.

On each page, you'll also find an historic event in technology—everything from Pascal's birthday to the day that Bill Gates thought of the name "Micro-Soft." There are many fun surprises, too, so keep your eyes peeled.

Leo Laporte's 2006 Technology Almanac is the computer book I've always wanted to read. I know you'll enjoy reading it as much as we've enjoyed writing it.

Leo
Leo Laporte

September 30, 2005

January 2006

January 2006

SUNDAY	MONDAY	TUESDAY	WEDNESDAY	THURSDAY	FRIDAY	SATURDAY
1 — 1942 United Nations created	**2** — 1920 Isaac Asimov born	**3** — 1952 *Dragnet* debuts	**4** — 1642 Isaac Newton is born	**5** — 1896 X-rays first discovered	**6** — 2001 Congress certifies George W. Bush winner of 2000 elections over Democrat Al Gore	**7** — 1985 General Motors Creates Saturn vehicle
8 — 1642 Astronomer Galileo dies in Italy	**9** — 1894 First motion picture copyrighted	**10** — 1946 First meeting of the United Nations	**11** — 1935 Amelia Earhart flies from Hawaii to California	**12** — 1896 First X-ray photo taken	**13** — 1910 Lee De Forest makes the first radio broadcast from the Metropolitan Opera	**14** — 1952 The *Today Show* debuts
15 — 1929 Martin Luther King, Jr., born	**16** — 1953 Chevrolet introduces the Corvette	**17** — 1949 TV's first sitcom, *The Goldbergs*, debuts	**18** — 1974 *Six Million Dollar Man* debuts	**19** — 1955 President Dwight D. Eisenhower becomes the first president to host news conferences on TV	**20** — 1981 Iran hostage crisis ends after 444 days	**21** — 1976 First Concorde flights leave simultaneously from London and Paris
22 — 1973 Roe v. Wade decision handed down by U.S. Supreme Court	**23** — 1977 *Roots* miniseries premieres on ABC, becoming the most watched program in American history	**24** — 1965 Winston Churchill dies	**25** — 1955 An atomic clock accurate to within one second in 300 years is developed at Columbia University	**26** — 1962 U.S. launches Ranger 3 to the moon, unsuccessfully	**27** — 1967 Apollo I fire kills three U.S. astronauts	**28** — 1956 Elvis Presley makes first TV performance
29 — 1896 Radiation first used to treat cancer patient	**30** — 1957 First pacemaker put into use	**31** — 1958 First U.S. satellite launched				

January 1, 2006

SUNDAY

THIS WEEK'S FOCUS
THE INTERNET

TESTING YOUR CONNECTION SPEED

Whether you're connecting to the Internet via a dial-up or broadband connection, there are sure to be days where you're convinced that your connection is slower than it should be. You're probably just imagining things, of course—but maybe not. Sometimes Internet connections *do* slow down, for a variety of different reasons.

While a seemingly slow connection could just be a slowdown at the particular site you're visiting, it's also possible that the connection from your ISP is experiencing difficulties. Maybe the problem is in the line from your ISP to your house; maybe the problem is in your ISP's connection to the Internet itself. Whatever the case, several sites let you test your absolute speed at any given time.

These sites work by downloading a text or picture file to your computer, then uploading back again. The time it takes to download/upload the file is measured and translated into kilobytes per second (Kbps). The higher your Kbps number, the faster your connection speed.

You can test your connection speed at the following sites: Bandwidth Place (www.bandwidthplace.com/speedtest/), Broadband Reports (www.dslreports.com/stest/), CNET (reviews.cnet.com/Bandwidth_meter/7004-7254_7-0.html), and TOAST.net (www.toast.net). Remember, your speed will vary from day to day—and sometimes from minute to minute!

ON THIS DAY: MILLENNIUM BUG DOESN'T BITE (2000)

After years of worrying about the possibly catastrophic effects of the so-called Y2K bug, the first day of the new millennium hit with virtually no impact whatsoever on computers anywhere. Whether this was due to effective fixing or the lack of a real problem to begin with is unknown, but IT workers worldwide breathe a sigh of relief.

BLOG OF THE WEEK: SLASHDOT

Slashdot (www.slashdot.com) promises "news for nerds," and it delivers. This blog is chock-full of technology news and comments, an indispensable resource for those truly interested in new technology of all sorts.

January 2, 2006

MONDAY

THIS WEEK'S FOCUS
THE INTERNET

BROADBAND INTERNET

If you're suffering through an outdated dial-up Internet connection, help is on the way—in the form of a faster broadband connection. There are actually three different types of broadband connections available, all of them at least six times faster than traditional 56Kbps dial-up.

The most popular type of broadband connection today is available from your local cable company. Broadband cable Internet piggybacks on your normal cable television line, providing speeds in the 500Kbps to 3Mbps range, depending on the provider. Most cable companies offer broadband cable Internet for $30-$50 per month.

Competing with cable broadband is DSL, which stands for digital subscriber line and is typically available from your local phone company. DSL service piggybacks onto your existing phone line, turning it into a high-speed digital connection. Not only is DSL faster than dial-up (384Kbps to 1.5Kbps, depending on your ISP), you also don't have to surrender your normal phone line when you want to surf; DSL connections are "always on." Most providers offer DSL service for $20-$50 per month.

Finally, if you can't get DSL or cable Internet in your area, you have another option: connecting to the Internet via satellite. Any household or business with a clear line of sight to the southern sky can receive digital data signals from a geosynchronous satellite at 500Kbps. The largest provider of satellite Internet access is Hughes Network Systems; its DIRECWAY system (www.direcway.com) enables you to receive Internet signals via a small dish that you mount outside your house or on your roof. Fees are around $60 per month.

ON THIS DAY: MICRO-SOFT NAMED (1975)

On the second day of January, 1975, Bill Gates and Paul Allen wrote a letter to MITS, the company that manufactured the Altair computer, offering a version of BASIC for the Altair 8800. In the letter, Gates and Allen referred to their company for the first time as "Micro-Soft"—which later got shortened to the more familiar "Microsoft."

FACT OF THE WEEK

The number of Internet users worldwide hit 935 million in 2004, and will top 1 billion by the end of 2005, reports the Computer Industry Almanac (www.c-i-a.com). The U.S. leads all nations with 185 million Internet users.

January 3, 2006

TUESDAY

THIS WEEK'S FOCUS
THE INTERNET

INTERNET EXPLORER KEYBOARD SHORTCUTS

You don't have to do all your web surfing with a mouse. Internet Explorer offers a bevy of keyboard shortcuts that can speed up your surfing, as long as you don't mind using your fingers. Here are some of my favorites:

- Return to previous page: BACKSPACE
- Enter a new URL: CTRL+O
- Add the current page to your Favorites list: CTRL+D
- Open the Favorites pane: CTRL+I
- Open the History pane: CTRL+H
- Open the Search pane: CTRL+E
- Find an item on the current page: CTRL+F
- Refresh the current page: CTRL+R
- Print the current page: CTRL+P
- Open a new IE window: CTRL+N
- Close the current IE window: CTRL+W

ON THIS DAY: *TIME* NAMES COMPUTER "MAN OF THE YEAR" (1983)

In its first issue of 1983, *TIME* magazine altered its annual tradition of naming a "Man of the Year," instead naming the computer as its "Machine of the Year" for 1982. *TIME* publisher John A. Meyers wrote that no human candidate for the honor "symbolized the past year more richly, or will be viewed by history as more significant, than a machine: the computer."

GADGET OF THE WEEK: CITIZEN TASER X26C PISTOL

Neighbors bugging you? Then take them out hard with this portable TASER pistol (we're kidding, of course). As you no doubt know, the TASER is a generally non-lethal weapon that uses electric bolts to temporarily override the target's central nervous system, reducing him or her to a heap of quivering Jell-O. The X26c is a handheld TASER gun with a range of 15 feet and a laser sight for more accurate targeting. Buy it for $999 at www.taser.com/self_defense/.

January 4, 2006

WEDNESDAY

THIS WEEK'S FOCUS
THE INTERNET

MOZILLA FIREFOX: A BETTER BROWSER

As popular as Internet Explorer is, it's not my favorite web browser. That honor goes to Mozilla Firefox, a modern update of the venerable Netscape browser.

Why is Firefox better than Internet Explorer? There are a lot of reasons. First, Firefox is a more streamlined program than IE. The program itself loads faster, and it also loads websites more quickly. It's not as bloated as IE has become.

Second, there's the interface. Firefox is simpler and cleaner than IE; it's a nicer-looking interface, if nothing else. in addition, Firefox offers tabbed browsing, which lets you open multiple pages within the same browser window, then switch between pages by selecting different tabs. It's a lot more elegant solution than opening several IE windows at a time.

Finally, and most important, Firefox is a much more secure browser. IE has holes galore, and an entire subindustry has sprung up to exploit IE's security flaws. Not so with Firefox; if you want a safer browsing experience, Firefox is the unquestionable choice.

If all this sounds attractive to you, navigate over to www.mozilla.org and download the latest version of Firefox. Did I mention that it's free?

ON THIS DAY: HP-35 CALCULATOR INTRODUCED (1972)

Hewlett-Packard introduced the world's first handheld scientific calculator on this date in 1972. The HP-35 was named for its 35 keys, weighed a whopping nine ounces, and sold for an even more whopping $395. It was the final nail in the coffin for the venerable (but less expensive) slide rule.

HARDWARE OF THE WEEK: LINKSYS WRT54GS WIRELESS ROUTER

Connect multiple PCs to a single broadband Internet connection with this Linksys wireless router. It's compatible with both 802.11a and 802.11g WiFi networks; SpeedBooster technology increases "real-world" network performance by up to 35%. It also features 128-bit encryption and a built-in firewall. Buy it for $99.99 at www.linksys.com.

January 5, 2006

THURSDAY

THIS WEEK'S FOCUS
THE INTERNET

FIREFOX TIPS AND TRICKS

One of the great things about Firefox is how customizable it is. Here are some of my favorite tips and tricks for personalize Firefox your way:

- **Create multiple home pages**—To have Firefox start with multiple preset pages loaded into its tabbed interface, simply open all the tabbed pages you want to see on startup, select Tools, Options, and then click the Use Current Pages button.

- **Open multiple bookmarks at once**—Firefox lets you group your favorite bookmarks in folders. To open all the bookmarked pages in a folder at once within the tabbed interface, just click the Bookmarks menu, right-click the desired folder, and then select Open in Tabs.

- **Bookmark RSS feeds**—When you visit a website that has an RSS feed, you'll see a Live Bookmark icon in the lower-right corner of the Firefox window. Click this icon to display a list of feeds from this site, and then click the feed you want. When Firefox asks where you want to store this feed, select Bookmarks Toolbar Folder. This creates a button for this feed on the Bookmarks toolbar; click the button to check the latest news and postings.

- **Faster URL entry**—Simply enter the middle part of the URL, and then press Ctrl+Enter. Firefox will automatically add the www. and .com for you! (BTW, this tip also works in Internet Explorer.)

ON THIS DAY: AMY JOHNSON DIES (1941)

Amy Johnson was a pioneering female British aviator who first achieved fame as a result of her 1930 attempt to fly solo from London to Australia. Although she failed to beat the record, she was the first woman to make the trip and was given a hero's welcome. She disappeared over the Thames on this date in 1941, while on a flying mission for the British Air Ministry.

SOFTWARE OF THE WEEK: WS_FTP

If you need to do a lot of file uploading and downloading, nothing beats the old FTP standard. Although you can use Internet Explorer to access FTP sites, the whole process is a lot easier with a dedicated FTP program, such as the venerable WS_FTP. Download the home version from www.ipswitch.com for $39.95—or just use the free evaluation version.

January 6, 2006

FRIDAY

THIS WEEK'S FOCUS
THE INTERNET

THE WAYBACK MACHINE: A BLAST FROM THE PAST

The Web is in constant flux. Most websites get updated on a weekly, daily, or even hourly basis, and a lot of good stuff gets thrown out to make way for the new. Wouldn't it be great if you could go back in time to see what a given website looked like last month, last year, or even 10 years ago?

Well, Sherman, what you need is the Wayback Machine. This is a cool directory that stores cached copies of other websites throughout history. Enter any URL and you'll get a list of cached pages dating back to 1996, if the site was actually around back then. Some sites are stored only a page or two deep; some are stored in their entirety.

I gotta tell you, browsing through the Wayback Machine is a real blast. Want to see what the Microsoft site looked like on October 20, 1996? (It's not a pretty sight, let me warn you.) Or how CNN.com was reporting the events of September 11, 2001? (Ten different versions are stored, from various times during the day.) Or how your own web page looked at some point in the past? Well, it's all stored in the Wayback Machine, thanks to the Internet Archive.

Check out the Wayback Machine for yourself at www.waybackmachine.org. I typically find myself spending hours at a time browsing through the archives, reacquainting myself with websites of the past. It's great that somebody's doing this—it's important to keep this kind of historical record. And its fun!

ON THIS DAY: PETER DENNING BORN (1942)

On January 6, 1942, Peter Denning was born. Denning developed the concept of virtual memory, where a computer can store more data than its physical memory can hold by using other storage devices (such as disk drives) to mimic traditional memory.

WEBSITE OF THE WEEK: THOMAS

Here's one of the few examples of the Federal government actually providing something useful to its citizens. The THOMAS site, run by the Library of Congress, provides access to all sorts of legislative information, from minutes of the Congressional Record to legislative details and committee reports. Check it out, all for free, at thomas.loc.gov.

January 7/8, 2006

SATURDAY/SUNDAY

THIS WEEK'S FOCUS
THE INTERNET

ANYBODY REMEMBER ARCHIE AND VERONICA?

No, I'm not talking about the comic book characters. The Archie and Veronica I'm referring to are two tools for searching the Internet, pre-World Wide Web. In a world before hyperlinks, searching wasn't quite as user-friendly as it is today.

When it came to pre-web searching, the first tool of choice was *Archie*, whose name was derived from the word "archive." The name made sense, as Archie was a tool used for searching anonymous FTP archives across the Internet for files to download. It wasn't very elegant, but it worked.

When you wanted to organize the files on a dedicated server, the tool of choice was *Gopher*, which was widely used on university sites across the country. Each Gopher server contained lists of files and other information, both at that specific site and at other Gopher sites around the world. Gopher worked similarly to a hierarchical file tree; you clicked folder links to see their contents and navigated up and down through various folders and subfolders.

This brings us to *Veronica*, which was a server-based tool used to search multiple Gopher sites for information. You used Veronica somewhat like you use one of today's web search engines—you entered a query and clicked a Search button, which generated a list of matching documents.

Finally, there was a search tool called *WAIS*, which stood for wide area information server. WAIS let you use the old text-based Telnet protocol to perform full-text document searches of Internet servers. WAIS was more powerful than Veronica but wasn't around long; it was quickly superseded by the then-newly developed World Wide Web and various web-based search tools.

ON THIS DAY: HOLLERITH TABULATING MACHINE PATENTED (1889)

On January 8, 1889, Herman Hollerith received a patent for his Hollerith Tabulating Machine, a punch-card system that went on to be used to compile the 1890 U.S. census. Thirty-five years later, the Hollerith Tabulating Machine company merged with two other firms to become International Business Machines—or what we now call IBM.

BLOG OF THE WEEK: BOING BOING

Boing Boing (www.boingboing.net) bills itself as "a directory of wonderful things." What it is is a collection of humorous and intelligent musings on a variety of matters, all of which somehow tend to center around technology topics. It's not a tech blog per se, but because it's written for and by techies, you'll find a lot of tech-related info. Always a fun read.

January 9, 2006

MONDAY

THIS WEEK'S FOCUS
DIGITAL MUSIC

HOW DIGITAL SAMPLING WORKS

The whole digital music thing starts in the recording studio, where a digital recording is made by creating digital samples of the original sound or piece of music. The way it works is that special software "listens" to the music and takes a digital snapshot of it at a particular point in time. The length of that snapshot (measured in bits) and the number of snapshots per second (called the sampling rate) determine the quality of the reproduction. The more samples per second, the more accurate the resulting "picture" of the original music.

For example, compact discs sample music at a 44.1kHz rate—in other words, the music is sampled, digitally, 44,100 times per second. Each sample is 16 bits long. When you multiply the sampling rate by the sample size and the number of channels (two for stereo), you end up with a *bit rate*. For CDs, you multiply 44,100 × 16 × 2, and end up with 1,400,000 bits per second—or 1,400Kbps.

All these bits are converted into data that is then copied onto some sort of storage medium. In the case of CDs, the storage medium is the compact disc itself; you can also store this digital audio data on hard disk drives or in computer memory.

The space taken up by these bits can add up quickly. If you take a typical three-minute song recorded at 44.1KHz, you end up using 32MB of disk space. Although that song can easily fit on a 650MB CD, it's much too large to download over a standard Internet connection or to store on a portable music player. This is where audio compression comes in. By taking selected bits out of the original audio file, the file size is compressed. Compression can be either *lossy* (meaning that some information is lost in the process) or *lossless* (which retains all the data from the original file). More on compression tomorrow.

ON THIS DAY: GALAXIES ARE ACCELERATING (1998)

On this day in 1998, two teams of scientists announced the discovery that galaxies are flying apart at ever faster speeds. This revelation was based on their observations of distant exploding stars. The discovery implies the existence of a mysterious, self-repelling property of space first proposed by Albert Einstein, which he called the cosmological constant.

FACT OF THE WEEK

In spite of all the buzz, Jupiter Research (www.jupiterresearch.com) isn't too bullish on the viability of digital music downloads. Their research indicates that 51% of all online adults think physical music is more valuable than digital, and that by 2009 digital music sales will represent just 12% of all consumer music spending.

January 10, 2006
TUESDAY

THIS WEEK'S FOCUS
DIGITAL MUSIC

LOSSY DIGITAL AUDIO FORMATS

Lossy compression works by sampling the original digital audio file and removing those ranges of sounds that the average listener can't hear. You can control the sound quality and the size of the resulting file by selecting different sampling rates for the data. The less sampling going on, the smaller the file size—and the lower the sound quality.

The problem with shrinking files in this manner is that by making a smaller file, you've dramatically reduced the sampling rate of the music. This results in music that sounds compressed; it won't have the high-frequency response or the dynamic range (the difference between soft and loud passages) of the original recording. To many users, the sound of the compressed file will be acceptable, much like listening to an FM radio station. To other users, however, the compression presents an unacceptable alternative to high-fidelity reproduction.

The most popular lossy compressed formats today are

- **AAC** (short for Advanced Audio Coding), the proprietary audio format used by Apple's iTunes and iPod. You can't play AAC files in Windows Media Player or other non-Apple programs.
- **MP3** (short for MPEG-1 Level 3), the most widely used digital audio format today, with a decent compromise between small file size and sound quality. Almost all media players are MP3-compatible.
- **WMA** (short for Windows Media Audio), Microsoft's digital audio format that offers similar audio quality to MP3 at half the file size—or better quality at the same file size.

ON THIS DAY: 45 RPM RECORD INTRODUCED (1949)

On this date in 1949, RCA introduced the "single," a 7" diameter 45 RPM record. A single could play up to eight minutes of sound per side, and along with the long-playing 33 1/3 RPM records introduced a year earlier, sounded the death knell for the older 78 RPM format.

GADGET OF THE WEEK: STIKAX MUSIC MIXER

The StikAx is a handheld device that you use to mix sound and images on your PC or Mac. It works with accompanying TrakAx software and a sample loop CD to let you create your own audio and video mixes—without touching a computer or musical keyboard. It's a great gizmo for wannabe DJs who don't want to learn how to use complicated mixing software. Buy it for $90 from www.stikax.com.

January 11, 2006

WEDNESDAY

THIS WEEK'S FOCUS
DIGITAL MUSIC

LOSSLESS DIGITAL AUDIO FORMATS

If you care about audio fidelity, lossy compression just doesn't cut it for your digital audio files. No matter how high the sampling rate or how good the compression algorithm, lossy files don't sound quite as good as the originals. (Remember that word "lossy"—you lose something in the translation!)

If you want to create a high-fidelity digital audio archive to play on a home audio system, a better solution is to use a lossless compression format. These formats work more or less like ZIP compression—redundant bits are taken out to create the compressed file, which is then uncompressed for playback. So what you hear has exact fidelity to the original, while still being stored in a smaller-sized file.

Of course, a lossless compressed file isn't nearly as small as a file with lossy compression. Whereas an MP3 file might be 10% the size of the original, uncompressed file, a file with lossless compression is typically about 50% of the original's size. This is why lossless compression isn't recommended for portable music players, where storage space is limited. If you're storing your CD collection on hard disk, however, it works just fine; you can easily store 1,000 CDs on a 300GB hard disk, using any lossless compression format.

The most popular llossless audio formats include ALAC (a lossless compression option available for use with Apple's iTunes and iPod), FLAC (an open-source lossless format that works on both Windows and Linux machines), and WMA Lossless (Microsoft's lossless format, available in Windows Media Player versions 9 and 10).

ON THIS DAY: SMOKING CAUSES CANCER (1964)

On this date in 1964, the U.S. Surgeon General's Report on smoking was released. The report stated that "cigarette smoking is a health hazard of sufficient importance in the United States to warrant appropriate remedial action." The following year Congress mandated printed health warnings on all cigarette packages.

HARDWARE OF THE WEEK: CREATIVE I-TRIGUE SPEAKER SYSTEM

In my humble opinion, the best music-oriented speaker system for your computer is Creative's I-Trigue L3500. The left and right speakers each feature two Titanium drivers and a lateral firing transducer, and they sound extremely neutral and open. The deep bass is provided by the separate subwoofer, and the result is a full-range system ideal for listening to your favorite digital music files. Buy it for $99.99 at www.creative.com.

January 12, 2006

THURSDAY

THIS WEEK'S FOCUS
DIGITAL MUSIC

ONLINE MUSIC STORES

One of the great things about digital music is that you can listen to just the songs you want. You're not forced to listen to an entire album; you can download that one song you really like and ignore the rest. Even better, you can take songs you like from different artists and create your own playlists; it's like being your own DJ or record producer!

Where do you go to download your favorite songs? You have a lot of choices, but the easiest is to shop at an online music store. These are websites that offer hundreds of thousands of songs from your favorite artists, all completely legal. You pay about a buck a song and download the music files directly to your computer's hard disk.

The most popular online music stores today include Apple iTunes Music Store (www.apple.com/itunes/store/), MSN Music (music.msn.com), and Wal-Mart Music Downloads (musicdownloads.walmart.com).

Of these sites, iTunes was the first and is far and away the most successful. In fact, iTunes often offers exclusive, non-album tracks from major recording artists that you'll find nowhere else (legally anyway). Know, however, that iTunes only offers music in AAC format, and the only players that are AAC-compatible are those in the Apple iPod family. If you have a non-Apple music player, you'll want to use one of the non-iTunes music stores—most of which offer downloads in WMA format.

ON THIS DAY: FIRST X-RAY PHOTO (1896)

On January 12, 1896, Dr. Henry Louis Smith took the first x-ray photograph. The x-ray showed the location of a bullet in the hand of a corpse. How the bullet got there is an interesting story; Smith obtained the corpse's hand and fired a bullet into it for the experiment.

SOFTWARE OF THE WEEK: ALBUM ART FIXER

If you download, trade, or rip a lot of music, you know that music player software doesn't always get the album art right. That's where Album Art Fixer comes in; it helps you find and replace the album art and other info for all your music downloads in Windows Media Player. The program's a shareware download at www.avsoft.nl/ArtFixer/.

January 13, 2006

FRIDAY

THIS WEEK'S FOCUS
DIGITAL MUSIC

FILE TRADING NETWORKS

Some of the best music on the Internet doesn't come from any website—it comes from other users. Over the past few years, the web has seen a profusion of file-trading networks, where you can swap music files with your fellow computer users. You connect your computer (via the Internet) to the network, which already has thousands of other users connected; when you find a song you want, you transfer it directly from the other computer to yours.

Of course, file trading works in both directions. When you register with one of these services, other users can download digital audio files from *your* computer, as long as you're connected to the Internet.

Most file-trading networks require you to download a copy of their software and then run that software whenever you want to download. You use their software to search for the songs you want; the software then generates a list of users who have that file stored on their computers. You select which computer you want to connect to, and then the software automatically downloads the file from that computer to yours.

The most popular of these file-sharing services include eDonkey (www.edonkey.com), Grokster (www.grokster.com), iMesh (www.imesh.com), LimeWire (www.limewire.com), Kazaa (www.kazaa.com), Morpheus (www.morpheus.com), and WinMX (www.winmx.com. In addition, BitTorrent (www.bittorrent.com) is a file-sharing program especially for large video files. Many people use BitTorrent to find and download their favorite television shows—great if you ever miss an episode!

Of course, in a vast majority of cases, file trading such as what's discussed here is illegal and violates the owners' copyrights. Perhaps more importantly, you could be arrested, fined and/or jailed for trading copyrighted files on the Internet. You have been warned.

ON THIS DAY: ADDING MACHINE PATENTED (1874)

On this day in 1874, the U.S. Patent Office issued a patent for the Spalding Adding Machine. This machine was a precursor of today's calculators and computers, capable of doing simple arithmetic.

WEBSITE OF THE WEEK: ALLMUSIC

Allmusic (AKA the All-Music Guide) is one of my favorite sites on the web, period. It offers an encyclopedic database of albums, songs, and artists, all meticulously cross-referenced. Look up an album, find out who played on it, and then click to see what other albums that artist is associated with it. It's all free, at www.allmusic.com.

January 15, 2006

SATURDAY/SUNDAY

THIS WEEK'S FOCUS
DIGITAL MUSIC

INTERNET RADIO

Many real-world radio stations—as well as web-only stations—broadcast over the Internet using a technology called *streaming audio*. Streaming audio is different from downloading an audio file. When you download a file, you can't start playing that file until it is completely downloaded to your PC. With streaming audio, however, playback can start before an entire file is downloaded. This also enables live broadcasts—both of traditional radio stations and made-for-the-web stations—to be sent from the broadcast site to your PC.

Internet radio can be listened to with most music player programs. For example, Windows Media Player has a Radio page that facilitates finding and listening to a variety of Internet radio stations. In addition, many Internet radio sites feature built-in streaming software or direct you to sites where you can download the appropriate music player software.

Here's a short list of sites that offer links to either traditional radio simulcasts or original Internet programming:

- LAUNCH (launch.yahoo.com/launchcast/)
- Live365 (www.live365.com)
- Radio-Locator (www.radio-locator.com)
- RadioMOI (www.radiomoi.com)
- SHOUTcast (yp.shoutcast.com)
- Web-Radio (www.web-radio.com)

ON THIS DAY: NCSA OPENS (1986)

Twenty years ago today, on January 15, 1986, the National Science Foundation opened the National Center for Supercomputer Applications (NCSA) at the University of Illinois. This is where Marc Andreesen invented the Mosaic web browser that formed the nucleus of his later Netscape browser.

BLOG OF THE WEEK: FLUXBLOG

Here's a new concept: a weblog that offers a variety of MP3 files for free downloading. Fluxblog (www.fluxblog.org) is one of the best MP3 blogs, with a constantly changing assortment of music for download. It's a great way to discover new music to listen to!

January 16, 2006

MONDAY

THIS WEEK'S FOCUS
YAHOO!

YAHOO! SEARCH SHORTCUTS

Yahoo! offers a number of very specific searches, all of which you can perform directly from the main search box. Here's some of the most fun ones:

- **Airport conditions and information**—Enter the airport name, followed by the keyword airport, like this: ohare airport.
- **Area codes**—Enter the keywords area code followed by the name of the city, like this: area code chicago. To find out what city an area code is in, just search on the three-digit code.
- **Conversions**—Convert any measurement into different units, enter the keyword convert followed by the original measurement, like this: convert 7 miles.
- **Dictionary definitions**—Find the definition of any word by entering the keyword define followed by the word you want defined, like this: define defenestrate.
- **Flight tracking**—Track the departure, arrival, and gate information for any flight by searching on the airline and flight number.
- **Postage tracking**—Track the status of any FedEx, UPS, or USPS shipment by searching on the tracing code. (For FedEx shipments, enter the keyword fedex before the number; for USPS shipments, enter the keyword usps before the number.)
- **Traffic conditions**—Find current traffic conditions by entering the name of the city followed by the keyword traffic, like this: newark traffic.

ON THIS DAY: DIAN FOSSEY BORN (1932)

Dian Fossey, born on this day in 1932, was the famous zoologist who made extensive studies of mountain forest gorillas. She founded and directed the Karisoke Research Center in Rwanda and wrote the book *Gorillas in the Mist*. She was killed in 1985, presumably by a band of gorilla poachers.

FACT OF THE WEEK

As of Q1 2005, Yahoo! had 8,023 full-time employees. The company was created in January 1994 and went public in April 1996. Since then, Yahoo! stock has split five times, the latest split occurring on May 11, 2004.

January 17, 2006

TUESDAY

THIS WEEK'S FOCUS
YAHOO!

NAVIGATING! YAHOO!

Here's a quick way to get from one Yahoo! site to another, without entering the full URL. Just enter an exclamation point (!), followed by the name of the site, into the Yahoo! search box. Here are some examples:

- ! address book
- ! auctions
- ! briefcase
- ! calendar
- ! chat
- ! fantasy sports
- ! finance
- ! geocities
- ! groups
- ! hot jobs
- ! mail
- ! maps
- ! messenger
- ! music
- ! my yahoo
- ! personals
- ! shopping
- ! weather
- ! white pages
- ! yahooligans
- ! yellow pages

ON THIS DAY: COMPUTER DISCOVERS NEW PLANETS (1996)

Ten years ago today, on January 17, 1996, Paul Butler and Geoffrey Marcy announced that they had discovered two new planets using a new computer technique to analyze spectrographic images of stars. The first planet they discovered using this method was orbiting the star 47 Ursae Majoris.

GADGET OF THE WEEK: SURVIVAL KIT IN A SARDINE CAN

This one is exactly as the name says: a survival kit crammed into a sardine can. You get 25 survival items, including a compass, hook and line, duct tape, matches, whistle, razor blade, fire-starter cube, signal mirror, safety pin, and the like. It's small enough to carry in your glove compartment or purse—just pop it open when an emergency arises! Buy it for 12 bucks at www.whistlecreek.com.

January 18, 2006

WEDNESDAY

THIS WEEK'S FOCUS
YAHOO!

YAHOO! MAPS

Yahoo! Maps (maps.yahoo.com) is one of my favorite mapping sites. I like it not only because you get maps and driving directions, but because you can also search for specific types of businesses in the general vicinity.

When you display a map, you see the SmartView section on the right side of the page. Click a particular type of business, select any filters, and see the nearest businesses displayed on the map. For example, you can display all nearby restaurants or fine-tune the display for only Chinese, Indian, or Italian. It's kind of cool.

You can also search for specific businesses by clicking the Find More Nearby link. This brings up a Yahoo! Yellow Pages page; perform your search, and those matching businesses closest to your original location will be listed.

By the way, you don't have to go to the Yahoo! Maps site to generate a map on Yahoo! If you enter a street name, city, and state into the standard Yahoo! search box, Yahoo! will display a small map of that location. Click the map to see a larger version on the Yahoo! Maps site.

ON THIS DAY: HARVARD MARK I OUTLINED (1938)

On this date in 1938, J.W. Bryce wrote a memo that formalized IBM's development of a new computing machine for Harvard University. This computer, the Harvard Mark I, was completed in 1944; it was the first fully automatic computer, capable of computing three additions or subtractions per second.

HARDWARE OF THE WEEK: LOGITECH MEDIAPLAY CORDLESS MOUSE

Logitech's MediaPlay is more than just a mouse; in addition to scrolling around your desktop, you can pick it up and use it like a television remote control. You get all the standard (and not-so-standard) mouse buttons, including tilt-wheel and page navigation buttons, along with special buttons to control your Windows Media Center PC. Buy it for $49.95 from www.logitech.com.

January 19, 2006

THURSDAY

THIS WEEK'S FOCUS
YAHOO!

YAHOO! SEARCH META WORDS

Here's a cool way to fine-tune your Yahoo! search: Use meta words to generate very specific types of results. Just enter the meta word followed by a colon, and then your normal search keywords. Here's some of what you can do:

- `site:`—Find all documents within a particular domain and its subdomains. For example, `site:leoville.com` finds all matching documents on my Leoville site.
- `link:`—Find all documents that link to a particular URL. For example, `link:http://www.leoville.com/` finds all pages that link to my Leoville site.
- `inurl:`—Find a specific keyword that is part of a URL. For example, `inurl:leo` finds all URLs that contain the word *leo*.
- `intitle:`—Find a specific keyword that is part of the page titles. For example, `intitle:leo` finds all documents that have the word *leo* in their title.

ON THIS DAY: APPLE LISA INTRODUCED

On this day in 1983, Apple introduced the Lisa computer. Although the Lisa ultimately flopped (it was taken off the market in April 1985), it paved the way for the much more popular Apple Macintosh—and included the first GUI and the first computer mouse. Why was the Lisa a failure? Look no further than the price, a whopping $9,995—and that's in 1983 dollars!

SOFTWARE OF THE WEEK: YAHOO! TOOLBAR

If you're a heavy-duty Yahoo! user, you need the Yahoo! Toolbar. The Toolbar gives you quick access to a slew of Yahoo! sites and functions, and also includes a built-in pop-up blocker. Download it for free at `toolbar.yahoo.com`.

January 20, 2006

FRIDAY

THIS WEEK'S FOCUS
YAHOO!

YAHOO! MY WEB

Here's a relatively new feature from Yahoo!: the capability to save your favorite web pages on the web itself, where you can retrieve them from any computer. Yahoo! My Web lets you build your own password-protected collection of favorite web pages, organize them into folders, and then search your saved pages using Yahoo! search. You can even email and share your favorite pages with friends and colleagues.

Here's something particularly neat about Yahoo! My Web. It doesn't just save a link to a page, it also saves a cached copy of that page. That means you can go directly to the page you viewed, even if that site is down or otherwise not available.

My Web also lets you import your previously saved Internet Explorer favorites. The first time you activate My Web, you'll be prompted to automatically import your bookmarks; you can also do the import thing from the My Web management page.

Yahoo! My Web works best with the Yahoo! Toolbar. You can access My Web from myweb.search.yahoo.com.

ON THIS DAY: BUZZ ALDRIN BORN (1930)

Astronaut Edwin "Buzz" Aldrin, Jr., was born on this date in 1930. For you youngsters in the audience, Buzz was the second man to walk on the moon. He currently runs Starcraft Boosters, Inc., a rocket design company, as well as the ShareSpace Foundation, a nonprofit organization devoted to opening the doors to space tourism for the masses.

WEBSITE OF THE WEEK: MY YAHOO!

Personalized start pages are all the rage, and my favorite remains My Yahoo! I've found that My Yahoo! offers the most content of any personalized page, along with a wide variety of customizable designs; it also lets you incorporate RSS feeds (RSS feeds enable you to publish content from other sites on your site, including news, headlines, event listings, and the like). Sign up today (for free) at my.yahoo.com.

January 21/22, 2006

SATURDAY/SUNDAY

THIS WEEK'S FOCUS
YAHOO!

YAHOO! BRIEFCASE

Here's a feature most people don't know about. Yahoo! provides a way for you to share your files with other users, thanks to Yahoo! Briefcase (briefcase.com). It's a lot easier than emailing files back and forth—or using the old sneakernet method.

Put simply, Yahoo! Briefcase is a way to store files online—on Yahoo!'s Web servers—and then access them from any PC with an Internet connection and a Web browser. You can even set selective access to your Briefcase files so that your friends, family, and co-workers can view, access, and download designated files. Yahoo! Briefcase lets you store up to 30MB worth of files.

To add a file to your Briefcase, just click the Add Files link and follow the onscreen instructions. You share the files or photos in your Briefcase by setting access levels for folders within your Yahoo! Briefcase. To change the access level for a folder within your Briefcase, click the Edit icon next to the folder on your Yahoo! Briefcase page, and then select the desired access level—Private, Friends, or Everyone.

If you want others to access your Yahoo! Briefcase page, they access a Web address that looks like this: briefcase.yahoo.com/yahooID/. Replace yahooID with your own Yahoo! ID to complete the URL. Anyone accessing a folder in your Yahoo! Briefcase can download any files in the folder.

ON THIS DAY: FIRST COMMERCIAL 747 FLIGHT (1970)

It's hard to believe, but 747s have been flying for more than 35 years now. On January 21, 1970, the first jumbo jet was put into service when a Pan American Airways Boeing 747 flew its first flight between New York's JFK and London's Heathrow airports. The 747 was twice as wide as the then-dominant 707 and could carry more than 400 passengers more than 5,500 miles.

BLOG OF THE WEEK: YAHOO! SEARCH BLOG

The Yahoo! Search Blog (www.ysearchblog.com) is the official blog of Yahoo! developers. This is where you find what's going on at Yahoo! before the general public knows.

January 23, 2006

MONDAY

THIS WEEK'S FOCUS
MICROSOFT WINDOWS

WHY YOU NEED SERVICE PACK 2

Microsoft is constantly updating its products—so much so that the Windows XP you buy today is quite a bit different from the Windows XP that went on sale back in 2001. If you're running an older version of XP, you can update it to the new version by installing Service Pack 2 (which followed on the heels—and integrates—the previous Service Pack 1). This software update upgrades your system to include all sorts of bug fixes and security patches, as well as adds the new features found on the version of XP currently distributed by Microsoft.

Several big changes are noticeable with Service Pack 2. First are those changes that were initially part of Service Pack 1, most of which let you change certain components of the operating system itself. These changes—mandated by the courts, as part of the government's anti-trust action against Microsoft—are designed to reduce Microsoft's monopoly of the computer desktop by letting you configure Windows XP to use various accessory programs of your choice, instead of Windows Media Player, Internet Explorer, and the like.

Service Pack 2 also includes additional changes that build on the SP1 update. Most of these new changes have to do with improved security—making Windows and your PC safer when you connect to the Internet.

There's also a new Wireless Network Setup Wizard for configuring your computers in a WiFi network, and some neat changes to Internet Explorer (including a very effective pop-up blocker).

If you've purchased a copy of Windows XP or a new PC since fall 2004, you probably have the latest version of Windows XP with Service Pack 2 already built in. If you have an older version of XP, you can upgrade it with SP2 going to www.microsoft.com/windowsxp/sp2/.

ON THIS DAY: INTEGRATED CIRCUIT CONCEIVED (1959)

Robert Noyce, co-founder and research director of Fairchild Semiconductor, first conceived of the integrated circuit on this day in 1959. In 1968 Noyce went on to found Intel with Gordon Moore and Andy Grove.

FACT OF THE WEEK

Windows XP has been around for four years now, yet corporate use still trails that of older versions of the operating system. AssetMetrix (www.assetmetrix.com) claims that Windows XP is found on just 38% of corporate PCs, while Windows 2000 is still used on 48% of those PCs. Come on, IT guys—get with the program and upgrade!

January 24, 2006

TUESDAY

THIS WEEK'S FOCUS
MICROSOFT WINDOWS

USING THE CONTROL PANEL

Most—but not all—of Windows XP's configuration settings are found somewhere within the Control Panel. You open the Control Panel by clicking the Start button and then clicking the Control Panel icon.

By default, the Control Panel includes a variety of standard Windows XP configuration utilities, such as Add New Hardware and User Accounts. In addition, many applications and utilities install their own Control Panel items. When you click an icon in the Control Panel, a dialog box opens that lets you configure settings specific to that item. For example, clicking the Date and Time icon opens the Date and Time Properties utility that enables you to set the date and time of the Windows XP system clock.

To launch a particular Control Panel applet, you start by picking a category. When the Pick a Task page appears, either click a task or click an icon to open a specific configuration utility. (When you click a task, the appropriate configuration utility is launched.)

If you want to bypass all the category and task steps, you can display all the Control Panel utilities at once—just like the way it used to be in Windows 9X/Me. To switch to the so-called "classic" view, click the Switch to Classic View link in the Control Panel's activity center. (That's the big pane on the left of the Control Panel window.)

ON THIS DAY: IBM DEDICATES THE SSEC (1948)

On this date in 1948, IBM dedicated the Selective Sequence Electronic Calculator (SSEC). This early mainframe computer was placed on public display near IBM's Manhattan headquarters so passers-by could watch it work, and thus influenced the public perception of what a computer looked like.

GADGET OF THE WEEK: GLOBAL PET FINDER

When Fido runs away, find him fast with the Global Pet Finder. This is a small GPS tracking device that attaches to your pet's collar; you can track your lost pet by accessing the Global Pet Finder website or by calling F-O-U-N-D from your cell phone. Two caveats: It's for dogs only (not cats), and you need to subscribe to the Global Pet Finder Service ($17.99/month). Buy it for $349.95 at www.globalpetfinder.com.

January 25, 2006

WEDNESDAY

THIS WEEK'S FOCUS
MICROSOFT WINDOWS

ENABLING CLEARTYPE

ClearType is a new display technology in Windows XP that effectively triples the horizontal resolution on LCD displays. (In other words, it makes things look sharper—and smoother.) If you have a flat-panel monitor or a portable PC, you definitely want to turn on ClearType.

To turn on ClearType, follow these steps:

1. From the Display Properties utility, select the Appearance tab.
2. Click the Effects button.
3. When the Effects dialog box appears, check the Use the Following Method to Smooth Edges of Screen Fonts option, and select ClearType from the pull-down list.
4. Click OK, and then click OK again.

Some computers (laptops, especially) come with additional ClearType fine-tuning. If you find a ClearType Tuner icon in your Control Panel, click it to display the ClearType Settings utility. From here you can run the ClearType Tuning Wizard, which takes you step-by-step through some basic ClearType calibration—similar to the "which looks better?" tests of a typical eye exam. (You can also download the ClearType Tuner utility from www.microsoft.com/windowsxp/downloads/powertoys/xppowertoys.mspx.)

ON THIS DAY: FLOURIDATED DRINKING WATER (1945)

On this day in 1945, Grand Rapids, Michigan, became the first U.S. city to add fluoride to its drinking water. With the intention of reducing tooth decay, one part per million of fluoride was added to the water supply. There were smiles all around.

HARDWARE OF THE WEEK: APC UNINTERRUPTIBLE POWER SOURCE

Provide full backup power to your PC with APC's Back-UPS battery-powered power source. The ES 725 Broadband is a 450 watt unit that offers a typical backup time of 4.3 minutes (14 minutes at half-load), long enough to get your system shut down properly during a power outage—or last you through a short brownout. It also functions as a power cleaner and surge protector, with four power outlets for your key devices. Buy it for $99.99 at www.apc.com.

January 26, 2006

THURSDAY

THIS WEEK'S FOCUS
MICROSOFT WINDOWS

USING A WEB PAGE AS A DESKTOP BACKGROUND

You probably already know how to change background images for your Windows desktop, but did you know you could display a real live web page as your desktop background? When you do this, your desktop background functions like a live web page whenever you're connected to the Internet, complete with live links and such. This is a great way to put your favorite web page or list of links at your fingertips.

Here's what you do:

1. Right-click anywhere on the desktop to open the Display Properties dialog box, and then select the Desktop tab and click the Customize Desktop button.
2. When the Desktop Items dialog box appears, select the Web tab.
3. To use your home page as the desktop background, select it from the web pages list.
4. To use another page as the desktop background, click the New button to display the New Active Desktop Item dialog box. Enter the URL of the page into the Location box, and then click OK.
5. To automatically update the content of this Web page, click the Properties button and select the Schedule tab. Check the Using the Following Schedules option, and then click the Add button. When the New Schedule dialog box appears, set a time for updating, check the If My Computer Is Not Connected option, and then click OK.
6. Click OK to close the remaining dialog boxes.

ON THIS DAY: INTERNAL COMBUSTION AUTO PATENTED

On this day in 1886, Karl Benz (name sound familiar?) patented the first automobile powered by an internal combustion engine. It was a single-cylinder engine with 984cc displacement, capable of 400 RPM and a top speed of 10 MPH. Wowzers!

SOFTWARE OF THE WEEK: THEME MANAGER 2005

With Windows XP, Microsoft actually took away a lot of the interface customization options that you had with Windows 98/Me. Fortunately, Theme Manager 2005 adds back a lot of that customization; you can change wallpapers, visual styles, desktop themes, cursors, login screens, and more. Best of all, Theme Manager 2005 is freeware, downloadable from www.stardock.com/products/thememanager/.

January 27, 2006

FRIDAY

THIS WEEK'S FOCUS
MICROSOFT WINDOWS

ADDING SPECIAL EFFECTS

All sorts of special effects are included with Windows XP. These effects are applied to the way certain elements look or the way they pull down or pop up onscreen. Some of these special effects can be changed from the Display Properties utility; others are changed from the System Properties utility.

You open the Display Properties utility by right-clicking anywhere on the desktop. You get to the settings you want by selecting the Appearance tab and then clicking the Effects button. From here you can make menus and ToolTips scroll or fade in and out; display drop shadows under menus; change the size of desktop icons; display the contents of windows while you're dragging them; and hide the underlined letters on menu items.

Even more special effects can be configured from the System Properties utility. You get there by opening the Control Panel, clicking the System icon, selecting the Advanced tab, and then clicking the Settings option (in the Performance section). When the Performance Options dialog box opens, select the Visual Effects tab.

Lots of effects are available here, some of which are extremely subtle. These include animating windows when minimizing and maximizing, drawing a gradient in window captions, fading in the taskbar, fading or sliding menus into view, showing translucent selection rectangle, sliding open combo boxes, smoothing edges of screen fonts, using drop shadows for icon labels on the desktop, and using Web view in folders. As you can see, some of these settings duplicate the settings in the Effects dialog box. Turning them off or on in one place turns them off or on in the other, as well.

ON THIS DAY: JIM CLARK LEAVES SILICON GRAPHICS (1994)

It was on this day in 1994 that Silicon Graphics co-founder Jim Clark left the company to start Mosaic Communications with Marc Andreesen. Mosaic would later become, of course, Netscape Communications Corporation.

WEBSITE OF THE WEEK: MICROSOFT

Well, we have to mention it somewhere. The Microsoft website offers a ton of free downloads, tips, and advice, as well as a the searchable Knowledge Base for when you have troubleshooting to do. Access it at the expected URL, www.microsoft.com.

January 28/29, 2006

SATURDAY/SUNDAY

THIS WEEK'S FOCUS
MICROSOFT WINDOWS

MOVING AND HIDING THE TASKBAR

By now I'm sure you're familiar with the Windows taskbar, that solid bar that typically sits at the bottom of your screen and holds buttons for any applications that are open at the time. The default position for the taskbar is horizontally across the bottom of the screen. If you don't like it there, however, you can move it to the top or to either side of the screen.

All you have to do is point to a part of the taskbar where no buttons appear. Then, click with the left mouse button and drag the taskbar to another edge of the screen, or even to the top of the screen. You'll see a shaded line that indicates the new position of the taskbar; release the mouse button to fix the taskbar in its new position.

If you're running Windows XP on a small display (640 × 480 resolution, for example), you might not want the taskbar taking up screen space all the time. For these situations, Windows XP lets you hide the taskbar and recall it only when you need it.

To hide the taskbar, right-click an open area of the taskbar and select Properties from the pop-up menu. When the Taskbar and Start Menu Properties dialog box appears, select the Taskbar tab. Check the Auto-Hide the Taskbar option, then click OK. This configuration automatically hides the taskbar so your applications can use the entire screen. To display the taskbar again, all you have to do is move your cursor to the very bottom of the screen. The taskbar then pops up for your use.

ON THIS DAY: FIRST GASLIT STREET (1807)

Way back on January 28, 1807, London's Pall Mall became the first street of any city in the world to be illuminated by gaslight. In 1809, Frederick Albert Windsor founded the Gas Light and Coke Company, England's first public gas company, with a primary interest in street lighting.

BLOG OF THE WEEK: LONGHORNBLOGS.COM

Interested in the next release of Windows, codenamed Longhorn? (Actually, Microsoft has now given it a real name: Windows Vista.) Check out LonghornBlogs.com, which features all the news that's fit to print about the upcoming operating system. Find it at www.longhornblogs.com.

January 30, 2006

MONDAY

THIS WEEK'S FOCUS
WINDOWS TWEAKS

SHOW ALBUM ART IN MY MUSIC FOLDERS

This tweak is actually fairly easy to do—no Registry editing requiring. It lets you find and display album art for all the music you have stored in your My Music folders.

First, you have to find the album art. You can use a third-party program for this (like my fav, Album Art fixer), or just do it yourself by searching and downloading the art from Amazon.com. (Google Image Search is also good for this, on the rare occasion Amazon doesn't carry the album you're looking for.)

When you find the album art, download it into the appropriate album folder within the My Music folder. But you're not done yet; you have to rename the graphics file folder.jpg. That's right, you create a separate folder.jpg file for each album folder.

Now it's a simple matter to display the artwork when you view the folders. Start by going to the parent My Music folder and selecting **View**, **Thumbnails**. Now do the same for each artist subfolder. The result will be mini album previews whenever you open each folder. Cool!

ON THIS DAY: DOUGLAS ENGLEBART BORN (1925)

On this day in 1925, Douglas Englebart was born. Just who is Douglas Englebart, you ask? Just the guy who invented the computer mouse is all. He first publicly demonstrated the mouse at a conference in 1968, where he also showed off his work in hypermedia and video conferencing. Not a bad legacy!

FACT OF THE WEEK

Recent research by the Stanford Institute for the Quantitative Study of Society shows that the average Internet user spends three hours a day online. Approximately a third of that online time occurs at work; not surprisingly, unemployed Internet users spend more time online than other users. (BTW, the Internet remains primarily a communications medium, with 57% of all time online devoted to email, instant messaging, or chat rooms.)

January 31, 2006

TUESDAY

THIS WEEK'S FOCUS
WINDOWS TWEAKS

REMOVE THE SHORTCUT ARROW FROM DESKTOP ICONS

Don't like those ugly little arrows that appear next to your desktop shortcut icons? Then delete them with this neat little Registry hack. Here's how it works:

1. Use RegEdit to open the Windows Registry.
2. Navigate to the following key: HKEY_CLASSES_ROOT\lnkfile.
3. Find the value IsShortcut and then delete it.
4. Now navigate the following key: HKEY_CLASSES_ROOT\piffile.
5. Find the IsShortcut value and then delete it.

That's it—no more ugly arrows! (You might need to reboot your PC for the changes to take effect.)

ON THIS DAY: FIRST PRIMATE IN SPACE (1961)

On this day in 1961, NASA launched a four-year-old male chimpanzee named Ham on a Mercury-Redstone 2 rocket. During his 16 1/2-minute suborbital flight, Ham experience about seven minutes of weightlessness, reached an altitude of 108 miles, and a speed of 13,000 MPH. He also performed some simple tasks, such as pulling levers when a light came on (for a reward of banana pellets). On re-entry, Ham was recovered safely 1,425 miles downrange.

GADGET OF THE WEEK: SCOTTEVEST CLOTHING

If you're a true gadget geek, you carry around a bunch of different gadgets on your person—a cell phone, PDA, iPod, GPS unit, digital camera, and who knows what else. The SCOTTeVEST line of technology-enabled clothing not only gives you lots of pockets for all your gadgets (30 or more per item, typically), but also has a series of holes and passages that let you string all your gadgets' wiring through the clothing's lining. Do your shopping at www.scottevest.com.

February 2006

February 2006

SUNDAY	MONDAY	TUESDAY	WEDNESDAY	THURSDAY	FRIDAY	SATURDAY
			1 1893 Construction of World's first movie studio begins in New Jersey	**2** Groundhog Day 1982 *Late Night with David Letterman* debuts	**3** 1966 U.S. launches first operational weather satellite, ESSA-1	**4** 1789 George Washington is elected first president of U.S.
5 1974 U.S. space probe Mariner 10 returns first close-up photos of the cloud structure surrounding Venus	**6** 1944 First successful fertilization of a human egg in a test tube	**7** 1964 Beatles arrive in U.S. for the first time	**8** 1928 First trans-Atlantic transmission of a televised image made between Purley, England and Hartsdale, New York	**9** 1964 Beatles makes first appearance on the *Ed Sullivan Show*	**10** 1961 The Styrofoam cooler is invented	**11** 1928 The La-Z-Boy recliner is invented, eventually being offered to the public in 1929
12 1941 First test-injection of penicillin on a human	**13** 1923 Chuck Yeager, first pilot to break the sound barrier, is born	**14** 1990 U.S. space probe Voyager I takes the first photograph of the entire solar system	**15** 1954 Scientists achieve an ocean exploration depth record of 13,287 feet (2 1/2 miles) in the Atlantic Ocean	**16** 1923 King Tutankhamen's tomb in Thebes, Egypt is opened	**17** 1869 Periodic Table of Elements is created by Dmitri Mendeleev	**18** 1977 First space shuttle orbiter, the Enterprise, is flight tested atop a 747 jet
19 1878 Phonograph patented by Thomas Edison	**20** President's Day 1986 Soviet Union launches space station Mir	**21** 1931 Alka Seltzer is introduced, much to the joy of over-indulgers everywhere	**22** 1857 Radio innovator Heinrich Hertz born	**23** 1997 First sheep cloned in Scotland, the first mammal ever successfully cloned from a cell from an adult animal	**24** 1991 Gulf War U.S. ground attack begins in Kuwait and Iraq	**25** 1913 The 16th Amendment is ratified, paving the way for income tax collection
26 1829 Levi Strauss is born	**27** 1900 Aspirin is patented in the U.S. by Felix Hoffman	**28** 1983 Last episode of M*A*S*H airs	**29**	**30**	**31**	

February 1, 2006

WEDNESDAY

THIS WEEK'S FOCUS
WINDOWS TWEAKS

CLOSING STUCK PROGRAMS IN XP

As robust as Windows XP is, especially compared to older versions, there still are occasions where a program freezes on you. This is typically a less serious situation than with older versions of Windows, when a stuck program could bring down your entire system. Now when a program freezes, it seldom affects anything else. All you have to worry about is closing the stuck program.

The first thing to do is press Ctrl+Alt+Del (what we old-timers call the "three-fingered salute") to display the Windows Task Manager. (You can also get to it by right-clicking the taskbar.) This is a souped-up version of the old Close Program dialog box found in previous versions of Windows, and it offers a lot more functionality.

When the Windows Task Manager opens, select the Applications tab, which displays a list of all programs currently running on your system. If a program is frozen, this might be indicated in the Status column. Or it might not. In any case, you want to select the frozen program, and then click the End Task button. Nine times out of ten, this should take care of your problem.

If this doesn't make the program close, go back to the Windows Task Manager and click the Processes tab. Find the file that's frozen, select it, and then click the End Process button. By the way, if you're forced to use the "end processes" method to shut down a program, you probably should restart Windows to clear up any loose ends still floating around system memory.

The worst-case scenario is that you can't close the frozen program no matter what you try, and it then starts to affect the rest of your system. If this happens to you, you need to restart Windows. If things get so bad you can't restart the operating system, you must reboot your computer by pressing Ctrl+Alt+Del twice.

ON THIS DAY: JAVA DEVELOPMENT STARTS (1991)

On the first day of February, 1991, Sun Microsystems initiated development of Java programming technology. The three members of the initial development team were Mike Sheridan, James Gosling, and Patrick Naughton.

HARDWARE OF THE WEEK: USB MINI-AQUARIUM

What you have here is a miniature fake aquarium, completely powered from your computer's USB port. The aquarium comes with the Plexiglas tank, two "fish," and a USB cable. Just add water to the tank, attach it to your computer via USB, and watch the fake fish swim around. Buy it for $19 at www.usbgeek.com.

February 2, 2006

THURSDAY

THIS WEEK'S FOCUS
WINDOWS TWEAKS

ADD A BACKGROUND GRAPHIC TO THE IE TOOLBAR

Here's a cool tweak that any Internet Explorer user can implement. It adds a graphic of your choosing (in BMP format) as the background for the Internet Explorer toolbar.

1. Use RegEdit to open the Windows Registry.
2. Navigate to the following Registry key: HKEY_CURRENT_USER\Software\Microsoft\Internet Explorer\Toolbar.
3. Add a new string value named BackBitmapIE5.
4. For the value of the new string, enter the name and location of the desired BMP file.

ENABLE WINDOWS XP FILE ENCRYPTION

Some users worry about their employers spying on the files stored on their work computers. It's not paranoia to worry about this; all companies have a legal right and, in many cases, an obligation to do so. For that reason, I recommend encrypting any data you want to keep private. (This is also a good idea for home computers you share with curious family members.)

If you're using Windows XP and your disk is formatted with NTFS, you can use the system's built-in encryption to keep others—even the system administrator—from reading your files. Here's how to do it:

1. Open My Computer and navigate to the folder you want to encrypt.
2. Right-click the folder and select Properties from the pop-up menu.
3. When the Properties dialog box appears, select the General tab and click the Advanced button.
4. When the Advanced Attributes dialog box appears, check the Encrypt Contents to Secure Data option, then click OK. This will keep prying eyes out of your private files!

ON THIS DAY: FIRST USE OF LIE DETECTOR (1935)

On this day in 1935, detective Leonard Keeler conducted the first use of his invention, the Keeler polygraph—otherwise known as the lie detector. Keeler used the lie detector on two criminals in Portage, Wisconsin, who were later convicted of assault when the results of the lie detector were introduced as evidence.

SOFTWARE OF THE WEEK: TWEAK UI

If you're serious about tweaking Windows XP, you need Microsoft's very own tweakers tool—the venerable Tweak UI. This official utility lets you make all sorts of little changes to Windows, without having to access the Registry. Download it for free at www.microsoft.com/windowsxp/downloads/powertoys/xppowertoys.mspx.

February 3, 2006

FRIDAY

THIS WEEK'S FOCUS
WINDOWS TWEAKS

ACTIVATING XP SPECIAL EFFECTS

Windows XP includes all sorts of visual special effects that affect the way certain elements look, or the way they pull down or pop up onscreen. Some of these special effects can be changed from the Display Properties utility. Others are changed from the System Properties utility.

You access the Display Properties utility by clicking the Display icon in the Control Panel. When the utility opens, select the Appearance tab and click the Effects button. Here you can choose to make menus and ToolTips scroll or fade in and out; display drop shadows under all Windows menus; display large icons on your desktop; display the contents of windows when they're being dragged; or hide those underlined letters on menu items.

The second set of special effects is available from the System Properties utility, which you get to by clicking the Systems icon in the Control Panel. When the utility opens, select the Advanced tab and click the Settings button (in the Performance section). When the Performance Options dialog box appears, click the Visual Effects tab and choose which effects you want. Here you can choose to animate windows when minimizing and maximizing; display a gradient color in window captions; enable per-folder type watermarks; fade in the taskbar; fade out menu items; fade/slide menus and ToolTips; slide taskbar buttons; smooth the edges of screen fonts; smooth-scroll list boxes; use Web view in folders; and much more.

Note that if you have a slower PC, enabling some of these special effects will make your PC run slower. If you experience a performance hit, experiment with turning off some of these effects. But if your system has the horsepower—especially enough RAM and a powerful enough video card—you should try turning on all these special effects. It really makes Windows XP a lot more fun to look at!

ON THIS DAY: FIRST U.S. WEATHER SATELLITE (1966)

On February 3, 1966, NASA launched its first operational weather satellite, ESSA-1, to provide cloud-cover photography to the U.S. National Meteorological Center. The satellite included two cameras; one camera could be pointed at some point on Earth every time the satellite rotated along its axis.

WEBSITE OF THE WEEK: TWEAKXP.COM

Dedicated Windows tweakers flock to TweakXP.com (www.tweakxp.com), a great site with tons of tweaks, tips, and downloads. Be sure you subscribe to the TweakXP newsletter, so you'll get the latest tweaks delivered directly to your email inbox.

February 4/5, 2006

SATURDAY/SUNDAY

THIS WEEK'S FOCUS
WINDOWS TWEAKS

CHANGE OWNER INFORMATION

When you first configured your Windows XP computer, you entered a username and organization. But what if you don't like that name—or want to give your old PC to someone else? Fortunately, a little Registry editing lets you change the owner and organization info on any Windows PC machine. Here's what to do:

1. Use RegEdit to open the Windows Registry.
2. Navigate to the following Registry key: HKEY_LOCAL_MACHINE\SOFTWARE\Microsoft\Windows NT\CurrentVersion.
3. Double-click the RegisteredOwner key and enter a new name.
4. Double-click the RegisteredOrganization and enter a new name.

ADDING PROGRAMS TO THE XP START MENU

The Start menu in Windows XP is a lot different than the Start menu you got used to in previous versions of Windows. It does a good job of hiding things you don't use that often, while keeping your most frequently used programs front and center.

If you don't like the way programs move on and off the Start menu, depending on when you last used them, you can choose to dock any icon to the Start menu—permanently. Just follow these steps:

1. From the Start menu, click the All Programs link to display the Programs menu.
2. Navigate to the program you want to add to the Start menu, and right-click that program.
3. From the pop-up menu, select Pin to Start Menu.

The program you selected now appears on the Start menu, just below the browser and email icons. To remove a program you've added to the Start menu, right-click its icon and select Unpin from Start Menu.

By the way, you can use this same method to add *any* program file to the Start menu, even if it doesn't appear on the Programs menu. Just use My Computer to navigate to an application file, and then right-click the file and select Pin to Start Menu from the pop-up menu.

ON THIS DAY: FIRST REMOVABLE SHIRT COLLAR (1825)

Okay, it seemed high-tech at the time. On February 5, 1825, housewife Hannah Lord Montague took her scissors and created the first detachable collar on one of her husband's shirts, in order to reduce her laundry load to the collar only. Her husband, Orlando, showed off his wife's invention to other men around town, and soon mass-market collars were being manufactured for sale around the world.

BLOG OF THE WEEK: TECHWHACK

TechWhack is a blog that assembles all manner of tech-related news and information. I especially like the Tips and Tricks section at the bottom of the home page. Check it out for yourself at www.techwhack.com.

February 6, 2006

MONDAY

THIS WEEK'S FOCUS
BLOGGING

WHAT'S A BLOG?

If you've been on the Internet for any length of time, you're probably familiar with message boards and discussion groups, where users post and respond to messages in organized threads. Well, a blog—short for "web log"—is like a message board, but a little more owner-driven. The typical blog is a personal website that is updated frequently with commentary, links to other sites, and anything else the author might be interested in. Many blogs also let visitors post their own comments in response to the owners' postings, resulting in a community that is very similar to that of a message board.

There are literally hundreds of thousands of blogs on the web, covering just about any topic you can think of. To find a particular blog, check out the following blog directories:

- Blog Search Engine (www.blogsearchengine.com)
- Blogwise (www.blogwise.com)
- Daypop (www.daypop.com)
- Feedster (www.feedster.com)
- Globe of Blogs (www.globeofblogs.com)
- Weblogs.com (www.weblogs.com)

And if you want to see a good example of a blog, look no further than my own Laporte Report blog, located at www.leoville.com/blog/. It's a daily assemblage of news and comments for the technology junkie—check it out!

ON THIS DAY: GOLF ON THE MOON (1971)

On February 6, 1971, Alan Shepard pulled out a six-iron and took aim at a couple of golf balls. No big deal, except he was walking on the moon at the time, as part of the Apollo 14 mission. The first ball landed in a nearby crater, while the second ball traveled "miles and miles and miles," thanks to the 1/6th lunar gravity.

FACT OF THE WEEK

Blogs had a big impact on the presidential election of 2004—and the views were fairly evenly divided. In the month prior to the election (September 27 through October 31, 2004), The Pew Internet & American Life Project (www.pewinternet.org) said 37% of blog postings were conservative in nature, 39% were liberal, and the balance were neutral.

February 7, 2006

TUESDAY

THIS WEEK'S FOCUS
BLOGGING

WHY BLOG?

Blogging takes time. You have to create the blog, and then keep it updated on a fairly frequent basis. Trust me, I run my own blog (at leo.typepad.com/tlr/) and I update it daily. It's a lot of fun, but it can also be a bit of a chore.

Some people view their blogs as a kind of personal-yet-public scrapbook—an online diary to record their thoughts for posterity. Even if no one else ever looks at it, it's still valuable to the author as a repository of thoughts and information they can turn to at any later date.

While some blogs are completely free-form, most blogs have some sort of focus. For example, my blog is all about technology because that's what I'm interested in. Other bloggers write about music, or politics, or cooking, or whatever *they're* interested in. Their blogs include their thoughts on the topic at hand, as well as links to interested news articles and websites.

Other people blog for a cause. Liberal blogs link to left-leaning stories and pages; conservative blogs contain commentary and links that reinforce their right-leaning viewpoints. There are blogs for every point on the political spectrum, and some you've never thought of.

HOW BLOGS ARE ORGANIZED

A blog is organized much differently than a normal web-site. Instead of the standard home page plus subsidiary page structure, a blog typically has just a single page of entries. This main page contains the most recent posts, and might require a bit of scrolling to get to the bottom. There's no introductory page; this main page serves as both introduction and primary content.

Older posts are typically stored in the blog archives. You'll normally find a link to the archives somewhere on the main page; there might be one huge archive, or individual archives organized by month.

The postings themselves are arranged in reverse chronological order. That means that the most recent posting is always at the top of the page, with older postings below that. Comments to a posting are typically in normal chronological order; you might have to click a link to see a separate page of comments.

ON THIS DAY: FIRST UNTETHERED SPACEWALKS (1984)

On this day in 1984, the first untethered spacewalks were made by *Challenger* astronauts Bruce McCandless II and Robert L. Stewart. The astronauts used manned maneuvering unit (MMUs) to check out equipment, maneuver within the cargo bay, and collect engineering data—while completely unattached from the shuttle itself.

GADGET OF THE WEEK: PRECISION SHOTS LASER SLINGSHOT

Combine stone-age weaponry with 21st-century technology and you get this fun little gizmo—a slingshot with a laser sight. The laser lets you accurately hit objects up to 150 feet away, which is pretty impressive. Buy the optional shotgun pouch and use it to shoot a spray of BBs! It's just $69.95, at www.catsdomain.com.

February 8, 2006

WEDNESDAY

THIS WEEK'S FOCUS
BLOGGING

CREATING YOUR OWN BLOG

If you want to create your own personal blog, the easiest way to go is to use a blog hosting community. These host sites offer easy-to-use tools to build and maintain your blog, and do all the hosting for you—often for free. Creating your own blog is typically as simple as clicking a few buttons and filling out a few forms. And once your blog is created, you can update it as frequently as you like, again by clicking a link or two.

If you want to create your own blog, consider Blogger (www.blogger.com), one of the first and the biggest bloghosting services on the Web. It's a good choice if you want to create your own blog; basic blogging is free. (Note that Blogger was recently acquired by Google.)

I would also be remiss if I didn't mention TypePad (www.typepad.com), which I use to host my own blog. Like Blogger, TypePad offers quick and easy blog hosting. Basic hosting is $4.95 per month, although the first month is free.

Many traditional Web hosting sites also offer blog hosting. For example, Tripod (www.tripod.lycos.com) offers a Blog Builder tool as part of all its hosting plans, including its free plan. Not a bad deal.

BEST SITES FOR BLOGGING

Whether you want to start up your own blog or read the blogs of others, there are several blog portals on the Web that deserve your attention. Most of these portals offer searchable directories of tens of thousands of individual blogs. Some of the best blog portals include

- Blog Universe (www.bloguniverse.com)
- BlogChalking (www.blogchalking.tk)
- Blogwise (www.blogwise.com)
- Eatonweb Portal (portal.eatonweb.com)
- Globe of Blogs (www.globeofblogs.com)
- Weblogs.Com (www.weblogs.com)

ON THIS DAY: HARVARD MARK I PATENT (1945)

On February 8, 1945, four engineers filed a calculator patent for the Automatic Sequence Control Calculator (ASCC), commonly known as the Harvard Mark I computer.

HARDWARE OF THE WEEK: AEROCOOL AEROBASE UFO GAMING PAD

This odd little accessory is a mousepad that looks like a flying saucer. It connects to your PC via USB, and it has 11 blue LEDs around its perimeter that flash in 15 different sequences. Buy it for $18 at www.aerocool.us.

February 9, 2006

THURSDAY

THIS WEEK'S FOCUS
BLOGGING

EXPLORING THE BLOGOSPHERE

When you take all the blogs on the web together, you get something called the *blogosphere*. It's important to think of the blogosphere as separate from the web, because of all the interlinking going on. Look at any blog, and you're likely to see a list of related blogs (sometimes titled "friends of..."). Bloggers like to link to other blogs that they like.

In fact, a lot of blogs are nothing more than links to interesting blog entries. (And sometimes relevant news articles.) The blogger finds something interesting, and then uses his own blog to draw attention to that other posting. In this way, bloggers are a lot like radio disc jockeys, "spinning" links and snippets the same way a DJ spins songs. In this way, you can use the blogosphere to keep up to date on important news and commentary.

The best bloggers, like the best DJs, do more than just repeat information you can find elsewhere. They not only sort through the blogosphere to find the most interesting articles, they also provide some background and organization to these postings, and in many cases add their own commentary. The best blogs have a definite point of view, no matter what content they're linking to.

The way to get the most efficient use of the blogosphere is to find one or two bloggers that you really like, and then use those blogs as a kind of guide to the rest of the blogosphere. Let the bloggers lead the way—and be prepared to spend some time jumping from link to link!

ON THIS DAY: DAVID WHEELER BORN (1927)

David Wheeler, programmer supreme, was born on this date in 1927. He is most famous for his work on assembly language, where he pioneered the so-called Wheeler Jump technique. For his work, Wheeler received the IEEE Computer Society Pioneer award.

SOFTWARE OF THE WEEK: NEWSGATOR

NewsGator is a news aggregator add-in that runs within Microsoft Outlook. Use it to subscribe to various RSS news feeds—including many blogs—and have news from those sites delivered directly to Outlook folders. Try it for free, or buy it for $29 from www.newsgator.com.

February 10, 2006

FRIDAY

THIS WEEK'S FOCUS
BLOGGING

POLITICAL BLOGS

The blogosphere is jam-packed with would-be political pundits. Whether you're left-wing, right-wing, or somewhere in the middle (is anyone in the middle anymore?), you can find dozens of blogs to either reinforce your preconceived opinions or challenge your mindset, as the case may be.

On the left side of the aisle, you should check out The Agonist (scoop.agonist.org), Daily Kos (www.dailykos.com), The Decemberist (markschmitt.typepad.com/decembrist/), Electablog (www.electablog.com), Eschaton (www.atrios.blogspot.com), Stage Left (www.stageleft.info), and Talking Points Memo (www.talkingpointsmemo.com).

On the right side of the aisle, some of the most popular blogs include AndrewSullivan.com (www.andrewsullivan.com), Ankle Biting Pundits (www.anklebitingpundits.com), Captain's Quarters (www.captainsquartersblog.com), INDC Journal (www.indcjournal.com), Instapundit (www.instapundit.com), Little Green Footballs (www.littlegreenfootballs.com/weblog/), and Power Line (www.powerlineblog.com).

Then there are those middle-of-the-road blogs that strive for a bit of independence from party politics—or aim to skewer both parties equally, such as PoliBlog (www.poliblogger.com) and the always fun and gossipy Wonkette (www.wonkette.com). And you can find a good listing of political blogs of all stripes at eTalkingheads (directory.etalkinghead.com), which organizes its directory by idealogy—conservative, liberal, libertarian, independent, and so on.

ON THIS DAY: WILHELM ROENTGEN DIES (1923)

Wilhelm Conrad Roentgen was the German physicist who discovered X-rays. He received a Nobel Prize for Physics (in 1901) for his discovery, and died on this date in 1923.

WEBSITE OF THE WEEK: THE WEBLOG REVIEW

The Weblog Review is a website that reviews weblogs. That's pretty simple; users submit their blogs for review, and the TWR editors review them. Check it out at www.theweblogreview.com.

February 11/12, 2006

SATURDAY/SUNDAY

THIS WEEK'S FOCUS
BLOGGING

MOBLOGS AND PHOTO BLOGS

Bloggers can post just about anything on their blogs, including text comments, links to other blogs, and even pictures. In fact, some blogs are nothing but pictures, often taken with camera phones. These blogs document the blogger's life in pictures, much as a standard blog documents things in words.

A picture blog that primarily uses digital photos from camera phones is called a *mobile blog* or *moblog*. Picture blogs devoted to quality digital photography are called *photoblogs*.

Most blog hosting services let you easily upload your digital photos as blog entries, so creating a moblog or picture blog is a fairly easy thing to do. If you want a picture-specific blog host, check out BusyThumbs.com (www.busythumbs.com) or Textamerica.com (www.textamerica.com). These sites also let you browse through other bloggers' photos, which is always fun.

LEO'S FAVORITE BLOGS

With literally millions of blogs on the Web, it's tough to choose just a handful of favorites. I'm up to the task, however, so here are my favorite blogs in a number of categories.

- **News blogs**—Big Blog (bigblog.com), Plastic (www.plastic.com), Unknown News (www.unknownnews.net)
- **Tech blogs**—Boing Boing (www.boingboing.net), Gizmodo (www.gizmodo.com), Slashdot (www.slashdot.org), and Techdirt (www.techdirt.com)
- **Sports blogs**—BadJocks (www.badjocks.com) and Fanblogs (www.fanblogs.com)
- **Movie blogs**—GreenCine Daily (daily.greencine.com), DVD Verdict (www.dvdverdict.com), filmfodder (www.filmfodder.com), and Tagline (www.tagliners.org)
- **Music blogs**—Fresh Tuneage (www.freshtuneage.com), MP3blogs (www.mp3blogs.org), and People Talk Too Loud (www.peopletalktooloud.com)

ON THIS DAY: BOSTON COMPUTER SOCIETY FOUNDED (1977)

On February 12, 1977, computing enthusiast Jonathan Rotenberg founded the Boston Computer Society. Four people attended the first meeting of the group; membership eventually reached several thousand, and the BCS became an extremely influential force in the burgeoning personal computer community.

BLOG OF THE WEEK: PHOTOBLOGS.ORG

Photoblogs.org is a text blog about photoblogging, if that makes sense. It's a great source of information about photoblogs, as well as a kind of directory to the best photoblogs on the web. Check it out at www.photoblogs.org.

February 13, 2006

MONDAY

THIS WEEK'S FOCUS
PODCASTING

ALL ABOUT PODCASTS

Despite the name, podcasts don't necessarily have anything to do with iPods. A podcast is essentially a homegrown radio program, distributed over the Internet, that you can play on any portable audio player—iPods included.

Anyone with a microphone and a computer can create their own podcasts. That's because a podcast is nothing more than an MP3 file posted to the Internet. Most podcasters deliver their content via an RSS feed, which enables users to easily find future podcasts by subscribing to the podcaster's feed. The podcasts are then downloaded to the listener's portable audio player and listened to at the listener's convenience.

What kinds of podcasts are out there? It's an interesting world, full of all sorts of basement and garage productions. Probably the most common form of podcast is the amateur radio show, where the podcaster assembles a mixture of personally selected music and commentary, just like we used to get on local radio stations before Clear Channel took over the world. But there are also professional podcasts by real radio stations and broadcasters, interviews and exposés, and true audio blogs that consist of running commentary and ravings. The variety is staggering, and the quality level ranges from embarrassingly amateurish to surprisingly professional.

How do you find podcasts? Fortunately, there are several podcast directories on the web, which I'll discuss later this week. In addition, Apple has added podcast subscriptions to its iTunes software, which makes things that much easier for iPod users.

ON THIS DAY: FRANCE GOES ATOMIC (1960)

The French joined the nuclear club on this date in 1960, detonating a plutonium bomb in what was then French Algeria, in the Sahara desert.

FACT OF THE WEEK

The Pew Internet and American Life Project (www.pewinternet.org) reports that 29% of Americans who own iPods or other MP3 players have listened to podcasts. (That's 6 million podcast listeners, if you do the math.)

February 14, 2006

TUESDAY

THIS WEEK'S FOCUS
PODCASTING

PREPARING TO PODCAST

What do you need to create your own podcasts? It's a short list, really.

First, you need a microphone. Any type will do, even the cheapie one that came with your PC. Second, you need a PC. Third, you need some sort of audio recording software. Fourth, it helps to have a set of headphones, although that isn't strictly necessary. Fifth, you need an Internet connection, so you can post your podcasts to the web. Finally, you need someplace to post them—some sort of audio blog host or RSS syndicator.

Before you start recording, it helps to make sure everything actually works. You'll want to connect the microphone to your PC, which is easy enough if you're using a standard low-end PC-type mic. If you're using a higher-end professional mic, you might need to invest in a mixer or microphone pre-amp and the appropriate interface to your PC. Obviously, you want to set volume levels before you press the Record button.

Next up, you need to prepare your recording software. I like Audacity, which I'll talk about later this week, but any audio recording program will do. Some really serious podcasters use professional recording/mixing software, such as Cubase, Cakewalk, or even ProTools, although this can be overkill. After all, you're going to be distributing your work in low-fi MP3 files, so anything too fancy gets lost in the mix.

When you go to save your podcast, you'll need to choose the sample rate and file format. For your original audio file, I recommend choosing the highest sample rate possible and to save to uncompressed WAV format. (Whether you record in mono or stereo depends on what type of podcast you're creating.) After you've made and edited your recording, you'll downmix to a lower bit rate MP3 file, but you want to start with the highest quality audio you can.

ON THIS DAY: ENIAC REVEALED (1946)

On Valentine's Day, 1946, John Mauchly and J. Presper Eckert unveiled the ENIAC computer at the University of Pennsylvania. The ENIAC could perform 5,000 calculations per second—a thousand times faster than previous computers.

GADGET OF THE WEEK: EDIROL R-1 PORTABLE MUSIC RECORDER

If you want to create your podcasts on the go, consider investing in a portable audio recorder, such as the Edirol R-1. This is a high-quality handheld audio recorder that records and plays back at any of nine quality levels, ranging from 64Kbps compressed MP3 format to uncompressed 24-bit linear WAV. Maximum recording time is 137 minutes (in 64Kbps MP3 mode), using the included 64MB memory card. Buy it for $450 at www.edirol.com.

February 15, 2006

WEDNESDAY

THIS WEEK'S FOCUS
PODCASTING

RECORDING A PODCAST

Making a podcast recording doesn't have to be any more complex than setting the volume levels, pressing the Record button, and then talking into the microphone. Naturally, you can stop and restart the recording as necessary; you can even go back and rerecord any section that you don't like. Most recording programs also let you edit your recordings, by snipping out unwanted sections and moving sections around as necessary.

If you want to incorporate music into your podcast, you'll need to import it into your master recording as a separate digital audio file. In most cases, it's not that difficult. Most programs let you mix multiple audio streams, so you have separate tracks for your voice, imported music, sound effects, and the like.

When you've finished your basic recording, you can then incorporate all manner of audio special effects wizardry. Depending on the program you're using, you might be able to add a little reverb to your voice (to give it more presence), do some equalization (for that professional FM announcer sound), and even edit out any pops, crackles, and noise.

Your original recording should be saved in high-quality WAV format, and you should stay in the WAV format throughout the editing process. When you have your podcast in its final form, you then export the file into MP3 format. If the podcast is voice-only, a relatively low bit rate (32 or 64Kbps) is fine. If the podcast has a lot of music, consider a higher bit rate, up to 128Kbps. Be sure you add the appropriate metatags for all the podcast info, and it's ready for distribution.

ON THIS DAY: MUSTARD ARRIVES (1758)

On this date in 1758, mustard was first advertised for sale in America, by Benjamin Jackson of Pennsylvania. In the *Philadelphia Chronicle*, Jackson claimed to prepare "the genuine Flour of Mustard-seed, of all Degrees of Fineness, in a Manner that renders it preferable to the European… and it will keep perfectly good any reasonable Time, even in the hottest Climates, and is not bitter when fresh made." Now pass the ketchup….

HARDWARE OF THE WEEK: GRIFFIN RADIO SHARK AM/FM RADIO/RECORDER

Even with all the music available on the Internet, sometimes you still want to listen to traditional AM or FM radio. Now you can add radio reception to your PC or Mac, thanks to Griffin's radio SHARK. It's an add-on radio antenna and receiver that connects to your computer via USB, and it also functions as a radio recorder (so you can time-shift your favorite radio programming). It's easy to use and, with it's Jaws-fin design, even easier on the eyes. Buy it for $69.99 from www.griffintechnology.com.

February 16, 2006

THURSDAY

THIS WEEK'S FOCUS
PODCASTING

PUBLISHING YOUR PODCAST

Once you've saved your podcast in MP3 format, you have to get it out on the Internet. This is a two-step process.

First, you have to upload the MP3 file to a server. If you have your own personal website, you can use that server to store your podcasts. You'll need a fair amount of storage space because audio files can get rather large, depending on the recording quality and length. For example, a 30-minute podcast saved at 64Kbps will be about 8MB in size. Use a higher bit rate and the file size goes up accordingly.

If you don't have your own server, consider using an audio blog hosting service, such as Audio Blog (www.audioblog.com), Liberated Syndication (www.libsyn.com), Ourmedia.org (www.ourmedia.org), and Podbus.com (www.podbus.com). You'll pay $5-$10 per month for file storage, and most of these sites will also help you with the RSS syndication of your podcasts.

Which leads us to the second step of the process, the RSS syndication. This is best accomplished via an audio blog hosting service blogging software. Most blogging software and services can generate an RSS feed, or you can use FeedBurner (www.feedburner.com) to do the work for you (for free). If you use FeedBurner, you'll have to create a link on your website to the FeedBurner file, so people can find the feed.

And that's that. All you have to do now is wait for users to find your podcasts and subscribe to your feed.

LEARN MORE ABOUT PODCASTING

Interested in learning more about Podcasting? Never fear. Our friends at Que Publishing have got you covered. Pick up a copy of *Absolute Beginner's Guide to Podcasting*, by George Columbo. Available at bookstores everywhere!

ON THIS DAY: FIRST MAN-MADE DIAMONDS (1953)

On this day in 1953, Erik Lundblad of Swedish company ASEA subjected graphite to 83,000 atmospheres of pressure at 2000° centigrade for an hour. The result was the first man-made diamonds, about the size of grains of sand.

SOFTWARE OF THE WEEK: AUDACITY

My favorite audio editing software is Audacity. Its interface is simple enough for any novice to figure out, yet it also offers advanced editing features and special effects—which makes it perfect for creating all types of podcasts. Audacity is available for Windows, Mac, and Linux; download it for free from audacity.sourceforge.net.

February 17, 2006

FRIDAY

THIS WEEK'S FOCUS
PODCASTING

WHERE TO FIND PODCASTS

Want to jump into the wild, wonderful world of podcasts? Then you need to browse through a podcast directory and see what's there for the listening. Some of the most popular podcast directories include

- All Podcasts (www.allpodcasts.com)
- Digital Podcast (www.digitalpodcast.com)
- iPodderX Directory (www.ipodderx.com/directory/)
- Podcast Alley (www.podcastalley.com)
- Podcast Bunker (www.podcastbunker.com)
- Podcast Directory (www.podcastdirectory.com)
- Podcast.net (www.podcast.net)
- Podcasting News Directory (www.podcastingnews.com/forum/links.php)
- Podcasting Station (www.podcasting-station.com)
- PodcastPickle.com (www.podcastpickle.com)
- Syndic8 Podcast Directory (www.syndic8.com/podcasts/)

And then there's the iTunes Podcast Directory, located at www.apple.com/itunes/store/podcasts.html. We'll talk more about this one tomorrow, so stay tuned!

ON THIS DAY: FIRST CAR STARTER (1911)

Before 1911, when you wanted to start your car you got out the crank. General Motors changed all that by installing the first self-starter on a Cadillac, and soon cranks were a thing of the past. The self-starter was designed and patented by Clyde Coleman and Charles Kettering, two GM engineers.

WEBSITE OF THE WEEK: IPODDER

iPodder is a great website for anyone interesting in podcasts. First, it's a blog that tracks new and interesting podcasts and related news. Then there's the iPodder directory of podcasts by category, which makes iPodder a great home base for whatever you're interested in listening to. Check it out at www.ipodder.org.

February 18/19, 2006

SATURDAY/SUNDAY

THIS WEEK'S FOCUS
PODCASTING

PODCASTS AND ITUNES

Podcasting hit the mainstream when Apple released version 4.9 of its iTunes software, which adds podcast capability to your iPod. Now you can browse for and subscribe to podcasts directly from iTunes, and then easily sync them to your iPod.

With iTunes 4.9 installed, your iPod gets a new Podcasts menu item. This lets you dial up all your stored podcasts, and then play them back in any order. Easy as pie, and it's nice how Apple chose to keep podcasts separate from regular music downloads.

Along with iTunes' podcast capability, Apple established the iTunes Podcast Directory (www.apple.com/itunes/store/podcasts.html). Browse or search through the available podcasts, download the ones you like, and subscribe to the ones you want to hear again. Most of the podcasts here are relatively professional, including programs form ABC News, ESPN, and podcast guru Adam Curry. Best of all, all the podcasts from iTunes are free. (This might be the only thing you'll ever get free from Apple!)

When you subscribe to a podcast, iTunes will automatically check for updates and download new episodes to your computer. Naturally, the new podcasts are transferred to your iPod when it's next connected and synced. It's the easiest way I've found to check out the whole podcast phenomenon—as long as you're an iPod owner, that is.

If you're running iTunes but haven't yet upgraded to version 4.9, do it today!

ON THIS DAY: VACUUM CLEANER PATENTED (1901)

On February 18, 1901, Hubert Cecil Booth received a U.K. patent for a dust-removing suction cleaner. Prior cleaning machines used compressed air to blow the dirt around; Booth's machine was the first to suck it up, by creating a vacuum inside the device. Booth build his machine on a horse-drawn cart, with an engine driving a pump and a long hose extending into the house to be cleaned.

BLOG OF THE WEEK: ADAM CURRY'S WEBLOG

Former MTV VJ (remember VJs?) Adam Curry is known as the father of podcasting, and this is his blog. It's about all things Adam Curry, with a definite emphasis on the world of podcasting. Check it out at www.curry.com. Even if you're too young to remember VJs, mullets, or Adam Curry (or Martha Quinn, my personal favorite!), there's still plenty here to keep you entertained.

February 20, 2006

MONDAY

THIS WEEK'S FOCUS
EMAIL

POP VS. WEB-BASED EMAIL

There are two primary types of email accounts: POP and web-based. POP stands for *post office protocol* and is the standard type of email account you receive when you sign up with an ISP. You're assigned an email account, given an email address, provided with an email program to install, and told which email servers to use in your program's configuration.

To use POP email, you have to use a POP email program, such as Outlook or Outlook Express or Eudora. That email program must be configured to send email to your ISP's outgoing mail server (with a name that looks like mail.domain.com), and to receive email from your ISP's incoming mail server. If you want to access your email account from another PC, you'll have to use a similar email program and go through the entire configuration process all over again on the second computer. (BTW, many ISPs maintain two email servers, one for incoming mail and one for outgoing mail, typically with different addresses.)

You're not limited to using the "hard-wired" POP email offered by your ISP, however; you can also send and receive email from special web-based email services. These services—such as Hotmail, Yahoo! Mail, and Google's Gmail—enable you to access your email from any computer at any time, using any web browser.

If you use a PC at work or on the road, web-based email is a convenient way to check your email at any time of day, no matter where you are. You don't have to go through the same sort of complicated configuration routine that you have with POP email; all you have to do is go to the email service's website, enter your user ID and password, and you're ready to send and receive messages.

ON THIS DAY: ELEVATOR PATENTED (1872)

On this date in 1872, Cyrus W. Baldwin received a patent for a vertical geared hydraulic electric elevator, which had been installed in the Stephens Hotel in New York City. Safety devices were included to prevent a fall if the suspending devices failed.

FACT OF THE WEEK

Commtouch (www.commtouch.com) did a little digging and discovered that 24.7% of all email spam comes from Europe, while 24.2% comes from either Japan or Korea. Next up is the U.S., from where 22.8% of all spam originates. China (including Taiwan and Hong Kong) accounts for 14.4% of all spam, while just 6% of spam comes from South America.

February 21, 2006

TUESDAY

THIS WEEK'S FOCUS
EMAIL

EMAIL ACRONYMS AND PROTOCOLS

As you go about configuring your POP email program, you'll run into a variety of acronyms. These acronyms describe specific protocols that define how email is stored and transmitted. Although it's not necessary to have a thorough understanding of these protocols, it is nice to know what all the acronyms mean.

SMTP (Simple Mail Transfer Protocol) is the general protocol for transmitting all email across the Internet. You send email from your PC using SMTP; email sent from one server to another also uses SMTP.

When it comes to retrieving the email stored on your ISP's server, two different protocols can be used. With POP3 (Post Office Protocol 3), your email messages are stored on the ISP's server only until your email program retrieves them, at which time they are deleted from the server. With IMAP (Interactive Mail Access Protocol), you don't actually download messages to your computer; instead, you view your email messages as they're stored on the ISP's server. Both POP3 and IMAP protocols are supported by most major email programs.

To keep things straight, just remember that you use SMTP to *send* messages, but you use POP3 or IMAP to *retrieve* messages waiting for you on your ISP's mail server. Got that?

ON THIS DAY: FIRST FLYING CAR (1937)

Believe it or not, we've had flying cars for almost 70 years now—just not for general use. The very first combination automobile-airplane was first tested on February 21, 1937, by the Westerman Arrowplane Corporation of Santa Monica, California. The vehicle was dubbed the Arrowbile, and the Studebaker Corporation, which supplied the 100-horsepower engines, eventually took delivery of five Arrowbiles. Obviously, however, the Arrowbile never made it to market.

GADGET OF THE WEEK: VOISEC REFRIGERATOR VOICE RECORDER

Need to leave a message for your spouse or your kids—or to yourself? Then check out the VoiSec digital voice recorder. It's a small button-shaped device that you can stick on the front of your refrigerator; press the button to record a message, or lift the lid for playback. It sells for about $40 at www.voisec.se/pages/voisec.html.

February 22, 2006

WEDNESDAY

THIS WEEK'S FOCUS
EMAIL

HTML EMAIL

Many email messages are just plain text—because that's all you're sending, in most cases. But virtually all email programs can also send and receive fully formatted text, with backgrounds and colors and embedded HTML hyperlinks. Unfortunately, although HTML email is pretty, there are several potential problems with it.

First, adding HTML code to an email message increases the size of the message. Add too much code—and a few graphics—and you end up with a large message that takes a long time to receive over a dial-up connection. Send an HTML email and your recipient might get annoyed at you.

Second, many spam messages are sent in HTML format, so you can see pretty pictures of the product you're being pitched. For this reason, many spam filters automatically block HTML email. Send an HTML-formatted message and you risk it being blocked.

Third, some viruses are designed to be embedded in HTML code. No surprise, then, that some antivirus programs automatically flag or block HTML email—or can be configured to display the HTML email without the HTML formatting. Again, send an HTML email and you risk it being blocked.

HTML email blocking is also an issue with some content filtering software—including AOL's kid-safe features.

Send an HTML email to a kid, and it will almost certainly get blocked.

Bottom line? HTML email is nice, but it might not always get delivered. Be prepared to send messages in plain text format if you run into problems.

ON THIS DAY: AMERIGO VESPUCCI DIES (1512)

Spanish astronomer Amerigo Vespucci, who realized that Columbus had discovered a new continent, died on this date in 1512. Because of his work, they named the new continent Vespuccia after him. (Oh, sorry, I meant *America*, not Vespuccia.)

HARDWARE OF THE WEEK: SEAGATE EXTERNAL HARD DRIVES

My favorite external hard drives are those from Seagate. What I like most about Seagate's drives is their form factor. These 3.5-inch drives are compact and lightweight; you can position them vertically or lay them horizontally. And if you go the horizontal route, they're stackable! Seagate offers capacities up to 400GB—one of the largest external drives on the market today. Prices range from $250 for a 200GB model to $350 for the big 400GB drive; find out more at www.seagate.com.

February 23, 2006

THURSDAY

THIS WEEK'S FOCUS
EMAIL

MANAGING MASS MAILINGS IN OUTLOOK EXPRESS

Most of your email will go to just one or a handful of recipients. However, you can use any email program to manage large lists of recipients, effectively creating your own mini-mass mailings that you can initiate with a single click of your mouse. Here's how you do it in Outlook Express.

The key to managing a mass mailing is to create a *mailing group* in your address book or contact list. Once this group is created, you can compose a single message, address it to the group, and then send it on its way. Even though you had to enter only one "address" in the To field (for the group), your message will automatically be sent to all the individual recipients you included when you created the group address.

To create a mailing group in Outlook Express, click the Inbox icon in the Folder List (or select Go, Inbox), and then click the Address Book button (or select Tools, Address Book). When the Address Book window appears, click the New Group button. When the New Group dialog box appears, enter the name of the group in the Group Name box. You can then add individual names to the group, by clicking the Select Members button, selecting names from the Address Book list, and then clicking the Select button. Click OK when done.

When you're ready to send your mass mailing, create a new message and compose the text as normal. Then click the Address Book button on the toolbar; when the Select Recipients dialog box appears, select the name of your mailing group, click the To: button, and then click OK. Separate copies of your message will now be sent to every member in your mailing group.

ON THIS DAY: FIRST CLONED SHEEP (1997)

On this date in 1997, scientists in Scotland announced they had succeeded in cloning a female lamb named Dolly. She was the first mammal ever successfully cloned from a cell from an adult animal.

SOFTWARE OF THE WEEK: MOZILLA THUNDERBIRD

Just as I'm a big fan of the Mozilla Firefox web browser, I also like the companion email program, Thunderbird. This is a faster, safer, and slightly easier to use program than Outlook Express or Eudora, and (like Firefox) it's completely free. Download it at www.mozilla.org/products/thunderbird/.

February 24, 2006

FRIDAY

THIS WEEK'S FOCUS
EMAIL

TROUBLESHOOTING EMAIL PROBLEMS

The most common problem with email is an incorrect address. It's easy to mistype an address or get the wrong address from someone. If you're having trouble getting a message to another user, try to verify the email address with the recipient personally.

You can sometimes track down the cause of a bounced message by carefully deciphering the error message that accompanies the returned message. You'll often find specific reasons why the message was returned, which can help you formulate your reaction.

Sometimes a really large message (or a message with a large attachment) can take so long to retrieve that it times out the email program. (Any message more than 1MB in size could cause this sort of problem on a dial-up connection.) The really bad thing is, if you can't download this message, it clogs up your message retrieval so you can't grab any other message after this one, either. If you run into this sort of problem, call your ISP's voice support line and ask them to remove the extra-large message from the queue so you can retrieve the rest of your message.

It's also possible that your ISP has a message size limit. Some ISPs won't let you send or receive messages larger than 3MB or so. Check the help files at your ISP's website to see whether you're affected by this sort of artificial constraint—and if you are, learn how to either compress or break up your attachments into smaller messages.

ON THIS DAY: STEVE JOBS BORN

Steve Jobs bounced into this world on February 24, 1955. If you don't know who Steve Jobs is, you've never used an Apple computer; Jobs was the guy who created Apple (along with his pal Steve Wozniak, of course).

WEBSITE OF THE WEEK: GMAIL

I've long been a fan of web-based email services, having used both Hotmail and Yahoo! Mail for several years now. But far and away the best web-based email service today is Google's Gmail. It's a completely free service, offers 2GB of storage, and has a terrific user interface. Sign up at gmail.google.com.

February 25/26, 2006

SATURDAY/SUNDAY

THIS WEEK'S FOCUS
EMAIL

EMAILING ANONYMOUSLY

Every email you send includes a header (often hidden by user-friendly email programs) that contains your email address, your ISP's email server address, and other similar information. This data is sent automatically whenever you send an email, and is easily read by anyone on the receiving end—which means just about anybody can find out who you are, even if you change the visible email name and address.

To send truly anonymous email, all of this technical information needs to be stripped from your message headers. The easiest way to do this is to use a *remailer* service. Remailers essentially resend your email messages to your intended recipients, stripping out the header info so that the email can't be traced back to you.

Some remailers operate with your standard POP email program; others require you to create your messages on their websites using their own web-based email systems. In addition, some are free services and some are subscription based. All are fairly effective at what they do.

Some of the more popular anonymous email services include Anonymous.To (www.anonymous.to), BiKiKii Anonymous Remailer (bikikii-remailer.noneto.com), HushMail.com (www.hushmail.com), Offshore Mixmaster Anonymous Remailer (www.anon-remailer.gq.nu), Riot Anonymous Remailer (riot.eu.org/anon/), SecureNym (www.securenym.net), and W3 Anonymous Remailer (www.gilc.org/speech/anonymous/remailer.html).

ON THIS DAY: SGI BUYS CRAY (1996)

On February 26, 1996, Silicon Graphics, Inc., acquired Cray Research for $767 million. This made SGI the leading supplier of high-speed computers in the United States.

BLOG OF THE WEEK: GAWKER

Okay, so Gawker supposedly focuses on Manhattan-based media news and gossip, and you don't live in Manhattan. Big deal; Gawker is still one of the smartest, funniest blogs on the web, and it's not all that NY-centric, anyway. Take a look at www.gawker.com.

February 27, 2006

MONDAY

THIS WEEK'S FOCUS
UPGRADING YOUR PC

BEFORE YOU UPGRADE

Before you dive headfirst into upgrading your personal computer, it helps to know what you're getting yourself into. Here's a short checklist to complete before you add any new device to your system:

- Determine whether the peripheral you want to add will work on your system
- Assemble all necessary tools (important if you're performing an internal upgrade)
- Print out a system hardware report from the Windows Device Manager, so you'll know exactly what hardware is installed on your system
- Make a note of your system's key configuration settings
- Gather your original Windows installation CD or disks—assuming you have them
- Make a backup of your important data files
- Set a System Restore point
- Read the instructions of the item you want to install—then read them again (never assume you know everything you need to know!)

ON THIS DAY: DAVID SARNOFF BORN (1891)

Television pioneer David Sarnoff was born on this date in 1891. Sarnoff was the first general manager of RCA and founded the television network NBC. He steered the company into the world of television broadcasting—first in black and white, and then with color programming.

FACT OF THE WEEK

According to ITtoolbox (www.ittoolbox.com), the primary reason (36%) for a company to upgrade its PCs is that new applications require the upgrade. Another 32% upgrade due to efforts to improve staff productivity, while 12% of companies upgrade because the old machines broke down. Another 11% upgrade in a hope to reduce service costs, while just 3% upgrade because their old units are susceptible to security threats or viruses.

February 28, 2006

TUESDAY

THIS WEEK'S FOCUS
UPGRADING YOUR PC

UPGRADING SYSTEM MEMORY

The easy part of increasing your computer's memory is performing the installation. The hard part is determining what type of memory you need to buy. That's because there are many different types of memory available; you need to figure out what type of memory your computer uses to buy the right RAM for your system.

All memory today comes on modules that contain multiple memory chips; the capacity of each chip adds up to the total capacity of the memory module. The memory modules plug into memory sockets located on your PC's motherboard; installation is actually quite easy.

Your computer could be using one of three different types of memory modules—SIMMs, DIMMs, or RIMMs. You'll need to consult your PC's instruction manual (or look up your PC's model number in a manufacturer's cross-listing) to determine the type of module your machine uses.

Not only do you have to specify what type of memory module your system uses, you also have to specify what type of RAM chip is installed on the module. There are five primary types of memory chips in use today: Synchronous Dynamic RAM (SDRAM), Double-Data-Rate SDRAM (DDR SDRAM), Rambus DRAM (RDRAM), Fast-Page-Mode DRAM (FPM DRAM), and Extended Data Out DRAM (EDO DRAM). The differences between these chip types are technical and not of interest to most users. The key thing is to recognize what type of chip is used in your particular PC, and get the right chip for your system.

ON THIS DAY: CORE MEMORY PATENTED (1956)

On February 28, 1956, MIT's Jay Forrester was awarded a patent for his design for a "multicoordinate digital information storage device," otherwise known as core memory. Core memory became the standard memory device for computers until it was supplanted by semiconductor-based RAM in the mid-1970s.

GADGET OF THE WEEK: ULTRA ANTISTATIC WRIST STRAPS

Whenever you're working inside your PC, you need to take precautions against static electricity, which can fry those sensitive components. That's why I recommend Ultra's Antistatic Wrist Strap, which helps ground you against pesky electrical discharges. Buy it for $6.99 at www.ultraproducts.com.

March 2006

March 2006

SUNDAY	MONDAY	TUESDAY	WEDNESDAY	THURSDAY	FRIDAY	SATURDAY
			1 1692 Salem Witch Hunt begins	**2** 1972 Pioneer 10 launched to Jupiter	**3** 1847 Alexander Graham Bell born	**4** 1997 Human cloning research funding banned in U.S. by President Bill Clinton
5 1616 Copernican Theory is ruled "false and erroneous" by the Catholic church	**6** 1886 First AC power plant begins operation in Great Barrington, MA	**7** 1897 World's first cornflakes served to patients at a mental hospital in Battle Creek, MI	**8** 1972 The Goodyear blimp is flown for the first time	**9** 1858 Mailboxes patented in Philadelphia, PA	**10** 1964 First Ford Mustangs produced	**11** 1989 Cops—the first reality-based television show—premieres on FOX
12 1894 First bottles of Coca-Cola sold	**13** 1781 Uranus discovered	**14** 1879 Albert Einstein born	**15** 44 B.C. Julius Caesar is murdered	**16** 1966 First U.S. manned docking of two spacecraft, the Gemini VIII and Gemini Agena	**17** 1958 U.S. launches its first man-made object into space—the Vanguard I satellite	**18** 1834 First U.S. railroad tunnel completed in Pennsylvania
19 1831 First recorded bank robbery in U.S. history occurs in New York	**20** 1727 Sir Isaac Newton dies	**21** 1980 U.S. boycotts the Olympics being held in Moscow	**22** 1895 First motion picture shown on a screen in Paris, France	**23** 1983 Ronald Reagan initiates "Star Wars" program	**24** 1955 First seagoing oil drill rig is placed into service	**25** 1970 The Concorde airplane makes its first sound barrier-breaking flight
26 1941 Italy attacks the British fleet at Suda Bay, Crete, using the first manned torpedoes in existence	**27** 1961 First mobile computer—essentially a UNIVAC Solid-State 90 computer filling an entire van—is used	**28** 1979 Three Mile Island nuclear accident	**29** 1974 U.S. satellite Mariner 10 takes the first close-up pictures of Mercury	**30** 1858 The first pencil with an attached eraser is patented	**31** 1932 Ford Motor Company unveils its "V-8" engine, the first of its kind	

March 1, 2006

WEDNESDAY

THIS WEEK'S FOCUS
UPGRADING YOUR PC

ADDING NEW PORTS

Let's say you want to add a second printer to your system. Most PCs come with only a single parallel port, to which you probably have your existing printer connected. The answer, of course, is to add a second parallel port to your system. (Another solution is to buy a USB printer and connect to a free USB port; let's not ignore easy fixes!)

Or maybe you've become enamored of USB devices and completely filled all your system's USB connections. Buy one more USB-compatible peripheral, and then what do you do? Here you have a choice: You can increase the number of connections available by adding an external USB hub, or you can add an extra USB card to your system.

Here's another one. You want to connect your new digital video camcorder to your PC to edit your home movies but your camcorder connects via FireWire, and your PC doesn't have a FireWire connector. The solution? Add a FireWire port to your system unit.

Adding ports of any kind to your system is a simple internal upgrade. Just head down to your local computer store and purchase the appropriate port expansion card. (All different types are available.) Then power down your PC, remove the case, and insert the new card into any open card slot. Put it all back together and boot it up, and you have new ports available to use. Easy!

ON THIS DAY: DIRECT-DIAL TRANSATLANTIC PHONE SERVICE (1970)

On this day in 1970, direct-dial transatlantic phone service was initiated between the U.S. and Britain. The service was a joint effort between America's AT&T and the British Post Office (which at that time also handled the British telephone system).

HARDWARE OF THE WEEK: SUNBEAM 20-IN-1 SUPERIOR PANEL

Sunbeam's 20-in-1 Superior Panel lets you add some extra connections to the front panel of your system unit. It takes up a normal 5.25-inch drive bay and offers a combination of card readers (all popular formats), FireWire, USB (2), SATA 2, RCA audio, and headphone connections, along with a fan controller and cool blue LCD display. Buy it for $34.99 at www.sunbeamtech.com.

March 2, 2006

THURSDAY

THIS WEEK'S FOCUS
UPGRADING YOUR PC

UPGRADING FOR DIGITAL VIDEO EDITING

Do you make your own home movies? Then you'll want to turn your PC into a video editing studio—which is especially easy if you have a digital video (DV) camcorder. You might need to beef up your hardware a bit, however, because video editing is one of the most demanding applications for a computer system.

Assuming you have a fairly powerful microprocessor to start with, the first thing you'll probably need to upgrade is your hard disk. Raw digital video takes up about 3.6MB for each second of footage; work with an hour-long movie, and you'll fill up two-thirds of a 20GB disk. So a big—no, make that a *huge*—hard disk is a necessity. Some users simply add a second hard drive to their system, dedicated solely to video editing. Whatever you do, make sure it's a *fast* disk; choose an IDE drive with a 7,200 RPM spin rate, or (if you're flush) a SCSI drive. And when you install the drive, format it with the NTFS file system if you can.

Memory is also important. Lots of it. Like 1GB worth, at a minimum. Any less and you'll find your system slows down considerably when processing all that digital video data. And while we're on the subject of memory, it helps to have a video card with lots of onboard memory. For best performance, look for a card with 64MB or 128MB video RAM.

Finally, consider going with a big monitor—the easier to view your movies with. In fact, you might think about a dual-monitor system, so you can edit on one screen and view your results on the other.

ON THIS DAY: FIRST PUSH-BUTTON PHONE (1959)

On this day in 1959, an experimental push-button phone was tested by the Southern New England Telephone Company in New Haven, Connecticut. The test was designed to see whether customers dialed fewer wrong numbers using the new push-button design.

SOFTWARE OF THE WEEK: ALOHA BOB PC RELOCATOR

When you're migrating from an old computer to a newer one, the most difficult task is moving all your old files and settings from one machine to another. The task is made a little easier with Aloha Bob PC Relocator, a nifty little upgrade/transfer program. Buy it for $29.95 from www.alohabob.com.

March 3, 2006

FRIDAY

THIS WEEK'S FOCUS
UPGRADING YOUR PC

UPGRADING YOUR HARD DISK

When you go shopping for a new hard drive, you need to understand the relevant specifications in order to make an informed buying decision—and get the right hard drive for your system. This information, typically available somewhere on the drive's packaging, tells you how much data the disk can hold and how fast it can access that data. Here are the important specs:

- **Capacity**—Disk drive size is typically measured in gigabytes (GB). The bigger the drive, the higher the price.
- **Access time**—This is the amount of time it takes for the heads to locate a specific piece of data on the hard drive. Manufacturers typically specify the "average access time," because the actual seek time varies depending on the location of the heads and where the next bit of data is stored.
- **Spin and data transfer rates**—The spin rate spec measures the speed at which the platters spin, in revolutions per minute (RPM); faster is better. The data transfer rate is the speed at which the system copies data from the hard drive to your computer.
- **Drive interface type**—Several different interfaces are available that control the communication between your hard drive to your PC. The two primary interfaces in use today are ATA (sometimes called IDE) and SCSI; ATA is more popular and lower-priced, while SCSI is a tad faster.

ON THIS DAY: FIRST MEETING OF THE HOMEBREW COMPUTER CLUB

On March 3, 1975, the Homebrew Computer Club held its first meeting in a garage in Menlo Park, California. Founders Fred Moore and Gordon French hosted about 30 fellow hobbyists, who spent their first meeting discussing the Altair kit computer.

WEBSITE OF THE WEEK: TOM'S HARDWARE GUIDE

When you're talking PC hardware, your first stop should be Tom's Hardware Guide, the premiere site for all manner of hardware news and reviews. If Tom can't answer your questions, you're out of luck! Check it out at www.tomshardware.com.

March 4/5, 2006

SATURDAY/SUNDAY

THIS WEEK'S FOCUS
UPGRADING YOUR PC

USB: THE EASY WAY TO UPGRADE

The most common type of external connector on PCs today is the USB port. USB is a great concept in that virtually every type of new peripheral comes in a USB version. Want to add a second hard disk? Don't open the PC case; get the USB version. Want to add a new printer? Forget the parallel port; get the USB version. Want to add a wireless network adapter? Don't bother with Ethernet cards; get the USB version.

USB is so popular because it's so easy to use. When you're connecting a USB device, not only do you not have to open your PC's case, you don't even have to turn off your system when you add the new device. That's because USB devices are *hot swappable*. That means you can just plug in the new device to the port, and Windows will automatically recognize it in real time.

The original USB standard, version 1.1, has been around for awhile and, if your PC is more than three or four years old, is probably the type of USB you have installed. The newer USB 2.0 protocol is much faster than USB 1.1 and is standard on all new computers. USB 2.0 ports are fully backward compatible with older USB 1.1 devices. You want to use newer USB 2.0 connections when you're installing devices that transfer a lot of data, such as external hard drives.

To connect a new USB device, all you have to do is find a free USB port on the back of your system unit and connect the new peripheral. Windows should automatically recognize the new peripheral and either install the proper device driver automatically or prompt you to provide a CD or disk containing the driver file. Follow the onscreen instructions to finish installing the driver, and you're done. (The only variation on this procedure is if the peripheral's manufacturer recommends using its own installation program, which you should use if it exists.)

ON THIS DAY: STAPLER PATENTED (1868)

On March 5, 1868, C.H. Gould of Birmingham, England, received a patent for his design of a stapling device. And what would we do without our red Swingline staplers? (Trivia note: When the movie *Office Space* was in production, Swingline actually didn't make a red stapler; the production designer of the movie had to spray-paint a standard black Swingline red.)

BLOG OF THE WEEK: TECH BLOG

Want to find the latest tech-related news? Then check out Tech Blog (www.techblog.org), which assembles all the latest headlines for your clicking and reading pleasure.

March 6, 2006

MONDAY

THIS WEEK'S FOCUS
CHOOSING A BIG-SCREEN TV

HOW BIG IS BIG ENOUGH?

When you're choosing a new big-screen TV, know that bigger isn't always better. If the screen is too big, picture flaws (including interface lines and the "screen door" effect you see on some LCD projectors) will be more noticeable. The optimal screen size depends on how far away from the screen you'll be sitting, as well as what video sources you're using.

One important thing to consider is the aspect ratio of the screen—especially if you're moving from an old 4:3 ratio screen to a new 16:9 screen. Because screen size is measured diagonally from one corner to another, a 36-inch 16:9 screen will actually be a little shorter, top to bottom, than a 36-inch 4:3 screen. If you're moving from a traditional 4:3 ratio set to a 16:9 model, you'll need a screen that measures about 25% wider (diagonally) to maintain the same screen area for 4:3 programming.

You should also consider how far off-axis various chairs are in your viewing room. If you're sitting too far off-axis, consider going with a slightly bigger screen or with a technology with good off-axis viewing characteristics (direct view CRT or plasma) and avoiding those technologies with poor off-axis viewing (CRT rear projection and all front projection systems).

So what's the right size? For analog cable or standard definition broadcast signals shown on a 4:3 ratio display, the ideal screen size is four times the number of feet between you and the screen, expressed in inches. For high-definition broadcasts shown on a 16:9 ratio display, the ideal screen size is six times the number of feet between you and the screen, again expressed in inches. So, for example, if you're watching high-definition programming and sitting 10 feet from the screen, the ideal screen size is approximately 60 inches (6 × 10 feet, in inches).

ON THIS DAY: MICHELANGELO VIRUS STRIKES (1992)

On this day in 1992, the dreaded Michelangelo computer virus was set to strike. Experts predicted that as many as five million PCs were in danger of contracting the virus, set to erase data on the March 6 anniversary of the artist's birth. In reality, Michelangelo was a dud, infecting only a few thousand machines worldwide.

FACT OF THE WEEK

Digital television sales are on the rise. Research firm iSuppli (www.isuppli.com) says that 5.8 million digital sets were sold in 2004 and estimates sales of 12 million in 2005 and 19.8 million in 2006. For 2005, 26% of these sets will be traditional CRT, 28% will be rear projection, and 45% will be flat panel (either plasma or LCD).

March 7, 2006

TUESDAY

THIS WEEK'S FOCUS
CHOOSING A BIG-SCREEN TV

CRT PROJECTION

When you want a larger display for a bigger room (anything over 40 inches diagonal), you need to step up from a traditional direct view set to a rear-projection television (RPTV). RPTVs present a good value in terms of price and performance and come in a variety of different display technologies.

The lowest-cost RPTVs are powered by three small (7- to 9-inch diagonal) cathode ray tubes (CRTs). The process involves separating the video signal into three separate colors—red, green, and blue—that are then fed to three separate CRTs, one for each of those primary colors. The three CRTs are aimed at a reflecting mirror, typically located toward the bottom of the set, which reflects the picture onto the television's screen. The separate red, green, and blue pictures combine into the single full-color picture that you see on the screen.

Compared to other types of rear projection sets, a CRT projector has a few unique advantages. The chief advantage, of course, is the price: CRT projectors typically cost $500 to $1,000 less than similar-sized DLP or LCD projectors. Also attractive is the picture, which has a film-like quality not yet duplicated by other technologies.

On the downside, CRT projectors are big and bulky, taking up a lot more floor space than competing microdisplay projectors. CRT sets also output less light than other types of RPTVs, and those three CRTs require careful (and constant) convergence to keep the three colors aligned. Off-angle viewing is also somewhat problematic; if you have a chair or two sitting off the side of the screen, this might not be the right type of TV for you.

ON THIS DAY: CORNFLAKES FIRST SERVED (1897)

On this date in 1897, Dr. John Kellogg served the world's first cornflakes to his patients at a mental hospital in Battle Creek, Michigan. In 1906, Kellogg's brother, Will Keith Kellogg, added sugar to the recipe and began marketing cornflakes as a breakfast food.

GADGET OF THE WEEK: COUCH POTATO TORMENTOR

Here's a gadget for the spouse of the ultimate home theater enthusiast. The Couch Potato Tormentor is the "anti-remote," a small device that interferes with your television viewing. It randomly changes the channel your spouse is watching, which leads to all manner of hilarity (and possible spousal conflict). Buy it for $14.95 at www.couchpotatotormentor.com.

March 8, 2006

WEDNESDAY

THIS WEEK'S FOCUS
CHOOSING A BIG-SCREEN TV

DLP REAR PROJECTION

When a CRT rear projector won't work, consider a set based on the newer digital light projection (DLP) technology. DLP projectors are remarkably small and lightweight, thanks to the compact size of the necessary optical and electronic components. The largest manufacturer of DLP RPTVs today is Samsung.

The DLP system starts with the light generated by a high-performance projector lamp. The light from the lamp is projected through a condensing lens and shone through a rotating color wheel, which is composed of red, green, and blue segments. The color wheel spins rapidly to sequentially generate 16.7 million different colors from that single point of light.

The colored light now shines onto the digital micromirror device (DMD), which is an array of hundreds of thousands of tiny, independently hinged mirrors. There is one micromirror for each pixel in the display, all controlled by the DMD's computer processor. The mirrors tilt on or off to reflect light through a projection lens and onto the display's screen. A typical single-chip DMD generates a picture composed of 1280 × 720 pixels, although some newer models deliver 1920 × 1080 resolution, for a true 1080p high-definition picture.

DLP projectors have the brightest pictures of all RPTVs, with excellent black levels and shadow details. On the downside, because it's a partly mechanical system, because of that color wheel, there are moving parts that can break. In addition, some viewers can see "rainbows" caused by the rapidly shifting colors; if you're susceptible to this visual effect, consider choosing another display technology instead.

ON THIS DAY: VON ZEPPELIN DIES (1917)

On this date in 1917, Ferdinand Count von Zeppelin passed away; he was 79 years old. von Zeppelin was a German engineer, known primarily as a builder of rigid dirigible airships that became known as zeppelins.

HARDWARE OF THE WEEK: ATI HDTV WONDER

Add high-definition television to your PC with the ATI HDTV Wonder video card. This is a low-priced, high-performance video card that not only handles HDTV video, but also records via DVR functionality. Buy it for $149 at www.ati.com.

March 9, 2006

THURSDAY

THIS WEEK'S FOCUS
CHOOSING A BIG-SCREEN TV

LCD REAR PROJECTION

If you're susceptible to DLP "rainbows," consider an RPTV powered by LCD technology. Because LCD sets don't use a color wheel, there are no "rainbows" to see—and no moving parts.

An LCD projector starts with a similar bright beam of light, but this beam is projected onto and through a series of dichroic mirrors. These are mirrors that let one color of light (red, blue, or green) pass through, while reflecting all other colors. This creates three separate red, blue, and green beams, which are passed through separate LCD display chips.

Each LCD chip consists of a layer of liquid crystal material sandwiched between two plates of glass. The hundreds of thousands of liquid crystals are arranged in a grid pattern, representing picture pixels. When an electric charge is applied to the chip, individual crystals rotate the plane of polarized light, effectively acting as an on/off switch for each pixel of the picture. The colored light beams pass through the LCD display panels to create the screen image, which is then projected onto the display's screen.

The advantages of an LCD projector are the lack of rainbows, good brightness and color reproduction, and slightly more natural colors than with DLP sets. The downsides to LCD technology include limited contrast and black levels and possible "screen door" effects (the visible lines between LCD pixels). The largest manufacturer of LCD RPTVs is Sony.

ON THIS DAY: POWEROPEN ASSOCIATION FORMED

On March 9, 1993, Apple, Motorola, IBM, and four other computer companies formed the PowerOpen Association. PowerOpen was intended to promote new computer chip technology in preparation for the release of the next generation of personal computers. Apple became the primary user of PowerOpen chips, until its decision to switch to Intel chips in 2006.

SOFTWARE OF THE WEEK: CINEMAR MAINLOBBY

If you're looking for a good front end for a home theater PC (and don't want to go with Windows Media Center), consider Cinemar's MainLobby and associated applications. MainLobby lets you use your PC to control all sorts of devices, including home lighting. Helper applications (such as DVDLobby, MusicLobby, TVLobby, and so on) let you perform specific entertainment operations. Buy MainLobby for $59.99 at www.cinemaronline.com.

March 10, 2006

FRIDAY

THIS WEEK'S FOCUS
CHOOSING A BIG-SCREEN TV

PLASMA FLAT-SCREEN DISPLAYS

As popular as rear projection sets are, what a lot of people want is a TV they can hang on the wall. Two different types of flat panel display are available today: plasma and LCD. Of these, plasma is by far the most popular today.

Most plasma display are only about six inches deep. It's that thin because it contains no picture tubes or projection devices. Instead, it sandwiches a layer of ionized xenon and neon gas between two thin layers of glass. The gas is contained in hundreds of thousands of tiny cells and made up primarily of uncharged particles. When an electrical voltage is applied to the gas, negatively charged particles rush toward the positively charged area of the plasma and positively charged particles rush toward the negatively charged area. The rapidly moving particles collide with each other, exciting the gas atoms in the plasma and releasing photons of energy—which we see as light. When enough sub-pixels light up in a pattern, a picture is created; color intensity is increased or decreased by varying the pulses of current flowing through the different cells.

Plasma displays are popular because they're becoming quite affordable. Plasmas also deliver the closest picture to that of a traditional CRT, with excellent off-axis viewing. But beware of some low-priced 42-inch plasmas that promise "enhanced definition" or EDTV picture quality; these displays can't reproduce high-definition signals.

The primary downside—and it's a big one—to plasma display technology is the danger of burn-in. If you leave a static image on the screen for an extended period of time, the phosphors will burn in to that image, leaving a ghost image on the screen after the fact. This is a big issue if you watch a lot of letterboxed movies or if you watch your 4:3 programming unstretched on a 16:9 display. (The pillars on either side of the picture burn in.) For this reason, many cautious videophiles avoid plasma displays.

ON THIS DAY: URANUS HAS RINGS (1977)

On March 10, 1977, astronomers discovered that the planet Uranus has rings—at least 11 of them. They can't be seen from Earth because the rings get lost in the planet's glare. The rings were discovered when Uranus passed in front of a star and scientists noted a dip in the brightness of the star. By the way, most of Uranus' rings are not quite circular, and most are not exactly in the plane of the equator.

WEBSITE OF THE WEEK: AVS FORUM

One of my favorite home theater/consumer electronics websites is the Audio Video Science (AVS) Forum, located at www.avsforum.com. This site has tons of different forums devoted to different types of audio/video equipment and specific brands; users typically are well versed in the technology and prone to posting very detailed reviews. This is definitely the site to go to when you're considering buying a new big-screen TV.

March 11/12, 2006

SATURDAY/SUNDAY

THIS WEEK'S FOCUS
CHOOSING A BIG-SCREEN TV

LCD FLAT-SCREEN DISPLAYS

If you're worried about plasma burn-in, consider an LCD flat-panel display instead. LCD technology—which is the same technology used for flat-panel computer displays—doesn't suffer from any burn-in or ghosting and is becoming a viable alternative to other display technologies.

Like a plasma display, a flat-panel LCD display uses transmissive light technology. A typical LCD display contains hundreds of thousands of individual pixels; each pixel has three subpixels with red, green, and blue color filters.

Every pixel in an LCD display contains a liquid crystal suspension sandwiched between two panels of polarized glass or plastic. The panels are polarized at right angles to each other. With no voltage applied, the individual liquid crystals are naturally twisted; light passes straight through the liquid crystal layer and the pixel appears to be transparent. When electric current is applied to an LCD pixel, the individual liquid crystals untwist and align in the same direction, which changes the angle of the light passing through the liquid crystal layer. The result is that no light passes through this pixel, making it darker than surrounding pixels. By varying the voltage applied to each pixel, the brightness of each individual pixel (and color subpixel) is thus controlled, resulting in a pattern of bright and dark pixels that creates the overall picture.

In addition to the benefit of not having a burn-in problem, LCD displays are also lighter and use less energy than comparable plasma displays. On the downside, LCD displays are still a little more expensive than plasmas (although that's changing), and have slightly lower contrast and black levels.

ON THIS DAY: LUDDITE RIOTS (1811)

On March 11, 1811, the Luddite riots began in Nottingham, England. The riots started when a group of disaffected laborers attacked a factory, destroying the machines they feared would replace them. The term Luddite has since come to symbolize those who are opposed to all technological change.

BLOG OF THE WEEK: HOME THEATER BLOG

Keep abreast of all the latest developments in the home theater field at the Home Theater Blog. Check it out at www.hometheaterblog.com.

March 13, 2006

MONDAY

THIS WEEK'S FOCUS
COMPUTER SECURITY AND PRIVACY

IDENTITY THEFT

Imagine someone stealing your life. The thief steals your name, your Social Security number, your bank accounts, your credit cards. In the eyes of many, the thief *becomes* you—and uses your personal information to commit all manner of fraud.

This type of theft, which starts with a simple theft of data, is known as *identity theft*—and can be very serious, indeed. If your identity has been stolen, you won't be able to cash checks, use credit cards, or get cash from an ATM. You'll have previously written checks bounce, creditors harass you about nonpayment on your accounts, and financial institutions refuse to issue you any new credit. You'll have your good name—and credit rating—sullied, and experience all manner of problems that could take *forever* to work out.

How can an identity thief get his hands on your personal information? There are a disturbingly large number of ways, including stealing your wallet or purse, stealing your postal mail, dumpster diving through your trash, tricking your company's human resources department into providing your personnel records, tricking you into providing the information via so-called phishing schemes, stealing the data from a large bank or clearing house, and flat-out buying the information—illegally, of course.

Notice that only a few forms of identity theft happen exclusively over the Internet. In fact, a greater threat comes from poor security procedures at your bank or credit card issuing company. However it happens, identity theft is no joke; it's a huge problem, especially if you have to try to rebuild your identity and credit after a theft.

ON THIS DAY: HENRY SHRAPNEL DIES (1842)

Three guesses as to what Henry Shrapnel is famous for. That's right, this former English soldier was the inventor of the Shrapnel shell, a spherical artillery shell designed to explode in midair and spread its content of small lead musket balls in a manner that inflicted injuries on enemy soldiers over a wide area. He died on this day in 1842.

FACT OF THE WEEK

Two-thirds of computer security experts expect that the U.S. will suffer at least one devastating attack to its national information network or power grid within the next 10 years. According to the Pew Internet and American Life Project (www.pewinternet.org), these experts outlined several types of threats, including physical attacks to central parts of the Internet's infrastructure and cyber-terrorist attacks on key utilities and industries.

March 14, 2006

TUESDAY

THIS WEEK'S FOCUS
COMPUTER SECURITY AND PRIVACY

PREVENTING IDENTITY THEFT

Knowing how big a problem identity theft is, how can you best protect your personal information? Here are some tips:

- Never provide personal information via email or instant messaging—which means you should never respond to phishing emails, no matter how legitimate they may appear
- Verify that you're talking to authorized personal before you provide personal information over the phone
- Deposit all your outgoing bill payments in public post office boxes, *not* in your personal mailbox
- If you plan on being away from home for an extended period of time, have your mail put on vacation hold until you return
- Shred all charge card receipts, account statements, and voided checks before you take out your trash
- Don't give out your Social Security number to anyone, unless absolutely necessary
- Once a year, order a copy of your credit report from each of the three major credit reporting agencies; check this report for any unexpected, unauthorized, or incorrect activities

ON THIS DAY: GIANT BRAIN COMPLETED (1955)

On March 14, 1955, AT&T Bell Laboratories announced the completion of the first fully transistorized computer. The computer, officially named TRADIC, was informally known as the "giant brain."

GADGET OF THE WEEK: TRIMTRAC GPS SECURITY LOCATOR

If your car ever gets stolen, you can find it fast with the TrimTrac security locator. This gizmo operates via GPS technology, which fixes its location and then sends its location via SMS text messages over the GSM network. It's small and battery operated, so you can mount it out of sight anywhere in your vehicle. Buy it for $499.99 at www.trimble.com.

March 15, 2006

WEDNESDAY

THIS WEEK'S FOCUS
COMPUTER SECURITY AND PRIVACY

DEALING WITH IDENTITY THEFT

If you find that someone has stolen your personal information—especially your Social Security and credit card numbers—you should take the following steps, *immediately*:

- Contact the fraud departments of the three major credit bureaus (Equifax, Experian, and Trans Union) and report that your identity has been stolen. Ask that a "fraud alert" be placed on your file and that no new credit be granted without your approval.
- For any credit card, loan, or banking accounts that have been accessed or opened without your approval, contact the security departments of the appropriate creditors or financial institutions. Close these accounts, and create new passwords on any replacement accounts you open.
- File a report with your local police department or with the police where you believe the identity theft took place. Get a copy of this report in case your bank, credit card company, or other institutions need proof of the crime at some later date.
- File a complaint with the FTC by calling 1-877-ID-THEFT (438-4338) or going to www.consumer.gov/idtheft/.

ON THIS DAY: ADOBE AND ALDUS MERGE (1994)

PageMaker used to be the big desktop publishing program, and it was published by Adobe. That's not entirely correct. You see, PageMaker was developed by Aldus Corporation, which merged with (or rather, into) Adobe Systems on this date in 1994.

HARDWARE OF THE WEEK: KEYSPAN USB MINI HUB

Here's a USB hub that's no bigger than a credit card—and only 1 centimeter thick. This four-port hub is a great way to add more USB peripherals to your PC, without sacrificing a lot of desk space. Buy it for $49.99 at www.keyspan.com.

March 16, 2006

THURSDAY

THIS WEEK'S FOCUS
COMPUTER SECURITY AND PRIVACY

PRIVACY IN THE WORKPLACE

There's no hiding from the man. When you're at work, every move you make is subject to some sort of surveillance. Some of that surveillance comes in the form of monitoring your physical activities (via hidden video cameras and tracking of pass card use); some comes in the form of monitoring your online activities.

If a company wants to, it can monitor virtually everything its employees do while seated at their computers. Keystroke logger software can track which keys you tap on your keyboard; email sniffers can examine the contents of your incoming and outgoing email messages; website sniffers can tell your boss which websites you visit—and what you do while you're there.

According to a recent report by the American Management Association, nearly 75% of all U.S. companies use some form of surveillance to spy on their employees. And the Center for Internet Studies, which studies Internet addiction in the workplace, reports that 60% of companies have disciplined employees for inappropriate Internet use—and 30% have had to terminate employees for Internet abuse.

Surprisingly, some employers are quite tolerant about how their employees use their computers and Internet access; other employers, however, view any such excursion as an inappropriate and unallowable use of company resources. In any case, these snooping activities are almost always legal—so there's nothing to keep a company from spying on its employees but its own corporate conscience. Surf at your own risk.

ON THIS DAY: FIRST LIQUID-FUEL ROCKET FLIGHT (1926)

On this day in 1926, whiz-kid Robert Goddard launched the first U.S. liquid-fuel rocket flight from a field in Auburn, Massachusetts. The rocket achieved a height of 41 feet before the rocket nozzle melted and it returned to Earth 2.5 seconds later.

SOFTWARE OF THE WEEK: NSCLEAN

One of the best ways to keep your web surfing private is to better manage the cookies that certain sites leave behind on your PC. To that effect, I recommend NSClean, a very effective cookie cleaner that works with all web browsers. Buy it for $39.95 at www.nsclean.com.

March 17, 2006

FRIDAY

THIS WEEK'S FOCUS
COMPUTER SECURITY AND PRIVACY

OLD POSTINGS CAN COME BACK AND BITE YOU

You might not realize it, but every public posting you make—in Usenet newsgroups, online message boards, chat rooms, blogs, and the like—becomes part of the undying fabric of the Internet. Once you put a message out there, it stays out there. There's no deleting something once it enters cyberspace.

Consider the Usenet archive available at Google (groups.google.com). This archive stores every single newsgroup posting from the start of Usenet to today. If you posted something nasty about your boss five years ago, that posting still exists. If you mentioned an affair you had with a neighbor back in the mid-'90s, that posting still exists. If you asked a question about a particular illness, or proffered an opinion about a particular make of car, or let slip where you live, that information is still available to a dedicated searcher.

Which means, of course, that the biggest threat to your privacy is *you*, and your inability to keep your yap shut when you're surfing.

It also reinforces the general warning that all veteran Internet users should know: Don't post anything in a newsgroup, message board, chat room, or blog that you wouldn't want your future boss—or spouse—to read.

ON THIS DAY: RUBBER BAND PATENTED (1845)

On this date in 1845, Stephen Perry of London received a patent for a rubber band. The idea came from Central and South America, where footwear, garments, and even bottles were made out of the milk of the rubber tree. Sailors carried the rubber to England, and the rubber band was destined to follow.

WEBSITE OF THE WEEK: ELECTRONIC FRONTIER FOUNDATION

The Electronic Frontier Foundation (EFF) fights on the front lines of the privacy wars. If you're concerned at all about your right to online privacy, this is the website to visit. Check it out at www.eff.org.

March 18/19, 2006

SATURDAY/SUNDAY

THIS WEEK'S FOCUS
COMPUTER SECURITY AND PRIVACY

PHISHING SCHEMES

Ever receive an email that purported to come from eBay or PayPal or your bank or credit card company, asking you to click a link to update your personal account information? If so, I hope you didn't get taken in and click the link; you were the recipient of a phishing scheme, which is the newest way for scammers to steal your personal information.

The term comes from the scammers fishing (or "phishing") for your information, by dangling that official-looking email as bait. The bait and switch comes if you click the link in the email. You won't be taken to your bank or eBay or wherever; instead, that fake URL will take you another site, run by the scammer, which will be tricked up to look like the official site. (And you can't tell a good site from a bad one just by looking at it; some of these phishers do a remarkable job of mocking up a fake site to look identical to the real thing.)

If you enter your personal information on this fake site, as requested, you're actually delivering it to the scammer—which means you're now a victim of identity theft. The scammer can use the information you provided to hack into your account, make unauthorized charges on your credit card, and maybe even drain your bank account.

It goes without saying that you should never respond to this type of email, no matter how official-looking it appears. If you want to make changes to your eBay or PayPal or credit card account, never do so from an email link. Instead, use your web browser to go directly to the official site, and make your changes there. No one from your bank or credit card company will ever ask you for this information via email. Be warned!

ON THIS DAY: FIRST U.S. RAILROAD TUNNEL (1834)
On March 18, 1834, the first U.S. railroad tunnel was completed between Hollidaysburg and Johnstown, Pennsylvania. The Staple Bend Tunnel was 901 feet long, 25 feet wide, and 21 feet high; it was designed for the Allegheny Portage Railroad, which was the first railroad to go west of the Alleghany Mountains.

BLOG OF THE WEEK: PRIVACY.ORG
Track all news stories related to online privacy and security at Privacy.org, a blog devoted to protecting privacy everywhere. Read all about it at www.privacy.org.

March 20, 2006

MONDAY

THIS WEEK'S FOCUS
COMPUTER MAINTENANCE

FIVE WAYS TO SPEED UP YOUR PC

You read the title, so let's get right to it:

- **Delete unnecessary files**—Your computer runs faster when you have some free hard disk space. The easiest way to free up more space is to delete unnecessary files, via Windows XP's Disk Cleanup tool.
- **Delete unused programs**—Another great way to free up valuable hard disk space. Use the Add/Remove Programs utility from the Windows Control Panel.
- **Defragment your hard disk**—Fragmented disks run slower, so you should run the Windows Disk Defragmenter utility at least once a month.
- **Fix hard disk errors**—Keep your hard disk spinning smoothly by occasionally running the ScanDisk utility. This will find and fix any errors on the hard disk.
- **Reboot**—Even though Windows XP appears to run smoothly for extended periods, you can still get memory leaks and unclosed utilities that will slow things down over time. Clear up all the loose ends by rebooting your machine occasionally, which starts everything fresh.

ON THIS DAY: THEORY OF RELATIVITY PUBLISHED (1916)

In 1916, Einstein's Theory of General Relativity was published as an academic paper in the *Annalen der Physik*. Einstein's theory showed that Newton's Law of Gravitation was only approximately correct, breaking down in the presence of very strong gravitational fields.

FACT OF THE WEEK

Jupiter Research (www.jupiterresearch.com) says that 32 million U.S. households—about half of all online households—had broadband access in 2004. This number is expected to grow to 78% (88 million) by the end of 2010.

March 21, 2006

TUESDAY

THIS WEEK'S FOCUS
COMPUTER MAINTENANCE

MAINTAINING YOUR SYSTEM UNIT

Your PC system unit has a lot of sensitive electronics inside—everything from memory chips to disk drives to power supplies. Check out these maintenance tips to keep your system unit from flaking out on you:

- Position your system unit in a clean, dust-free environment. Keep it away from direct sunlight and strong magnetic fields. In addition, be sure your system unit and your monitor have plenty of air flow around them to keep them from overheating. (And would it kill you to dust the thing every once in a while?)
- Hook up your system unit to a surge suppressor to avoid deadly power spikes.
- Avoid turning on and off your system unit too often; it's better to leave it on all the time than incur frequent "power on" stress to all those delicate components. However...
- Turn off your system unit if you're going to be away for an extended period—anything longer than a day or two.
- Check all your cable connections periodically. Make sure that all the connectors are firmly connected and all the screws properly screwed—and make sure that your cables aren't stretched too tight or bent in ways that could damage the wires inside.

ON THIS DAY: TEACHING OF EVOLUTION PROHIBITED (1925)

On this date in 1925, the Butler Act became state law in Tennessee. This was the law that prohibited the teaching of evolution in all the state's public schools and universities, and eventually led to the famous Scopes "monkey trial." (Eighty years later, the science of evolution is still being fought as a "theory" by some narrow-minded creationists.)

GADGET OF THE WEEK: TOMTOM NAVIGATOR 2004

The TomTom Navigator is a cool little gizmo that turns your PDA into a full-fledged GPS navigation system. It offers all the features you'd find in a handheld GPS device but routes all the information and maps through your PDA via Bluetooth wireless technology. Buy it for $299 at www.tomtom.com.

March 22, 2006

WEDNESDAY

THIS WEEK'S FOCUS
COMPUTER MAINTENANCE

MAINTAINING YOUR KEYBOARD

Even something as simple as your keyboard requires a little preventive maintenance from time to time. Check out these tips:

- Keep your keyboard away from young children and pets—they can get dirt and hair and Silly Putty all over the place, and they have a tendency to put way too much pressure on the keys.
- Keep your keyboard away from dust, dirt, smoke, direct sunlight, and other harmful environmental stuff. You might even consider putting a dust cover on your keyboard when it's not in use.
- Use a small vacuum cleaner to periodically sweep the dirt from your keyboard. Alternatively, you can use compressed air to *blow* the dirt away. Use a cotton swab or soft cloth to clean between the keys. If necessary, remove the keycaps to clean the switches underneath. (And make sure you do this with your PC turned off; fiddling with a live keyboard can put all sorts of nonsense on your computer screen!)
- If you spill something on your keyboard, disconnect it immediately and wipe up the spill. Use a soft cloth to get between the keys; if necessary, use a screwdriver to pop off the keycaps and wipe up any seepage underneath. Let the keyboard dry thoroughly before trying to use it again.

ON THIS DAY: PENTIUM CHIP SHIPS (1993)

Before 1993, Intel's chips were numbered—80236, 80386, 80486, and so on. Intel abandoned the numbering when it released the Pentium microprocessor, which sharp-eyed readers will recognize as relating to the number "five," as in 80586—the natural successor to the 80486. In any case, March 22, 1993 is when the first Pentium chip was shipped.

HARDWARE OF THE WEEK: IOGEAR MINIVIEW MICRO KVM SWITCH

When you need to share USB devices between two PCs (or between a PC and Mac Mini), you need a KVM switch. (KVM stands for "keyboard, video, and mouse.") IOGEAR's MiniView Micro offers connections for audio, USB keyboard and mouse, and video monitor; just flip a switch to make all these peripherals work on one or another PC. Buy it for $69.95 from www.iogear.com.

March 23, 2006

THURSDAY

THIS WEEK'S FOCUS
COMPUTER MAINTENANCE

MAINTAINING YOUR MOUSE

If you're a heavy Windows user (and we all are), you probably put thousands of miles a year on your mouse. Just like a car tire, anything turning over that often needs a little tender loving care. Check out these mouse maintenance tips:

- Periodically open up the bottom of your mouse and remove the roller ball. Wash the ball with water (or perhaps a mild detergent). Use a soft cloth to dry the ball before reinserting it.

- While your mouse ball is removed, use compressed air or a cotton swab to clean dust and dirt from the inside of your mouse. (In extreme cases, you might need to use tweezers to pull lint and hair out of your mouse—or use a small knife to scrape packed crud from the rollers.)

- Always use a mouse pad—they really do help keep things rolling smoothly; plus, they give you good traction. (And while you're at it, don't forget to clean your mouse pad with a little spray cleaner—it can get dirty, too.)

Of course, if you have an optical mouse, you can skip these steps; there aren't any mouse balls to clean. However, you still might need to keep pieces of lint from clogging the optical sensor. Wiping off the bottom of your mouse makes sense, no matter what type of mouse you have!

ON THIS DAY: COLD FUSION CLAIMED (1989)

On this date in 1989, two Utah scientists claimed they had produced fusion at room temperature—so-called "cold fusion." Unfortunately, other scientists were never able to replicate their work, making this claim somewhat spurious.

SOFTWARE OF THE WEEK: PC CERTIFY

Generate all sorts of diagnostic tests and reports with PC Certify. This utility is based on a suite of high-level tests and utilities originally developed for Intel, and will test every piece and part of your system, from memory to joysticks. Buy the Power User version for $119 from www.pc-diagnostics.com.

March 24, 2006

FRIDAY

THIS WEEK'S FOCUS
COMPUTER MAINTENANCE

MAINTAINING YOUR PRINTER

Whether you own an inkjet or a laser printer, what you have is a complex device with a lot of moving parts. Follow these tips to keep your printouts in good shape:

- Use a soft cloth, mini-vacuum cleaner, and/or compressed air to clean the inside and outside of your printer periodically. In particular, make sure you clean the paper path of all paper shavings and dust.
- If you have an inkjet printer, periodically clean the ink jets. Run your printer's cartridge cleaning utility, or use a small pin to make sure they don't get clogged.
- If you have a laser printer, replace the toner cartridge as needed. When you replace the cartridge, remember to clean the printer cleaning bar and other related parts, per the manufacturer's instructions.

And, whatever you do, *don't* use alcohol or other solvents to clean any rubber or plastic parts—you'll do more harm than good!

ON THIS DAY: TUBERCULOSIS BACILLUS ANNOUNCED (1882)

On this date in 1882, German scientist Robert Koch announced that he had discovered the bacillus responsible for tuberculosis. For his work, Koch was awarded the Nobel Prize in medicine in 1905.

WEBSITE OF THE WEEK: PC PITSTOP

Give your PC an online checkup at PC Pitstop. The site offers a variety of free tests that will almost always make your PC run a little bit faster. I like this site so much I visit it at least once a month. Check it out at www.pcpitstop.com.

March 25/26, 2006

SATURDAY/SUNDAY

THIS WEEK'S FOCUS
COMPUTER MAINTENANCE

MORE BASIC MAINTENANCE

Here are some more common-sense things you should be doing to keep your system in tip-top shape:

- **Back up your data**—There's no excuse for losing valuable data if your system crashes. Buy a cheap external hard drive and configure it to automatically back up all your data on a regular basis.

- **Update your virus and spyware definitions**—New viruses and spyware gets released every day. Your system won't be protected unless the definitions are up to date—so download them!

- **Run Windows Update**—Yeah, I know it's a pain, especially if you're on a slow dial-up connection, but these regular updates are necessary to patch security holes and fix bugs. There's no good excuse *not* to do it.

- **Organize your files**—Okay, so this one won't make your system run any faster, but it will help you more easily find things. Anytime you have more than 10 files of a certain type, create a new subfolder for them! Don't get stuck with folders so overcrowded that finding that one file you need is practically impossible.

ON THIS DAY: TITAN DISCOVERED (1655)

On March 25, 1655, Christiaan Huygens discovered Titan, Saturn's largest satellite. Titan wasn't named until almost two centuries later, however, when Sir John Herschel, discoverer of Uranus, assigned names to the seven moons of Saturn that were known at that time.

BLOG OF THE WEEK: DOC SEARL'S IT GARAGE

This is a great blog for the technically minded, especially those who have technical problems. Doc Searl is business editor for *Linux Journal*, and his blog covers technical issues from the IT user's and developer's perspective. Find out more at www.itgarage.com.

March 27, 2006

MONDAY

THIS WEEK'S FOCUS
WINDOWS MEDIA CENTER

WHY YOU NEED A PC IN YOUR LIVING ROOM

Why exactly would you want to put a PC in your living room? Simple—a Media Center PC performs the functions of several different audio/video components, all in one box. When you hook up a Media Center PC to your home theater system, you can do away with your CD player, DVD player, and hard disk recorder—and gain additional functionality, to boot. Here are just some of the uses for a Media Center PCs:

- Recording television shows, via the PC's built-in tuner and hard disk, just like TiVo and similar DVR devices
- Storing and playing digital music, either by ripping your CD collection to hard disk or by downloading digital music from the Internet
- Playing music from CDs, using the built-in CD drive
- Recording CDs, using the built-in CD burner drive
- Playing movies from DVDs, using the built-in DVD drive
- Recording DVDs, using the DVD burner drive included with most Media Center PCs—which lets you burn DVDs of the television shows you record
- Listening to Internet radio, via the PC's Internet connection
- Viewing digital photos, which you can store on the PC's hard drive

ON THIS DAY: ALASKA EARTHQUAKE (1965)

On March 27, 1965, south central Alaska was rocked by a giant earthquake. The quake measured between 8.3 and 8.5 on the Richter scale and released more than twice the energy of the 1906 San Francisco earthquake. The death toll was only 131, but property damage was very high.

FACT OF THE WEEK

Digital Connect magazine (www.digitalconnectmag.com) reports that one million Media Center PCs will be sold in 2005, with four million projected to be sold in 2006. Media Center PC owners use their PCs to create or burn CDs and DVDs (72%), listen to digital music (69%), play video games (68%), watch DVDs (63%), edit photos (45%), watch TV (45%), record TV (42%), create slideshows (38%), and edit videos (37%).

March 28, 2006

TUESDAY

THIS WEEK'S FOCUS
WINDOWS MEDIA CENTER

BUYING A MEDIA CENTER PC

What makes a Media Center PC a Media Center PC? All Media Center PCs have the same components as a typical desktop PC, but with a few items fine-tuned for home theater use. In particular, most HTPCs include the following components:

- Big hard drive, ideally 200GB or more, to store lots of television programs as well as a large CD collection
- TV tuner—at least one, sometimes two or three (including, in some cases, an ATSC tuner for HDTV)
- CD/DVD drive (typically a burner drive)
- Windows XP Media Center Edition operating system, for the living room interface

Most Media Center PCs come in a form factor that isn't quite what you're used to with a desktop PC. The majority of Media Center PCs aren't vertical, they're horizontal, typically about 17 inches wide and a few inches tall, in a cabinet that closely resembles that of most audio components. The goal is to have the PC look at home in a rack of audio and home theater components—complete with the black metal or brushed aluminum faceplate.

In addition, many Media Center PCs go to some lengths to reduce the noise typically associated with personal computers. That means fewer, quieter fans—or, in some instances, no fans at all, with either water cooling or large heat sinks to dissipate the normal heat buildup.

ON THIS DAY: THREE MILE ISLAND ACCIDENT (1979)

On March 29, 1979, a nuclear accident occurred at the Three Mile Island nuclear power plant outside Harrisburg, Pennsylvania. The incident was caused by a combination of human and mechanical errors, when a cooling system malfunctioned and permitted a partial meltdown of the reactor's core. Fortunately, no deaths were recorded.

GADGET OF THE WEEK: PHILIPS RC9800I WIFI REMOTE CONTROL

The RC9800i is a unique universal remote control. First, it's activity-based rather than component-based, so it uses a lot of multiple-device macros. Second, it doesn't just control your audio/video equipment, it can also control a variety of Universal Plug-and-Play devices on your PC system. Third, it works via WiFi, in addition to traditional infrared. And fourth, because of the WiFi connection, it displays its own electronic program guide. Buy it for $499 from www.homecontrol.philips.com.

March 29, 2006

WEDNESDAY

THIS WEEK'S FOCUS
WINDOWS MEDIA CENTER

BUILDING YOUR OWN MEDIA CENTER PC

If you don't want to buy a pre-built Media Center PC, you can always build your own. When you go this route you're not limited to the Windows Media Center Edition operating system; in fact, many home theater PC boxes run Linux, which is perhaps a little more efficient than Windows for this type of dedicated purpose.

Building your own home theater PC is just like building your own desktop PC, with careful consideration for the following components:

- One or more large hard disks, often employed in a RAID configuration.
- High-end video card.
- One or more television tuner cards with recording capability, such as those by AccessDTV, Hauppage, MyHD, and Pinnacle Systems. For state-of-the-art performance, consider a card with HDTV capability.
- Sound card with surround-sound digital audio output.
- Combination CD/DVD player/recorder drive.
- System cabinet, in either a slim desktop/media center or small form factor "shuttle" design.
- Appropriate home theater PC operating software, such as MainLobby, MediaPortal, Meedio Essentials, SageTV, and Snapstream Beyond Media and Beyond PC.

ON THIS DAY: COCA COLA CREATED (1886)

On this date in 1886, the first batch of Coca Cola was brewed over a fire in a backyard in Atlanta, Georgia. Dr. John Pemberton had created the concoction as a cure for hangovers, stomach aches, and headaches, and advertised it as a "brain tonic and intellectual beverage." And here's the fun part: Coke contained cocaine as an ingredient until 1904, when the drug was banned by Congress.

HARDWARE OF THE WEEK: DENALI EDITION MEDIA CENTER

My favorite Media Center PC is one hog of a machine. The Denali Edition Media Center stands an impressive 8.25 inches high, while its 17.5-inch width matches all your other system components. Operation is absolutely silent due to the total absence of cooling fans; instead, the entire case acts as a giant heat sink to disperse the normal heat build up. You also get four TV tuners (two for HDTV), up to 1 terrabyte of hard disk space, and wireless keyboard, mouse, and remote control. Prices start at $4,600 and go a lot higher; learn more at www.niveusmedia.com.

March 30, 2006

THURSDAY

THIS WEEK'S FOCUS
WINDOWS MEDIA CENTER

USING MEDIA CENTER AS A DVR

Windows Media Center is a fully featured audio/video operating environment. One of the most important features of this environment is the capability to watch live television broadcasts—and record programming to the PC's hard disk, using the PC as a TiVo-like digital video recorder (DVR).

There are two ways to record a program to hard disk in Media Center. First, all you have to do is watch a program; the hard disk automatically stores a short cache of the selected program, so you can pause and rewind the "live" program. Second, you can set up Media Center to record programs in advance, using the built-in electronic program guide (EPG).

Using the EPG to schedule a recording is a snap. Just press the Guide button on the remote control or go to My TV, Guide; this displays the EPG's program grid. Scroll down to view more channels, or scroll right to see more of the schedule. (You can also search for specific programs to record—go to My TV, Search, Title and enter the program's name.) Highlight the program you want to record, press OK on your remote, and then select Record from the onscreen menu.

To play back a recorded program, go to My TV, Recorded TV. This displays a list of all recorded programs; select a program from the list and press OK on your remote. Playback will start automatically.

ON THIS DAY: METER DEFINED (1791)

On this day in 1791, responding to a proposal by the Acadèmiè des Sciences, the French National Assembly decreed that a meter (or what the French called a *metre*) would be a 1/10,000,000 of the distance between the North Pole and the equator. Now you know.

SOFTWARE OF THE WEEK: TWEAK MCE

I've always been a big fan of Microsoft's PowerToys utilities, and now there's one for Windows XP Media Center Edition called Tweak MCE. This utility provides access to systems settings and options not normally visible, including remote control settings, user interface options, and the like. Download it for free from www.microsoft.com/windowsxp/downloads/powertoys/mcepowertoys.mspx.

March 31, 2006

FRIDAY

THIS WEEK'S FOCUS
WINDOWS MEDIA CENTER

MEDIA CENTER MY MUSIC

I primarily use my Media Center PC as a digital music jukebox. I've ripped my entire CD collection to hard disk and now can play back any album in any order at any time, using Windows Media Center for navigation. It's like having a giant iPod in my living room—but with true CD-quality sound.

The first thing you need to do is tell Media Center to store your music in WMA Lossless format. Using any other form of compression will dramatically affect the audio quality of the recordings; with WMA Lossless, a CD only takes up about 250MB of disk space and you retain the original audio quality. Unfortunately, you can't make this selection from within Media Center. You have to exit the Media Center interface, start up the Windows Media Player software, and make this selection from the Tools, Options menu.

After you have WMP configured, you can start ripping your CDs. It takes a little less than five minutes to rip each CD in WMA Lossless format, so if you have a large music collection this could take some time. About 98% of the time Media Center downloads the correct track information and album art, which means every now and then you'll have to do a little manual tweaking. To retype album and artist info, select the album in My Music and press the Info button on your remote; to retype individual track info, select the track and press Info. To change album art, you'll have to use a third-party utility, such as Album Art Fixer (www.avsoft.nl/artfixer/).

Playing back your music is easy. You can select music by album, artist, or genre, or create your own playlists. Unfortunately, creating a playlist from within Media Center is tedious and somewhat counterintuitive; it's easier to exit Media Center and create your playlists from within Windows Media Player, instead.

ON THIS DAY: DAYLIGHT SAVINGS TIME BEGINS (1918)

On the last day of March 1918, the United States first observed Daylight Saving Time. The concept had already been introduced in Great Britain as a fuel-saving measure during World War I; the idea was introduced in the U.S. by the Daylight Savings Association and New York Senator William M. Calder.

WEBSITE OF THE WEEK: THE GREEN BUTTON

If you're a Windows Media Center user, you have to visit The Green Button, a website devoted to everything Media Center. Here you can find answers to all your questions, as well as links to some of the best Media Center add-ins and utilities. Bookmark it at www.thegreenbutton.com.

April 2006

April 2006

SUNDAY	MONDAY	TUESDAY	WEDNESDAY	THURSDAY	FRIDAY	SATURDAY
						1 April Fool's Day — 1889 First dishwashing machine offered for sale in Chicago, IL
2 1827 First lead pencils are manufactured	**3** 1860 Pony Express mail service begins	**4** 1983 Space Shuttle Challenger makes its maiden voyage	**5** 1994 Nirvana vocalist, Kurt Cobain, commits suicide in Seattle, WA	**6** 1930 Hostess Twinkies invented	**7** 1827 First matches go on sale in England	**8** 1862 First aerosol dispenser is patented
9 1959 NASA names first seven astronauts in history	**10** 1949 Safety pins patented in New York	**11** 1986 Halley's Comet makes last approach to Earth until 2061	**12** 1961 Russian cosmonaut Yuri Gagarin becomes the first human in space, orbiting the Earth in Vostok I	**13** 1970 Apollo 13 disaster; all three astronauts miraculously survived	**14** 1912 The Titanic strikes an iceberg in the North Atlantic and sinks, killing about 1,500 people	**15** 1923 Insulin first becomes available for diabetic patients
16 1947 First television camera zoom lens demonstrated	**17** 1790 Benjamin Franklin dies	**18** 1775 The famous ride of Paul Revere and William Dawes begins	**19** 1995 Oklahoma City, OK, Federal Building bombed, killing 168 people	**20** 1940 First electron microscope is demonstrated in Philadelphia, PA	**21** 753 B.C. Rome founded	**22** Earth Day — 1915 Chemical weapons ("mustard gas") first used in war by German troops
23 Administrative Professionals' Day (Secretaries Day) — 1564 William Shakespeare born	**24** 1886 Oil is first discovered in the Middle East, on the Egyptian shore of the Red Sea	**25** 1719 *Robinson Crusoe* is published	**26** 1986 Chernobyl nuclear accident	**27** 1981 First computer to use a mouse (Xerox STAR 8010) is introduced	**28** 1930 Eclipse of the sun first filmed	**29** 1923 Zipper patented in Hoboken, NJ
30 1939 Commercial TV debuts						

April 1/2, 2006

SATURDAY/SUNDAY

THIS WEEK'S FOCUS
WINDOWS MEDIA CENTER

MEDIA CENTER ADD-INS

Want to enhance the basic Windows Media Center experience? Then check out some of these cool third-party add-ins and widgets:

- **Comics for Media Center** (www.mcesoft.nl), displays seven days' worth of comics, including *Garfield*, *Peanuts*, *BC*, *Big Nate*, *Dilbert*, and more
- **MCE Customizer 2005** (www.mcecustomizer.com), similar to Tweak MCE, lets you reconfigure various "hidden" Media Center settings
- **MCE Outlook** (www.mce-software.com), lets you access your Microsoft Outlook contacts and email via Media Center
- **mceAuction** (www.cbuenger.com/mceauction/), for managing your eBay auctions via Media Center
- **mceWeather** (www.cbuenger.com/mceweather/), complete with Weather Channel maps and forecasts
- **Media Center Playlist Editor** (www.microsoft.com/windowsxp/downloads/powertoys/mcepowertoys.mspx), a better way to create and edit playlists from within the Media Center environment
- **My Netflix** (www.unmitigatedrisk.com/mce/), for managing your NetFlix rental queue

ON THIS DAY: FIRST PHOTOGRAPH OF THE SUN

On April 2, 1845, the first photograph of the sun was taken by H.L. Fizeau and J. Leon Foucault. Fortunately, they didn't look directly at their subject, and thus didn't go blind.

BLOG OF THE WEEK: MATT GOYER'S WINDOWS MEDIA CENTER BLOG

This is the personal blog of Matt Goyer, a program manager on Microsoft's Media Center team. This is a great place to get the inside scoop on Media Center use and development; check it out at mediacenter.mattgoyer.com.

April 3, 2006

MONDAY

THIS WEEK'S FOCUS
WACKY WEBSITES

SKELETAL SYSTEMS

This week I'm going to share some of my favorite weird and wacky websites, the kinds of sites that really don't serve a useful function but are fun and unusual, anyway. Like, for example, today's wacky site, the Skeletal Systems page on Michael Paulus' website.

Michael Paulus is an artist and illustrator based in Portland, Oregon. His "regular" work is a bit odd in and onto itself, full of strange and humorous images. But it's his Skeletal Systems work that really gets my attention.

The Skeletal Systems page is a character study of 22 famous cartoon characters—on the inside. That is, these sketches imagine the skeletal structures of the cartoon characters, big heads and all. It's pretty wacky stuff, with disturbingly funny skeletal renditions of Fred Flintstone, Charlie Brown, Betty Boop, Tweety Bird, Marvin the Martian, Pikachu, and the Powerpuff Girls, among others. Probably the most disturbing thing is that these sketches are all extremely detailed, and would pass for anatomically accurate if we weren't talking about cartoon characters.

You'll get a kick out of this one, believe me. Check it out for yourself at www.michaelpaulus.com/gallery/character-Skeletons/.

ON THIS DAY: FIRST MOBILE PHONE CALL (1973)

The very first mobile phone call was placed by inventor Martin Cooper on this date in 1973. The mobile phone itself was anything but mobile; it measured 10 inches high, 3 inches deep, and 1.5 inches wide and weighed a studly 30 ounces. Imagine trying to talk on this bad boy while negotiating bumper-to-bumper traffic!

FACT OF THE WEEK

Google (www.google.com) reports that **britney spears** was the number-one search term for all of 2004. Number-two was **paris hilton**, followed by **christina aguilera** (3), **pamela anderson** (4), and **chat** (5). The rest of the top 10 searches for the year were **games** (6), **carmen electra** (7), **orlando bloom** (8), **harry potter** (9), and **mp3** (10). It's good to know we're putting the power of Google to good use!

April 4, 2006

TUESDAY

THIS WEEK'S FOCUS
WACKY WEBSITES

THE DEATH CLOCK

Not to be a downer, but have you thought about death lately? If you have, then you'll probably like the Death Clock website, which answers the always amusing question, "When am I going to die?"

Here's how the Death Clock works. Input your date of birth, gender, mode (pessimistic, optimistic, sadistic), Body Mass Index, and smoking status, then click the Check Your Death Clock button. The Death Clock then uses various actuary tables and databases to determine exactly when it is that you're likely to kick the bucket. You'll get a target death date, as well as an attractive clock that counts down the precise number of seconds you have to live.

To be fair, the Death Clock is not meant to be funny or fun. The site includes links to a comprehensive health library, as well as the expected links to wills, obituaries, and similar services. Use it as you will.

For what it's worth, I'm slated to shuffle off this mortal coil on November 26, 2030. Find out when you should start writing your own obituary at www.deathclock.com.

ON THIS DAY: NETSCAPE FOUNDED (1994)

On April 4, 1994, Marc Andreesen and Jim Clark founded Mosaic Communications Corporation—later renamed Netscape Communications Corp. Andreesen developed the original Mosaic web browser while he was a student at the University of Illinois. Mosaic morphed into the groundbreaking Netscape browser, which remains the inspiration behind today's Mozilla Firefox browser.

GADGET OF THE WEEK: MOTOROLA OJO PERSONAL VIDEOPHONE

Our friends at Motorola bill the Ojo as a personal videophone, which describes exactly what it is. The Ojo has a built-in video camera to take pictures of you as you talk, and an LCD display to show pictures of whomever it is you're talking to. You use the Ojo for VoIP Internet phone communication; you can talk to (and see) anyone with a similar Ojo device and VoIP connection. Buy it for a paltry $799.99 at www.motorola.com/ojo/.

April 5, 2006

WEDNESDAY

THIS WEEK'S FOCUS
WACKY WEBSITES

DUMB AUCTIONS

Want to find the most off-beat, disturbingly strange, and downright dumb current auctions on eBay? Then turn to the Dumb Auctions site (www.dumbauctions.com), which searches out the weirdest and wackiest eBay auctions for your perusal and amusement. (Heck, you might even be interested enough to bid.)

What kinds of auctions am I talking about? How about a young girl selling her first baby tooth? (She apparently didn't think the tooth fairy paid enough…) Or a large letter "O" to hang on your living room door? (Great for Oprah, not so great for the rest of us…) Or a chicken nugget in the shape of a seahorse? (Huh?) You get the idea.

By the way, Dumb Auctions is just part of the Dumb Network. Other sites include Dumb Bumpers (www.dumbbumpers.com), which displays unusual bumper stickers; Dumb Criminals (www.dumbcriminalacts.com), which celebrates the most incompetent bad guys out there; Dumb Facts (www.dumbfacts.com), a database of truly useless information; Dumb Laws (www.dumblaws.com), which lists some extremely odd and outdated laws still on the books; and Dumb Warnings (www.dumbwarnings.com), a compendium of blatantly obvious signs and product labels. They're all fun, and worth checking out.

ON THIS DAY: OPPENHEIMER WINS FERMI AWARD (1963)

On this date in 1963, the United States Atomic Energy Commission gave the Fermi Award to John Robert Oppenheimer for his research in nuclear energy. Oppenheimer was the chief scientist of the Manhattan Projection during World War II, out of which was born the atomic bomb. (To his credit, Oppenheimer later opposed the more destructive hydrogen bomb—which caused the government to revoke his security clearance.)

HARDWARE OF THE WEEK: ACTIONTEC INTERNET PHONE WIZARD

ActionTec's Internet Phone Wizard is a broadband phone adapter you can use with the Skype service. It connects your PC or wireless router to any standard telephone; you can then dial other Skype users over the Internet at no charge. What's nice about the Internet Phone Wizard is that it lets you use the same phone for both traditional and Internet calling; it seamlessly integrates both services, automatically switching to the correct service depending on who you're calling. Buy it for $79.99 at www.actiontec.com.

April 6, 2006

THURSDAY

THIS WEEK'S FOCUS
WACKY WEBSITES

UGLYDRESS.COM

Not that I've ever been a bridesmaid, but I've been to enough weddings in my time to know that the bridesmaids get the worst of it—particularly when it comes to clothing. I mean, the groom and the groomsmen get to wear fancy tuxedos and the bride gets a gorgeous white gown, but the poor bridesmaids are stuck in the most hideous dresses you've ever seen.

Which is where today's wacky website comes in. UglyDress.com celebrates the worst bridesmaid dresses of all time. These are fug-ugly dresses, submitted by site visitors (primarily former bridesmaids) to document the horrors they've been subjected to. There are categories for Bad Patterns, Bad Color, Bad Shape, and Out of Date dresses; additional sections focus on Ugly Weddings, Ugly Shoes, and Ugly Tuxedos. There's even a contest you can enter to submit your own ugly dress pictures. (Man, some of these are extremely barf-inducing!)

Check it out, if you can stand it, at www.uglydress.com.

ON THIS DAY: WINDOWS 3.1 RELEASED (1992)

The 3.1 version of Microsoft Windows was released on this date in 1992. Windows 3.1 introduced support for sound cards and MIDI, Super VGA (800 × 600) monitors, scalable fonts, and 9600bps modems. (Woo!) It was also the first version of Windows to utilize the three-finger salute (Ctrl+Alt+Del) for rebooting the system—which was good, because you had to reboot Windows 3.1 *a lot*.

SOFTWARE OF THE WEEK: MARINE AQUARIUM SCREEN SAVER

You've seen this screen saver in lots of places; it displays a cool virtual saltwater aquarium on your computer screen, complete with bubbles and changing lighting. It's without a doubt the most realistic aquarium simulation I've seen. Buy it for $19.95 from www.serenescreen.com.

April 7, 2006

FRIDAY

THIS WEEK'S FOCUS
WACKY WEBSITES

MUSEUM OF BAD ALBUM COVERS

The days of artistic album covers are long over; fancy artwork just doesn't have the same impact on a small CD case as it does on a large LP cover. That said, there were some great album covers back in the day. And some really bad ones, too, as celebrated on the Museum of Bad Album Covers website.

The Museum of Bad Album Covers is just what it says—a collection of the worst album covers of all time. This site, run by Zonicweb, has assembled more than 100 really lousy covers, sorted them into categories, and displayed them for your viewing pleasure. There's even a constantly updated Top Ten list, as voted on by site visitors. (On my most-recent visit, the number-one worst album cover was the Scorpions' *Virgin Killer*, a decidingly disgusting piece of kiddie porn.)

For some reason, many of these bad album covers fall into the Christian music category, mostly by small-time gospel singers from the 1960s and 1970s with apparently low graphic-design budgets. But bad covers aren't limited to religious singers; there are a fair number of truly awful covers by heavy metal bands, funk singers, and jazz artists—and even a few Beatles covers. Awfulness knows no bounds.

You can browse through the Museum of Bad Album Covers for yourself at www.zonicweb.net/badalbmcvrs/. Take special note of the warning on the home page: "Zonicweb.net accepts no responsibility whatsoever for any nausea, vomiting, and/or retinal damage caused by viewing these album covers." Sweet!

ON THIS DAY: IBM ANNOUNCES SYSTEM/360 MAINFRAMES (1964)

On this date in 1964, IBM announced the release of its latest family of mainframe computers, the System/360. It was called the "360" because it was meant to address all possible sizes and types of customers with one unified software-compatible architecture. (It was a good idea; IBM eventually generated more than $100 billion in revenue from the System/360 line.)

WEBSITE OF THE WEEK: TODAY'S FRONT PAGES

Here's a fun website that is much more useful than wacky. The Newseum presents the front pages of daily newspapers from around the country and around the world. Click on any location to view today's image of that city's newspaper; click further to read the papers online. Check it out at www.newseum.org/todaysfrontpages/flash/.

April 8/9, 2006

SATURDAY/SUNDAY

THIS WEEK'S FOCUS
WACKY WEBSITES

DAILY ROTTEN

When you want a consistent source for the weirdest news on the Web, turn to the Daily Rotten. This is an excellent compendium of wacky news stories, compiled from legitimate news organizations everywhere. If it's oddball news you want, this is the place for it.

I particularly like the "Today in Rotten History" column. It's like the On This Date feature of this book, except it focuses on the most unusual events in history. It's a fun way to amuse your friends with odd and somewhat irrelevant facts.

The Daily Rotten is part of the Rotten.com website, which hosts extremely odd (and extremely bad) poetry about various world events. I'm not sure I quite understand this, but it's deeply disturbing and as such might appeal to some of you readers.

You can read the Daily Rotten at www.dailyrotten.com. Rotten.com is located at, naturally, www.rotten.com. Have fun!

Parents beware: While Daily Rotten is amusing for adults, it's not a child friendly site by any stretch. You've been warned.

ON THIS DAY: PRESPER ECKERT BORN (1919)

Presper Eckert, along with John Mauchly, was one of the chief designers of ENIAC, the world's first electronic digital computer. Eckert and Mauchly went on to found the Eckert-Mauchly Computer Corporation, which they sold to Remington-Rand in 1950. Eckert was born on April 9, 1919, and passed away on June 3, 1995.

BLOG OF THE WEEK: DAVE BARRY'S BLOG

That's right, this is the official blog of humorist Dave Barry. It's just as funny as his weekly column appearing in *The Miami Herald* and other newspapers, and updated more frequently! Here's the URL: weblog.herald.com/column/davebarry/.

April 10, 2006

MONDAY

THIS WEEK'S FOCUS
EASTER EGGS

DVD EASTER EGGS

Easter eggs are hidden bits in DVDs, video games, and so on that pop up when you press certain buttons or menu combinations. Easter egg content can include outtakes, hidden scenes, creator credits, extra features, and so on—but you have to know how to find them! Here are some of my favorite DVD Easter eggs:

- *Austin Powers: The Spy That Shagged Me.* Go to the main DVD menu, select Special Features, and then do nothing for about a minute. That's when the Dr. Evil rocket ship comes up from the bottom of the screen and reveals a new icon; click the icon to view a Dr. Evil special feature.
- *Die Another Day.* On the second disc of the two-disc special edition, select the image database, choose Locations and Sets, and then scroll through the images until you come to the one with Halle Barry coming out of the water. Press the Up bottom on your remote and you'll see a hidden Halle Barry menu; you can now view Ms. Barry coming up out of the water from different camera angles, which is not a bad thing.
- *The Incredibles.* On disc two, go to the main menu and wait 30 seconds. When the small picture of the Omnidroid appears, select it and press Enter to play a short clip. There are similar clips hidden on almost all the submenus, including a sock puppet version of the movie.
- *Star Wars Episode 1: The Phantom Menace.* Select Options from the main menu, and then punch 1 1 3 8 on your remote, pausing between each number. This displays a credit sequence that includes outtakes from the film.

ON THIS DAY: FIRST SYNTHETIC RUBBER (1930)

On this date in 1930, Dr. Arnold M. Collins isolated a chemical called chloroprene, the world's first synthetic rubber. Chloroprene was later marketed under the trade name of Du Prene, by the Du Pont company.

FACT OF THE WEEK

A survey commissioned by Yahoo! (www.yahoo.com) indicates that 72% of American adults have considered starting their own business, and a like number (75%) say that the Internet has made it easier to launch a small business.

April 11, 2006

TUESDAY

THIS WEEK'S FOCUS
EASTER EGGS

CD EASTER EGGS

Many recording artists put Easter eggs on their CDs, typically in the form of hidden tracks, backward masking, and other audio tricks. (Here's a tip: When listening to a CD, always let the last track play until it skips back to the first track; often, artists will hide a song here.) Here are a few good CD Easter eggs:

- *The Beatles: Sgt. Pepper's Lonely Hearts Club Band.* This is a famous Easter egg. On the LP version, the run out groove featured a high-pitched dog whistle, well above the range of human hearing, put there by John Lennon for all the dogs to hear.
- *Jimi Hendrix: Are You Experienced?* Play the cut "Third Stone from the Sun" at about twice normal speed and the formerly incoherent vocals will turn into a discussion between two aliens approaching Earth.
- *Dave Matthews Band: Before These Crowded Streets.* There's a hidden song at the end of the cut "Spoon," which is actually the reprise to "The Last Stop."
- *Alanis Morisette: Jagged Little Pill.* Let the last track (an alternate take of "You Oughta Know") play out and you'll hear the hidden track, "Your House."
- *Santana: Supernatural.* Play through the last track and you'll hear a hidden song—Ritchie Havens' "Freedom."
- *Todd Snider: Songs for the Daily Planet.* Look for a very funny hidden song after the end of the final track, "Joe's Blues."
- *Frank Zappa and the Mothers of Invention: We're Only In It for the Money.* Play the track "Hot Poop" backward and you'll hear the original lyrics to the song "Mother People" that the record label had forced Frank to change.

ON THIS DAY: APOLLO 13 LAUNCHED (1970)

On this day in 1970, the ill-fated Apollo 13 moon mission was launched from Cape Canaveral. Two days later, disaster struck when a liquid oxygen tank exploded, causing mission commander James Lovell to exclaim, "Houston, we've had a problem." The lunar landing was aborted, and the crippled spacecraft began a long, cold journey back to Earth—where it landed safely on April 17th. All the astronauts survived the harrowing adventure.

GADGET OF THE WEEK: EVA SOLO MAGNETIMER KITCHEN TIMER

This is a strangely appealing gadget. The Magnetimer is a simple round timer. It's magnetized, so you can stick it to any metal surface. When it's time to time, just slap it to the stainless steel fronts of your fancy-schmancy high-end appliances, and you're ready to start the countdown. If you ever watch *Good Eats* on the Food Channel, you'll notice host Alton Brown using a similar timer. Buy it for $35 at www.evasolo.com/products-minutur.html.

April 12, 2006

WEDNESDAY

THIS WEEK'S FOCUS
EASTER EGGS

GAME EASTER EGGS

Game designers love their Easter eggs, as this short list will attest:

- **Age of Empires**—In the chat box, type `jack be nimble`. This will cause the catapults to start catapulting men and cows at enemies.

- **Halo 2**—In any multiplayer level, make sure you have an empty rocket launcher and an energy sword. Take the rocket launcher out and aim it at a person. When the circle turns red, press the "Y" button to switch to the energy sword, and then simultaneously press the trigger button as if you're firing the rocket. This launches the master chief like a rocket with a sword in his hand!

- **King's Quest VIII**—In the Frozen Reaches, go to the outside of Queen Freesa's castle and click on the guard with your weapons cursor. Then click all the guards in the throne room, going counter-clockwise from the entrance. Next, face Queen Freesa and turn to the left wall, and then click the block on the wall that is down three and over seven from the top left-hand corner. You'll find out where Waldo is—and watch Freesa and Connor have a funny conversation.

- **Roller Coaster Tycoon 2**—Create a scenario with Giga Coaster, Hyper Twister Coaster, or Inverted Impulse building capability, and then create a line leading out of the entrance. As a guest is about step into the line, name that guest Elissa White. When Elissa steps into the line, she'll say "I'm so excited—it's an Intamin ride!" (In case you didn't know, Intamin is the designer of real-world Giga Coasters, Hyper Twister Coasters, and Inverted Impulse coasters—and Elissa White is one of the Roller Coaster Tycoon game designers.)

ON THIS DAY: FIRST INTERNET SPAM (1994)

This is where it all started. On this day in 1994, Laurence Canter and his wife Martha Siegel flooded Usenet newsgroups with a notice for his law firm's "Green Card Lottery."

HARDWARE OF THE WEEK: USB CAFÈ PAD

The USB Cafè Pad isn't a mouse pad, it's a drink warmer. Just plug it into your PC's USB port, set your coffee mug on top, flip the switch, and your hot beverage will stay relatively hot. Buy it for $15 from www.usbgeek.com.

April 13, 2006

THURSDAY

THIS WEEK'S FOCUS
EASTER EGGS

SOFTWARE EASTER EGGS

Software programmers apparently have a lot of free time, because they like to add little games and credit screens as Easter eggs to their programs. Here's a list of some popular ones:

- **Final Cut Pro**—In version 4.X, select Tools, Button List. In the resulting entry box, enter Bruce; this will appear as "ShiftBruce." Use the Backspace key to delete the Shift symbol, and then drag the resulting icon to any window, and click the button. Bruce the Wonder Yak will now appear!
- **mIRC**—Select Help, About, then click the end of Khaled's nose. A "squeek" sound will result and the little dot from the "i" will start bouncing around.
- **Mozilla Firefox**—Open a blank window and type about:credits into the address bar. This will display a hidden credits page.
- **Paint Shop Pro**—Pull down the Help menu and select About Paint Shop Pro. When the About dialog box appears, hold down Ctrl+Shift and click anywhere in the splash screen. Release Ctrl+Shift and type either paintshoppro (for versions 5 and 6) or jascsoftware (for version 7). This reveals the hidden Atomic Dog splash screen.
- **PowerPoint**—In PowerPoint 97, 98, or 2000, select Help, About Microsoft PowerPoint. When the dialog box appears, click anywhere on the graphic to the left, and the box displays developer credits.

ON THIS DAY: FIRST ELEPHANT IN AMERICA (1796)

On April 13, 1796, the first elephant was brought to the United States from Bengal, India. The two year-old elephant was exhibited at the corner of Beaver Street and Broadway in New York City.

SOFTWARE OF THE WEEK: COFFEE BREAK WORM

No, this isn't the kind of worm you find in computer viruses. It's a game that you can play on your coffee break; the objective is to make the onscreen worm eat as many fruits as it can. Download it for free from www.gamejamboree.com/download-Coffee-Break-Worm.html.

April 14, 2006

FRIDAY

THIS WEEK'S FOCUS
EASTER EGGS

LITERARY EASTER EGGS

Easter eggs aren't limited to digital products; many famous authors have written hidden gems into their manuscripts. Here are a few of the most famous ones:

- Stephen King likes to make sly references to his other works in the books he writes. The most unusual reference came in *Gerald's Game*, which was published in 1992 but made reference to events in his book *Dolores Claiborne*, which wasn't published until 1993! Check out pages 166–167 of the first edition of *Gerald's Game* to find the reference to events on pages 183–184 of *Dolores Claiborne*.
- Umberto Eco paid tribute to some of his favorite writers when naming two key characters in *The Name of the Rose*. The character named William of Baskerville refers both to Sir Arthur Conan Doyle's famous Sherlock Holmes story, *The Hound of the Baskervilles*, and to English philosopher William of Occam. The character named Jorge de Burgos is named after Argentine writer Jorge Luis Borges.
- Edgar Allen Poe liked to play word games in his poems. For example, check out Poe's poem "An Enigma." If you take the first letter of line one, the second letter of line two, and so on, it spells out Sarah Anna Lewis—a fellow poet whose work Poe had reviewed.

- Anne McCaffrey pays tribute to her late friend John Greene by including similarly named characters in all her books, such as Jay Greene in *Freedom's Landing*, Johnnie Green in *Power Lines*, Jayge in *Renegades*, and Jon Green in *Crisis on Doona*.

ON THIS DAY: CHINA'S GREAT SOFTWARE PURGE BEGINS (1995)

There have been lots of purges in Chinese history; this is the first one to involve computer software. The goal of this purge, which was undertaken with the assistance of U.S. software companies, was to remove illegally copied software from Chinese workspaces. It all began on this date in 1995.

WEBSITE OF THE WEEK: THE EASTER EGG ARCHIVE

Where do you find all these wonderful Easter eggs? At the Easter Egg Archive, of course. Go on your own Easter egg hunt at www.eeggs.com.

April 15/16, 2006

SATURDAY/SUNDAY

THIS WEEK'S FOCUS
EASTER EGGS

SEXUALLY EXPLICIT EASTER EGG

In most instances, Easter eggs in software and DVDs are relatively tame—typically extra scenes, outtakes, and credits. But that isn't always the case, as evidenced by the hidden sex scenes in the video game *Grand Theft Auto: San Andreas*.

It seems that the game developers at Rockstar Games decided to add a little nasty action to the best-selling game (above and beyond the well-known violence of the game itself), and created some rather explicit sexual action in the game's code. The average game player will never see the scenes, however, because they can't be accessed by traditional Easter egg methods; no manner of button pushing will get you to the dirty stuff, nor can you stumble across it accidentally.

Instead, you have to download a special third-party mod, called Hot Coffee, that unlocks these hidden scenes—actually, mini-games that let you engage in intimate relations with GTA's many "girlfriends," as well as show the girls without clothes in various parts of the main game. You probably heard something about this in the news in mid-2005. You can download Hot Coffee for free at www.gtagarage.com/mods/show.php?id=28; use it at your own discretion.

Parental warning: If your young children are playing GTA—with or without the Hot Coffee mod—we strongly urge you to consider curtailing your child's exposure to this game. While debatably suitable for adults, GTA is absolutely inappropriate for children. We don't endorse this game, nor the Hot Coffee mod. (And don't expect the Hot Coffee mod to last for long; Rockstar has released a new version of the game without the hidden material, and they're taking a lot of flack from parents' groups about the whole thing.)

ON THIS DAY: LISP LANGUAGE UNVEILED

On April 16, 1959, the LISP programming language was unveiled to the public. Developed by John McCarthy, LISP offered programmers a flexibility in organization particularly useful in the development of artificial intelligence programs.

BLOG OF THE WEEK: ENGADGET

Discover all the latest tech gadgets at Engadget, a blog for gadget lovers. Check it out at www.engadget.com.

April 17, 2006

MONDAY

THIS WEEK'S FOCUS
EBAY

EBAY NEWS AND ANNOUNCEMENTS

eBay is constantly changing. There's something new practically every day, and it's tough to keep track of all the changes. Fortunately, eBay keeps you up to date on all the latest goings-on via official announcements. There are three ways to access these almost-daily updates.

First, you can view them on the My Summary page of My eBay. Second, you can view them on the eBay General Announcements board (www2.ebay.com/aw/marketing.shtml). Finally, you can have eBay send you all the latest announcements via email. To sign up, click the Community link at the top of any eBay page, and when the next page appears, click the eBay Groups link in the People section. When the Groups page appears, go to the News & Events section and click the Announcements link. When the next page appears, click the eBay Announcements link, and when the next page appears, click the Join Group link in the top-right corner. That's harder than it needs to be, but you only have to do it once.

Of course, just reading the official announcements won't tell you all there is to know about eBay. For that you need a dispassionate third-party source, such as AuctionBytes (www.auctionbytes.com). This site not only features the latest eBay news, but also a ton of tips and other information about eBay and other online auction sites. You can also subscribe to the free *Update* and *NewsFlash* online newsletters to have current news delivered directly to your email inbox.

AuctionBytes isn't the only entity offering eBay-related newsletters. Other useful newsletters include the Auction Guild's *TAGnotes* (www.auctionguild.com) and Auction KnowHow's *Auction Gold* (www.auctionknowhow.com/AG/). eBay also has its own official newsletter, *The Chatter*, which you can read online at pages.ebay.com/community/chatter.

ON THIS DAY: BENJAMIN FRANKLIN DIES (1790)

Printer, publisher, inventor, scientist, diplomat, patriot. All these words describe the great Benjamin Franklin, who died on this date in 1790. He lived a long and healthy life, having been born on January 17, 1706.

FACT OF THE WEEK

Pierre Omidyar launched eBay on Labor Day of 1995. As the story goes, Pierre's then-girlfriend commented to him about how great it would be if she were able to collect PEZ dispensers using the Internet. Pierre did her a favor and developed a small PEZ-dispenser trading site, originally dubbed Auction Web, that quickly became a much larger site and morphed into the eBay of today. The company went public in 1998.

April 18, 2006

TUESDAY

THIS WEEK'S FOCUS
EBAY

USING FEEDBACK

eBay's feedback system is how users get graded on their performance. A buyer generates good feedback by paying promptly; a seller generates good feedback by offering quality merchandise and shipping soon after being paid.

One of the primary uses of feedback is to help you decide which eBay users to deal with, or not. That's because the higher one's feedback rating, the more trustworthy that user is—in most cases. For that reason, experienced eBay users zealously guard their feedback status, and strive for as close to a perfect score (that is, the fewest negatives) possible.

You can learn even more about a particular user by taking the time to read the individual feedback comments left by their trading partners. Read the positive comments to see if they're tempered in any way. Read the negatives to see what they're for, and if (and how) the user responded. For example, if a user gets negatives for slow shipping, that might not be a big deal if you're not in a hurry.

It also helps to check out the feedback of those users leaving negative comments. Some users have a history of handing out negatives, and you might need to discount their remarks. In short, look for the story behind the numbers to get a better picture about a user's true performance.

ON THIS DAY: ALBERT EINSTEIN DIES (1955)

Albert Einstein was perhaps the greatest scientific mind of the twentieth century; his many theories proposed entirely new ways of thinking about space, time, and gravitation. He was born on March 14, 1879, and died on April 18, 1955.

GADGET OF THE WEEK: SPORTVUE MC1 HEADS-UP DISPLAY

SportVue's MC1 adds a heads-up display to any motorcycle helmet, so you can see key readouts—speed, RPM, and gear—without taking your eyes off the road. The display is projected onto the helmet's windscreen via the MC1 device, which mounts onto the front of your helmet; it's the same technology used by fighter pilots. Buy it for $329 from www.sportvue.net.

April 19, 2006

WEDNESDAY

THIS WEEK'S FOCUS
EBAY

BIRDDOGGING FOR BARGAINS

Some eBay users are natural bargain hunters. So why reinvent the wheel? Find out what the best bargain hunters are bidding on, and you might be able to score a deal, too.

The term "birddogging" refers to the act of following the auctions of another eBay bidder. Find a user who always gets great merchandise, and then birddog his or her other auction activity. Chances are you'll find something you like, and then you can get in on the bidding, too.

To birddog another member, you first have to find him—which is as easy as looking at the high bidder in a particular auction. Once you have a member identified, it's time to do the actual birddogging, using eBay's search function. Go to the Search page and click the Items by Bidder link. When the Items by Bidder page appears, enter the bidder's user ID and click the Search button. (For the most possible results, you should also select the Even if Not the High Bidder option.) The results page lists all the other auctions your subject is bidding on, which amounts to your own personal shopping list. Bid away!

And here's something else to keep in mind: Once you find a seller who offers really good deals on merchandise you like, add that seller to your Favorite Sellers list. (Just click the Add to Favorite Sellers link on any of that seller's auction pages.) You'll then receive a weekly update of that seller's current auctions, and the seller will be added to the list of Favorite Sellers displayed in My eBay.

ON THIS DAY: FIRST FORTRAN PROGRAM (1957)

On this day in 1957, researchers at Westinghouse ran the first FORTRAN program. The first attempt at running the program produced a missing comma diagnostic; the second attempt was successful.

HARDWARE OF THE WEEK: LOGITECH DINOVO CORDLESS DESKTOP

When every mouse and keyboard starts to resemble every other mouse and keyboard, Logitech's diNovo Cordless Desktop stands out from the crowd; this mouse and keyboard combo features hip styling and the latest cordless technology. The unique part of the diNovo is a separate piece that docks to the keyboard, called the MediaPad, which you can use to control all your digital media. Buy the diNovo desktop for $149.95 at www.logitech.com.

April 20, 2006

THURSDAY

THIS WEEK'S FOCUS
EBAY

EBAY POWERSELLERS

One of the signs of a successful eBay seller is the achievement of PowerSeller status. PowerSellers, quite simply, are those sellers who generate the most revenue, month in and month out. You can't choose to be a PowerSeller—eBay chooses you, based on your past sales performance (at least $1,000 average sales per month). If you're chosen, you don't have to pay for the privilege; membership in the PowerSellers program is free. (Learn more about the PowerSeller program here: pages.ebay.com/services/buyandsell/welcome.html.)

Some sellers find that sales go up when they display the PowerSeller logo; some find that it actually drives away some potential buyers. (These buyers apparently prefer buying from "little guys"; the PowerSeller logo makes a seller look like a bigger player.) Most, however, agree that the PowerSeller logo means more to newbie buyers. These buyers are looking for some reassurance when they bid, and appreciate the good reputation conveyed by the earning of PowerSeller status.

There are other benefits to achieving PowerSeller status, of course. PowerSellers qualify for priority customer support from eBay. Bronze-level PowerSellers get dedicated 24/7 email support, while higher levels get honest-to-goodness live telephone support—something regular sellers can only dream of. Of course, this support is only worthwhile if you actually need the help, but still.

In addition, qualified PowerSellers can take advantage of eBay's Healthcare Solutions, a medical insurance plan provided by Marsh Advantage America. There's also a rebate when you use eBay's Direct Pay billing, and access to the PowerSellers Entrepreneur Resource Center, a one-stop site for third-party business services. Not bad for free.

ON THIS DAY: WHIRLWIND COMPUTER ON TV (1951)

On the evening of April 20, 1951, MIT project director Jay Forrester demonstrated the Whirlwind computer on Edward R. Murrow's *See It Now* television program. The Whirlwind used 4,500 vacuum tubes and 14,800 diodes and took up a total of 3,100 square feet.

SOFTWARE OF THE WEEK: EBAY TOOLBAR

Many frequent eBay operations can be accessed directly by the eBay Toolbar, an add-on toolbar that attaches to the Internet Explorer web browser. With the eBay Toolbar installed, you're a button click away from doing pricing research, checking on your auction bidders, and going directly to your My eBay; it also includes an Account Guard function that warns you if you accidentally navigate a spoofed eBay site. Download it for free at pages.ebay.com/ebay_toolbar/.

April 21, 2006

FRIDAY

THIS WEEK'S FOCUS
EBAY

FINDING STUFF TO SELL

When you're first starting out, you can find lots of stuff to sell just sitting around the house. Sooner or later, however, you run out of stuff. That's when you get serious about this online selling business and go looking for more merchandise to sell on eBay.

Many eBay sellers get their merchandise from professional liquidators. These are companies that purchase surplus items from other businesses, in bulk. These items might be closeouts, factory seconds, customer returns, or overstocked items—products the manufacturer made too many of and wants to get rid of. Some popular liquidators include AmeriSurplus (www.amerisurplus.com), Liquidation.com (www.liquidation.com), and My Web Wholesaler (www.mywebwholesaler.com). And don't forget Overstock.com (www.overstock.com), which offers good values on clearance and overstocked merchandise.

Another alternative for obtaining new merchandise is to purchase from a wholesale distributor. The distributor purchases their merchandise direct from the manufacturer, who in many cases doesn't deal directly with retailers, and then resells it to retailers and other third parties. You can find directories of wholesalers at Buylink (www.buylink.com), Wholesale Central (www.wholesalecentral.com), and Wholesale411 (www.wholesale411.com).

You also might want to consider using a supplier that offers drop shipping services. Drop shipping is where the supplier inventories the merchandise for you and ships directly to your customer when you make a sale. Check with your supplier to see what services are available.

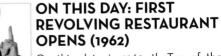

ON THIS DAY: FIRST REVOLVING RESTAURANT OPENS (1962)

On this date in 1962, the Top of the Needle restaurant opened in Seattle, Washington. This was the nation's first revolving restaurant and was opened by remote control by President John F. Kennedy from Palm Beach, Florida. The Needle remains one of the most famous and recognizable buildings in the Seattle skyline.

WEBSITE OF THE WEEK: ÁNDALE

Lots of companies offer third-party tools to help you manage your eBay auctions, but Ándale (www.andale.com) offers the most variety of tools—everything from listing creation to post-auction checkout to detailed sales and research reports. All their tools are available on an ala carte basis, so you only have to pay for what you need.

April 22/23, 2006

SATURDAY/SUNDAY

THIS WEEK'S FOCUS
EBAY

EBAY TRADING ASSISTANTS

Selling your own merchandise is only one way to make money on eBay. One of the newest types of eBay businesses exists to sell other people's merchandise. These so-called eBay resellers or drop-off stores—officially known as eBay Trading Assistants—help non-eBay users sell their goods on eBay by taking them on consignment and then reselling the items via normal eBay auction. The Trading Assistant handles the entire auction process, from taking photos and listing the item to handling payment and shipping the merchandise. The client doesn't have to bother with the whole eBay process, and the Trading Assistant earns a percentage of the final selling price.

To become a Trading Assistant, you have to have sold at least 4 items in the past 30 days and have a feedback rating of at least 50 with a percentage of at least 97%. To join up, just go to the Trading Assistants Hub (pages.ebay.com/tahub) and click the Become a Trading Assistant link.

Because many Trading Assistants work out of their home, this requires them to either pick up items from their clients' homes or have their clients drop off items at their home. A more professional (and more expensive) option is to rent your own retail storefront for merchandise drop-offs. The downside, of course, is you have additional costs—the rent and utilities for your store, plus signage and the like. It's your choice.

You can also make the whole process a lot easier by buying into a Trading Assistant franchise. If you don't mind sharing your profits with the head office, some of the more popular eBay drop-off franchises include Auction Mills (www.auctionmills.com), iSold It (www.i-soldit.com), and QuikDrop (www.quikdrop.com).

ON THIS DAY: WILHELM SCHICKARD BORN (1592)

Wilhelm Schickard, the father of the calculator, was born on April 22, 1592. Schickard created an early adding machine, called the "calculating clock," that used wooden gears to add and subtract up to six-digit numbers.

BLOG OF THE WEEK: OUTRAGEOUS EBAY AUCTIONS

I love this one. It's a blog devoted to the oddest, funniest, most outrageous current auctions on eBay. See what I mean at outrageous-ebay.blogspot.com.

April 24, 2006

MONDAY

THIS WEEK'S FOCUS
IPODS

IPODS RULE!

The most popular portable music player today is the Apple iPod. Apple moves a million or so of these puppies every month; for many teenagers, an iPod is a required accessory—even more than the latest camera phone. You have to hand it to Steve Jobs. He's done a good job of marketing this little white slab of plastic with some of the most inventive and entertaining television commercials ever produced.

Apple sells several different versions of the iPod. The big daddy of the iPod line is, quite simply, the hard-disk iPod. Shrink the case and use flash memory instead of a hard drive, and you get the new iPod Nano. Go with an even smaller case (and remove the display, for some unknown reason), and you get the iPod Shuffle.

Probably the best thing about the iPod and the iPod Nano is the Control Wheel. This is one area where Apple really got it right, and competitors still haven't come close. Operating an iPod is as easy as moving your thumb around the outside of the wheel to scroll through the menu system and then clicking the center button to select a menu item. It does exactly what you expect it to, no surprises.

As good as the iPod is, it has one big drawback: It only plays digital audio files in the AAC and MP3 formats. (AAC is somewhat proprietary to Apple, no surprise.) It does *not* play files in Microsoft's WMA format. So you won't be able to play WMA files downloaded from Napster, Musicmatch, and similar sites on your iPod device; if you have a big library of WMA files, choose another brand of player.

ON THIS DAY: APPLE IIC INTRODUCED

Hey kiddies, Apple doesn't just make iPods, it also makes computers! Back on April 24, 1981, Apple introduced the Apple IIc, which was a portable machine with the same capacity as the standard IIe model. It came with a whopping 128KB of RAM and a state-of-the-art floppy disk drive. Wowzers!

FACT OF THE WEEK

Our pals at the Pew Internet and American Life Project (www.pewinternet.org) tell us that 19% of folks age 18-28 own an iPod or other MP3 player. Of those age 29-40, 14% are iPod/MP3 player owners; of those 41-50, 11% own a player. Of oldsters between 51 and 69, only 6% own some sort of audio player. It really is an age thing!

April 25, 2006

TUESDAY

THIS WEEK'S FOCUS
IPODS

USING YOUR IPOD WITH A WINDOWS PC

The iPod works great with Macs, and you can also use it with Windows-based PCs. Of course, you first have to install Apple's iTunes software; iTunes is a music player program, like Windows Media Player, that is fine-tuned to work with both the iPod and with Apple's iTunes Music Store. You use iTunes to transfer music files to your iPod, as well as to purchase and download digital music from the ITunes Music Store site on the web.

Connecting the iPod to your PC is as simple as connecting a cable from the iPod to a USB or FireWire port on your computer. When the iPod is connected, your PC automatically launches the iTunes software and downloads any new songs and playlists you've added since the last time you connected.

By the way, while both USB and FireWire are equally fast, connecting via FireWire has the added benefit of recharging your iPod while connected to your PC. Otherwise you have to connect the iPod to the standard charger to recharge the battery.

Know, however, that the iPod won't be able to play back all the digital music you have stored on your PC. That's because the iPod only plays digital audio files in the AAC and MP3 formats. (AAC is somewhat proprietary to Apple.) It does *not* play files in Microsoft's WMA format, so you won't be able to play WMA files downloaded from Napster, Musicmatch and similar sites on your iPod device.

ON THIS DAY: INTEGRATED CIRCUIT PATENTED

On this date in 1961, Robert Noyce received a patent for the integrated circuit he developed. It was a controversial patent, however, as competitor Jack Kilby had invented a germanium version of the IC, while Noyce developed the silicon IC—which ultimately became the more accepted design.

GADGET OF THE WEEK: GRIFFIN BLUETRIP AUDIO HUB

Griffin's BlueTrip lets you connect your iPod to your home audio system, using Bluetooth wireless technology. The BlueTrip connects to your audio receiver via standard R/L RCA jacks or optical digital connection; it can transmit up to 30 feet away. Buy it for $149.99 from www.griffintechnology.com.

April 26, 2006

WEDNESDAY

THIS WEEK'S FOCUS
IPODS

CREATING PLAYLISTS FOR YOUR IPOD

Most users like to assemble collections of their favorite songs, in the form of playlists. The iTunes software provides two different ways to create playlists for use on your iPod.

The easiest type of playlist to create is the Smart Playlist. This is a playlist that iTunes automatically assembles from information you provide. You can create Smart Playlists based on artist, album, genre, file size, beats per minute, and other attributes. All you have to do is select File, New Smart Playlist. This displays the Smart Playlist dialog box; select the criteria you want, click OK, and iTunes finds all matching songs. For example, to create a playlist of all the Coldplay songs in your Library, choose Artist from the first drop-down list and enter Coldplay in the right-hand box.

You can also create playlists manually. Just select File, New Playlist to add a new playlist to the Source pane, and then click the new playlist twice to edit the title. You add songs to the new playlist by dragging songs from the Library onto the new playlist icon. Once you've added songs to a playlist, you can view the contents of the playlist by selecting it as your source. Then you can rearrange the songs by dragging them into the order you prefer.

The next time you connect your iPod to your computer, all your new or changed playlists will be transferred to the iPod.

ON THIS DAY: CHERNOBYL NUCLEAR DISASTER (1986)

Russia's Chernobyl nuclear plant exploded on this day in 1986. It was the world's worst civil nuclear catastrophe (to date), killing 31 people immediately (and thousands more over the long term) and sending a cloud of radioactive dust over Europe. It was the result of an experiment went wrong that caused the fourth reactor in the complex to explode and melt down.

HARDWARE OF THE WEEK: APPLE AIRPORT EXPRESS

Here's something a lot of people don't know: You can use Apple's AirPort Express as a budget digital media hub. You don't get a display, which means you need to queue things up on your computer, but it's a quick and easy digital media solution—especially if you're an iTunes user. The AirPort Express works with both PCs and Macs and connects to your PC via 802.11g WiFi wireless connection. Buy it for $129 at www.apple.com.

April 27, 2006

THURSDAY

THIS WEEK'S FOCUS
IPODS

IPOD SPEAKERS

When you want to listen to your iPod without earphones—or share the music with a roomful of friends—you need a set of powered speakers. Plug the speakers into the iPod's earphone jack, and you have a full-featured digital audio system. Here are some of the more popular iPod speaker systems:

- **Cube Travel Speakers**—The Cube is a fold-up speaker system powered by four AAA batteries, ideal for travel use. Buy it for $39.99 at www.pacrimtechnologies.com.
- **Macally PodWave**—The PodWave is a mini-speaker that attaches to the top of your iPod. It's battery powered, so it won't drain your iPod's battery. Buy it for $39 at www.macally.com.
- **Altec Lansing inMotion**—Here's another battery-powered iPod speaker system, but with better sound. The base of the unit incorporates a highly efficient Class D power amplifier and two 1-inch microdrivers. Buy it for $149.95 at www.alteclansing.com.
- **JBL OnStage**—This is an AC-powered speaker system just for iPods. Insert your iPod into the center cradle, and you get room-filling sound from the four built-in drivers. Buy it for $159.95 from www.jbl.com.

- **Bose SoundDock Speaker System**—Even better sound can be had from Bose's SoundDock. The shielded speakers deliver big Bose sound; all you have to do is dock your iPod at the front of the unit, then you can control the whole shebang with the included wireless infrared remote control. Buy it for $299 from www.bose.com.

ON THIS DAY: MICROSOFT PURCHASE OF INTUIT BLOCKED (1995)

Microsoft was on the acquisition path in 1995 and wanted to buy Intuit, the maker of Quicken. On April 27, however, the Justice Department stepped in to block the purchase, saying the deal could lead to higher software prices and diminished innovation. Microsoft eventually dropped the deal, deciding to compete directly against Quicken with its Microsoft Money software.

SOFTWARE OF THE WEEK: iART

iArt is a shareware program that lets you download album art for your iTunes library—and to display on any color iPod screen. You can use iArt to search Amazon or Google for the appropriate album art. Download it for $10 at www.ipodsoft.com.

April 28, 2006

FRIDAY

THIS WEEK'S FOCUS
IPODS

COOL IPOD ACCESSORIES

There's a huge business in selling iPod accessories, so let's look at some of the most popular ones:

- **Belkin iPod Voice Recorder**—This gadget features a built-in microphone and lets your iPod double as a digital audio recorder. You can record hundreds of hours of conversations in mono WAV format, and listen back via the built-in speaker. Buy it for $30 from www.belkin.com.

- **Belkin TunePower**—TunePower is a rechargeable external battery pack that gives you 8-10 hours of extra play time and recharges via your iPod's FireWire recharger cable. Buy it for $99.99 from www.belkin.com.

- **Solio Solar Charger**—Solio is an add-on battery pack for your iPod that recharges via three solar panels or standard AC power. Buy it for $89.99 from www.solio.com.

- **Griffin BlueTrip**—Griffin's BlueTrip lets you connect your iPod to your home audio system, using Bluetooth wireless technology. Buy it for $149.99 from www.griffintechnology.com.

- **naviPlay Remote**—The naviPlay is a remote control system for your iPod that uses Bluetooth wireless technology. The remote unit includes a headphone jack; tuck the remote somewhere on your person and hide your iPod up to 30 feet away. Buy it for $199 from www.tentechnology.com.

ON THIS DAY: KURT GÖDEL BORN (1906)

On this day in 1906, mathematician Kurt Gödel was born. He's best known for his proof of Gödel's Incompleteness Theorems, which showed that in any axiomatic mathematical system there are propositions that cannot be proved or disproved within the axioms of the system. (Want to learn more? Read Douglas Hofstadtler's book *Gödel, Escher, Bach: An Eternal Golden Braid*.)

WEBSITE OF THE WEEK: ILOUNGE

The iLounge (formerly known as the iPod Lounge) is the Web's premiere source for iPod news, reviews, forums, and more. It's a great place to find out about the latest iPod accessories and to download third-party iPod utilities. It's located at www.ilounge.com.

April 29/30, 2006

SATURDAY/SUNDAY

THIS WEEK'S FOCUS
IPODS

MORE COOL IPOD ACCESSORIES

Without further ado, here's another batch of cool accessories for your iPod:

- **Griffin iBeam**—Now this is a cool little gadget, a small laser pointer/flashlight that snaps onto the top of your iPod. Buy it for $19.99 at www.griffintechnology.com.

- **Belkin Digital Camera Link**—This gadget lets you use your iPod to store hundreds of digital photos, transferred directly from your digital camera. Buy it for $79.99 from www.belkin.com.

- **Belkin iPod Media Reader**—Belkin's Media Reader connects directly to your iPod and reads the following digital media formats: CompactFlash I, CompactFlash II, SmartMedia, Secure Digital, Memory Stick, and MultiMediaCard. Buy it for $99.99 from www.belkin.com.

- **H2O Audio Underwater iPod Housing**—Want to take your iPod boating or swimming? Then tuck it into H2O Audio's waterproof housing, and you won't do any damage. It's submersible up to 10 feet and lets you control your iPod from the outside. Buy it for $149.95 from www.h2oaudio.com.

- **Denision ice>Link Plus**—The ice>Link Plus kit lets you connect your iPod to your car's existing CD changer port—no FM transmitter needed. The sound is better than you get with an FM kit, plus you can control your iPod from your car radio controls. Buy it for $199 from icelink.denisionusa.com.

ON THIS DAY: GEORGE STIBITZ BORN (1904)

On April 30, 1904, computer pioneer George Stibitz was born. In 1937, Stibitz used a box of spare parts to construct the "Model K" (for "kitchen table") breadboard digital calculator. The Model K was the genesis for the later Complex Number Calculator, which was used for the first instance of remote job entry in 1940. Stibitz died on January 31, 1995.

BLOG OF THE WEEK: IPODITUDE

It's all iPod all the time at the iPoditude blog. This blog is chock full of news and comments about anything remotely iPod-related. See what I mean at www.ipoditude.com.

May 2006

May 2006

SUNDAY	MONDAY	TUESDAY	WEDNESDAY	THURSDAY	FRIDAY	SATURDAY
	1 Space Day — 1884 Construction begins on the first skyscraper	**2** 1933 Infamous Loch Ness Monster first sighted in Loch Ness, Scotland	**3** 1968 First successful heart transplant is performed	**4** 2003 First cloned mule is born at the University of Idaho	**5** 1961 U.S. makes its first manned space flight aboard the capsule Freedom 7	**6** 1937 The Hindenburg burst into flames, killing 36 passengers and crewmembers
7 1992 Space shuttle Endeavor makes its maiden voyage	**8** 1847 Rubber tires first patented	**9** 1960 First birth-control pill approved by the FDA	**10** 1954 First silicon transistor announced by Texas Instruments	**11** 1997 IBM computer Deep Blue defeats world chess champion Garry Kasparov	**12** 1816 First printing press is patented in Philadelphia, PA	**13** 1981 Pope John Paul II is shot in Rome's St. Peter's Square
14 Mother's Day — 1973 First U.S. space station, Skylab One, was launched	**15** 1940 Nylon stockings first offered for sale in Wilmington, DE	**16** 1995 First dual-processor PC introduced by Del	**17** 1973 Televised Watergate hearings begin	**18** 1980 Mount St. Helen's volcano eruption	**19** 2005 *Star Wars: Revenge of the Sith* opens in U.S. theaters	**20** Armed Forces Day — 1830 Fountain pen is patented in Reading, PA
21 1980 *Star Wars: Episode V: The Empire Strikes Back* released	**22** 1990 Windows 3.0 released	**23** 1785 Bifocal glasses invented by Benjamin Franklin	**24** 1938 Parking meters patented in Oklahoma City, OK	**25** 1992 Jay Leno's first appearance as the regular host of *The Tonight Show*	**26** 1930 Scotch tape invented in St. Paul, MN	**27** 1937 Golden Gate Bridge opens in San Francisco, CA
28 2003 First cloned horse is born in Cremona, Italy	**29** Memorial Day — 1992 First digital TV test by Zenith and AT&T	**30** 1971 Mariner 9 space probe is launched for Mars	**31** 1990 *Seinfeld* premieres			

May 1, 2006

MONDAY

THIS WEEK'S FOCUS
CHAT AND INSTANT MESSAGING

HOW INSTANT MESSAGING WORKS

Instant messaging works a little differently from most Internet applications. Email, Usenet, IRC, the Web—all these parts of the Internet operate via a client/sever model, with most of the heavy lifting done via a network of dedicated servers. Your email is stored on and managed by an email server, Usenet articles are stored on and propagated by Usenet servers, IRC chats are hosted on a network of IRC servers, and all the pages on the Web are hosted on millions of individual web servers.

Instant messaging, however, doesn't use servers at all. When you send an instant message to another user, that message goes directly to that user; it's not filtered by or stored on any dedicated servers. The technical name for this type of connection is *peer-to-peer (P2P)* because the two computers involved are peers to each other.

All instant messaging needs to work is a piece of client software (one for each computer involved, of course) and the IP addresses of the individual computers. The communication goes directly from one IP address to another, with no servers in the middle to interfere or introduce lag. (Naturally, the data must still make its way through numerous routers to get to the other PC, but that's part and parcel of any Internet-based application.)

Because of the way that it works, you use instant messaging in different ways than you use online chat or email. Instant messaging is ideal for very short, very immediate messages. (In fact, most instant messaging systems limit the length of the messages you can send through their systems.) Online chat is better for longer discussions and for group discussions. Email is better than either chat or instant messaging in communicating longer, more complex, and more formal messages.

ON THIS DAY: FIRST COMPUTER TIMESHARING (1964)

On the first day of May 1964, Dartmouth professors Thomas Kurtz and John Kemeny launched a computer timesharing system, which allowed several users to run their programs simultaneously on a single processor. The system used the BASIC programming language.

FACT OF THE WEEK

The Pew Internet and American Life Project (www.pewinternet.org) reports that 53 million Americans (42% of all Internet users) use instant messaging. Roughly 13 million people use IM on any given day, and 24% of all users say they use IM more frequently than they do email.

May 2, 2006
TUESDAY

THIS WEEK'S FOCUS
CHAT AND INSTANT MESSAGING

INSTANT MESSAGING CLIENTS

Instant messaging lets you communicate one on one, in real time, with your friends, family, and colleagues. It's faster than email and less chaotic than public chat rooms. It's just you and another user—and your instant messaging software.

There are several big players in the instant messaging market today, including

- AOL Instant Messenger (www.aim.com)
- ICQ (web.icq.com)
- Windows Messenger—also known as MSN Messenger (messenger.msn.com)
- Yahoo! Messenger (messenger.yahoo.com)

Unfortunately, many of these products don't work well (or at all) with each other; if you're using Windows Messenger, for example, you won't be able to communicate with someone running AOL Instant Messenger. That means you'll be messaging only with other users of the same program you're using.

Unless, of course, you go with a universal IM client, which lets you connect to multiple IM systems simultaneously, using the same program and interface. Some of the more universal IM programs include GAIM (gaim.sourceforge.net), Miranda (www.miranda-im.org), and Trillian (www.trillian.cc).

ON THIS DAY: MICROSOFT'S TWO-BUTTON MOUSE (1983)

On this day in 1983, Microsoft introduced the two-button Microsoft Mouse, to accompany its new Microsoft Word program. Microsoft built 100,000 of these rodent wonders, but only sold about 5,000; a much improved model, introduced in 1985, was more successful.

GADGET OF THE WEEK: TIVOLI MODEL ONE

The Tivoli Model One is the AM/FM table radio that the radio pros use. Designed by audio legend Henry Kloss, it's constructed from the highest-quality components and built to exact quality requirements. It's also the height of simplicity, with just three knobs to its name, and sounds terrific. Buy it for $99.99 from www.tivoliaudio.com.

May 3, 2006

WEDNESDAY

THIS WEEK'S FOCUS
CHAT AND INSTANT MESSAGING

GROUP CHATS IN AOL INSTANT MESSENGER

In addition to private conversations, AOL Instant Messenger also enables you to hold group conversations via the Buddy Chat feature of the AIM software. These Buddy Chats are private conversations among three or more people, unlike the public conversations in normal online chat.

To initiate a group chat, highlight a name in your Buddy list, and then click the Send Buddy Chat Invitation button. (Alternatively, you right-click the buddy's name in your Buddy List and select Send Chat Invitation from the pop-up menu.) When the Buddy Chat Invitation dialog box appears, select which screen name to invite, enter your message in the Invitation Message box, and then click the Send button.

If the user accepts your chat request, a new Chat Room window will open on your desktop. From within the Chat Room window, enter your messages in the bottom message box, and click Send to send your messages to the chat session. Messages from all users in the group chat will appear in the main message pane.

To add other buddies to this Buddy Chat, pull down the People menu, select Invite a Buddy, and send an invitation as described previously. If you receive a group chat request from another user, a flashing Chat icon will appear in the AIM window. (If your window is closed, a flashing icon will appear in your Windows Taskbar tray.) To accept the chat request, double-click on the flashing icon, and then select Accept.

ON THIS DAY: FIRST COMIC BOOK (1934)

On this day in 1934, *Famous Funnies*, the very first comic book ever, hit newsstands in the U.S. It was published by Delacourte and sold out its print run of 35,000 copies. (Learn more about the surprisingly interesting history of comic book publishing in Gerard Jones' book, *Men of Tomorrow: Geeks, Gangsters, and the Birth of the Comic Book*.)

HARDWARE OF THE WEEK: LOGITECH QUICKCAM ORBIT

Logitech's QuickCam Orbit is a webcam that doesn't just sit there and stare, it actually rotates and zooms and follows you as you move around the room. That's right, this camera features automatic face-tracking software and a mechanical pan-and-tilt mechanism with digital zoom to keep the camera focused on your face, no matter where your face might be. Buy it for $129.99 from www.logitech.com.

May 4, 2006

THURSDAY

THIS WEEK'S FOCUS
CHAT AND INSTANT MESSAGING

SENDING FILES VIA INSTANT MESSAGING

All instant messaging programs can be used to send files from one user to another, without having to bother with FTP or email attachments. It's a quick and convenient way to send data back and forth in real time.

To send a file using Windows Messenger (or its kissin' cousin, MSN Messenger), all you have to do is right-click the contact's name and select Send a File. When the Send a File dialog box appears, select the file you want to send, and then click the Open button. The file is now sent to the recipient you indicated.

In AOL Messenger, you select a name from your Buddy List, and then select People, Send File to Buddy. When the Send File To dialog box appears, click the File button to select which file to send, enter a message in the large text window, and then click the Send button.

Same thing with Yahoo! Messenger. Select a friend from your list and select Message, Send a File. When the Send a File dialog box appears, click the Browse button to display the Select the File to Send dialog box; select a file and click Open. Enter a message to accompany the file (in the Enter a Brief Description box) and click the Send button. That's all there is to it!

ON THIS DAY: COMMODORE SOLD (1995)

Venerable computer manufacturer Commodore was sold to German electronics company Escom AG on this date in 1995. Escom paid $10 million for the rights to the Commodore name, patents, and other intellectual property.

SOFTWARE OF THE WEEK: MIRC

Before there were chat rooms and instant messaging clients, there was Internet Relay Chat. IRC is still around, and the best IRC client is the venerable mIRC. Download the latest version for free at www.mirc.com.

May 5, 2006

FRIDAY

THIS WEEK'S FOCUS
CHAT AND INSTANT MESSAGING

TIPS FOR CHAT AND IM

Here are some tips to ensure a safe and enjoyable chat or instant messaging experience:

- Don't feel like you have to jump right in and participate in every single chat room discussion. It's okay to sit back, take your time, and watch the conversations flow before you decide to add your two cents' worth.

- Don't assume you're really talking to the person you *think* you're talking to. It's pretty easy to hide behind a nickname and create a totally different persona online.

- If you want to get personal with someone you meet in a chat channel, send a private message. Don't subject everyone in a room to your private conversations.

- Don't give out any personal information (name, address, phone number, or Social Security number) in any chat session—period. When you send personal information via chat or IM, you're opening yourself up for identity theft.

- Be very careful about accepting files sent to you during a chat or IM session. If you're not vigilant, you'll find yourself mindlessly accepting any file sent your way—which is a surefire way to infect your system with a computer virus. Carefully examine any file sent to you in a chat session, and reject anything that is in the least bit suspicious. Better safe than sorry!

ON THIS DAY: SCOPES ARRESTED FOR TEACHING EVOLUTION (1925)

On May 5, 1925, John T. Scopes was arrested for violating a state law against the teaching of Darwin's theory of evolution in a Tennessee public school. The next month he was found guilty of that heinous crime, at what became known as the Scope's Monkey Trial, and fined $100. Fortunately, we've since progressed beyond this ignorant nonsense—right?

WEBSITE OF THE WEEK: CYBERMOON STUDIOS

While Cybermoon Studios' site includes a bevy of laugh-inspiring material, the 8-bit D&D (that's Dungeons & Dragons) animation is enough to make any child of the '80s roll around on the floor laughing. If you ever played D&D in its heyday, be sure to check out the oh-so-cheesy 8-bit rendition of a typical D&D game. See for yourself at www.cybermoonstudios.com/8bitDandD.html.

May 6/7, 2006

SATURDAY/SUNDAY

THIS WEEK'S FOCUS
CHAT AND INSTANT MESSAGING

ONLINE CHAT COMMUNITIES

If you're serious about online chat, you might want to check out some of the other major chat communities on the Web. These sites include

- Internet TeleCafe (www.telecafe.com)
- MSN Chat (chat.msn.com)
- Talk City (www.talkcity.com)
- Yahoo! Chat (chat.yahoo.com)

If you're a subscriber to America Online, you can also access AOL's proprietary chat rooms. These are some of the busiest chat rooms on the Internet, and they're reserved exclusively for AOL members—which means you have to subscribe to AOL (and use the AOL interface) to join in the fun.

Of course, you can find chat rooms on lots of other websites. Many sites have their own private chat communities (as well as message forums), reserved for site visitors. If you're a fan of a specific site, click around and see if there's a way to chat in real-time with other visitors.

ON THIS DAY: HINDENBURG DISASTER (1937)

At 7:25 p.m. on May 6, 1937, the dirigible *The Hindenburg* exploded and burned while landing at the naval air station at Lakehurst, New Jersey. The zeppelin crashed into the ground 32 seconds after the flame was first spotted; 36 people died in the inferno. To this day, the cause of the fire remains the subject of much speculation.

BLOG OF THE WEEK: JAKE LUDINGTON'S MEDIA BLAB

Jake is a long-time digital media guru, and his blog covers just about everything related to the digital lifestyle. Read it at www.jakeludington.com.

May 8, 2006

MONDAY

THIS WEEK'S FOCUS
HDTV

DIGITAL TELEVISION

Our current NTSC television system is more than a half-century old. What was state-of-the-art in the 1950s is woefully inadequate today; today's larger television screens and higher-resolution source material require a better way to reproduce picture and sound than what we've grown used to over the years.

The first step to improving the quality of our television broadcasts is to move from analog to digital transmission. Just as digital compact discs replaced analog records, digital television will replace our outdated analog broadcasting system. Digital signals don't deteriorate as analog signals do; a digital broadcast looks every bit as good in your home as it does in the broadcast studio.

The primary advantage of digital broadcasting is that it slices the traditional analog television signal into a series of digital bits, which are then recombined to reproduce an exact copy of the original broadcast. These digital signals don't weaken with distance, as analog signals do. As long as the signal can be received, the picture is perfect, with no degradation or ghosting. Because digital signals are comprised of binary bits, a 1 is always a 1 and a 0 is always a 0. Because of this exact end-to-end reproduction, digital means better picture and sound quality, no matter what is broadcast.

It's important to note that digital television is not the same as high-definition television. While all HDTV is digital, not all digital broadcasts are high definition. (It's kind of like the old pasta analogy—while all spaghetti is pasta, not all pasta is spaghetti.) Digital technology is used in a variety of media, including DVDs, direct broadcast satellite (DBS), digital cable, and the new HDTV format.

ON THIS DAY: METRIC SYSTEM ESTABLISHED (1790)

On this day in 1790, following a motion by Charles Maurice Talleyrand, the French National Assembly decided on the creation of a decimal system of measurement called the metric system. The metric system was a base 10 system that adopted Greek prefixes for multiples and Latin prefixes for decimal fractions.

FACT OF THE WEEK

As of the end of 2004, 14 million U.S. households owned an HDTV-capable television, up from 8.7 million at the end of 2003. Strategy Analytics (www.strategyanalytics.com) forecasts that by 2008, 37 million households will be able to receive HDTV programming.

May 9, 2006

TUESDAY

THIS WEEK'S FOCUS
HDTV

COMPARING DIGITAL TELEVISION FORMATS

Digital programming comes in many shapes and sizes; not all digital broadcasts have the same resolution or aspect ratio. Here's a quick guide to all the possible format:

- **SDTV**—Standard definition digital television has the same resolution and aspect ratio as analog television but is transmitted in digital fashion. The SDTV picture is 480 × 640 pixels, with interlaced scanning.

- **EDTV**—Enhanced definition television has the same resolution as SDTV but with progressive scanning for a smoother overall picture. EDTV can be in either the 4:3 or 16:9 aspect ratio, for a resolution of either 480 × 640 or 480 × 720 pixels.

- **HDTV (720p)**—The first of the two current high-definition television formats features 720 × 1280 pixel resolution with progressive scanning. This format is ideal for programming with lots of movement, such as sporting events.

- **HDTV (1080i)**—The second current high-definition format features greater resolution (1080 × 1920 pixels) but with interlaced scanning. Both the 720p and 1080i formats feature a 16:9 aspect ratio and Dolby Digital 5.1 surround sound.

- **HDTV (1080p)**—This is the ultimate high-definition format, with 1080 × 1920 pixel resolution and progressive scanning. Because of the high bandwidth requirements, this format is not yet in use, although some newer HDTV sets can display this format—and upconvert 1080i programming to 1080p.

ON THIS DAY: BIRTH CONTROL PILL APPROVED (1960)

In 1960, the sexual revolution began when the Food and Drug Administration approved a pill safe for birth control use. The original contraceptive pill contained five times the estrogen that similar pills do today and ten times the progestin. The pill is the most common form of birth control today, and is about 99% effective in preventing pregnancy.

GADGET OF THE WEEK: MONSTER POWERCENTER

If you've spent a few grand or more on a quality audio/video system, you don't want to plug your stuff straight into a standard wall outlet. Monster's PowerCenter is a line conditioner that serves to filter out line noise as well as isolate noise between components. The HTS3500 (other models are available) filters 10 AC outlets, 3 coaxial connections, and 1 phone line connection. Buy it for $399.95 at www.monstercable.com.

May 10, 2006

WEDNESDAY

THIS WEEK'S FOCUS
HDTV

WHY HDTV IS BETTER

High-definition television is just one form of digital broadcasting—but it's the best one. The HDTV format combines several different technologies to offer a bigger, higher-resolution picture and sound field. Here's what you get when you go high-def:

- **Sharper picture**—True HDTV (720p or 1080i formats) delivers 1 million pixels or more of information. Standard TV and DVDs deliver approximately 300,000 pixels. That translates into more than three times as much detail as what you've been used to watching.

- **Less flicker**—The increased number of scan lines per frame (and, in the case of the 720p format, progressive scanning) translates into less-visible scan lines—which means you can sit closer to the screen without seeing flicker in the picture.

- **Widescreen picture**—When you display a widescreen movie on a standard 4:3 television, the movie is either letterboxed (with black bars at the top and bottom of the screen) or panned and scanned so that only a portion of the movie is displayed onscreen. With HDTV's 16:9 aspect ratio, you can see more of the widescreen movie with less letterboxing.

- **Better sound**—The HDTV format utilizes the Dolby Digital 5.1 surround-sound format, which is much improved over the NTSC stereo standard.

ON THIS DAY: TRANSCONTINENTAL RAILROAD COMPLETED (1869)

On May 10, 1869, the first transcontinental railroad was completed. The final spike joining the Central Pacific and Union Pacific railroads was driven at Promontory, Utah, with the engines No. 119 and Jupiter practically touching noses.

HARDWARE OF THE WEEK: TVIX DIGITAL MOVIE JUKEBOX

This is a promising device for those of you who want to store and stream digital movies. The TViX unit lets you rip CDs and DVD to the unit's hard drive, then play back movies and music through your home theater system. You can order the unit with anywhere from 120GB to 300GB of hard disk storage; prices range from $275 to $400. Buy it at www.tvix.co.kr/eng/.

May 11, 2006

THURSDAY

THIS WEEK'S FOCUS
HDTV

HOW HDTV WORKS

Images viewed on television screens are made up of small picture elements known as *pixels*. Each pixel is comprised of three closely spaced "dots" of color—red, green, and blue—arranged in rows and columns. The pixels are so close together that, from the proper distance, they appear connected—and create a complete picture.

The easiest way to increase picture resolution is to stuff more of these pixels onto the screen, by using smaller pixels. In the 1080i format, 18 HDTV pixels can fit in the same space occupied by 4 pixels used in traditional television sets.

HDTV pixels are not just smaller than SDTV pixels, they're also squarer. This enables an HDTV display to resolve finer details and hold smoother curves. The square pixels also remove some of the image distortion seen on older televisions.

By using these smaller, squarer pixels, an HDTV picture can contain up to seven times the total number of pixels found in a standard definition picture. The 1080i format uses more than 2 million individual pixels—compared with just 300,000 pixels in a traditional television screen. And *that's* why HDTV looks so much better than regular standard definition television!

LEARN MORE ABOUT HDTV

You can learn more about HDTV and home theater in two new books published by Que: *How HDTV and Home Theater Work*, by Michael Miller (the co-author of this book) and illustrator Michael Troller, and *Absolute Beginners Guide to HDTV and Home Theater*, by Steve Kovsky. Find both books at your favorite bookstore or online retailer!

ON THIS DAY: VISICALC DEMONSTRATED (1979)

The first public presentation of VisiCalc was given on this day in 1979. VisiCalc, a spreadsheet program for the Apple II, was developed by Dan Bricklin and Robert Frankston, and it helped usher in the personal computer revolution. Of course, most of us use Microsoft Excel for our spreadsheets these days, but VisiCalc paved the way.

SOFTWARE OF THE WEEK: MOVIE LABEL 2006

Movie Label 2006 is a shareware program that helps you organize your DVD movie collection on your PC. You can also use the program to download cover art and key information about your movies. Buy it for $39.95 from www.codeaero.com.

May 12, 2006

FRIDAY

THIS WEEK'S FOCUS
HDTV

PROGRESSIVE SCANNING

Normal television pictures are composed of 480 horizontal lines that scan across the screen in less than the blink of an eye. The conventional way of putting the lines onscreen is to display two sequential fields of 240 lines each; each odd and even field is scanned in 1/60th of a second each. This *interlaced scanning* process is used for all analog broadcasts, as well as for the 1080i HDTV format. (The 1080i format utilizes two sequential fields of 540 lines each.)

For SDTV broadcasts, interlaced scanning produces 60 fields or 30 complete frames every second—which is where the problems start. You see, this slower refresh rate results in some flickering of the picture, especially with larger screen sizes. (You're more apt to see this flickering the closer you sit to your TV.)

A more visually appealing approach is to scan all the horizontal lines in a single pass. This *progressive scanning* process flashes the complete picture onscreen in 1/60th of a second, as opposed to the half-picture fields displayed with interlaced scanning. The result is improved picture quality; even though the resolution (number of pixels) remains the same, the picture is refreshed twice as fast. Progressive scanning is used for the 480p EDTV format (which is what you get from a progressive scan DVD player) and for the 720p and 1080p HDTV formats.

Why spring for progressive scanning? Progressive scanning produces 60 complete frames every second—twice as many as an interlaced picture. This translates into a more accurate display when you have fast action onscreen, such as with sporting events. It's also why I recommend going with a progressive scan DVD player, which converts the standard 480i picture into a much more appealing 480p picture.

ON THIS DAY: DVORAK KEYBOARD PATENTED (1936)

On this date in 1936, the Dvorak typewriter keyboard was patented in the U.S. by efficiency experts August Dvorak (cousin of the famous composer) and William Dealey. This new keyboard maximized efficiency by placing common letters on the home row and made the stronger fingers of the hands do most of the work. Unfortunately, the less-efficient QWERTY keyboard was so well entrenched that the Dvorak keyboard never caught on. Take a look at your keyboard: the first row of letters, starting at the left, reads "QWERTY."

WEBSITE OF THE WEEK: HDTVOICE.COM

Want to learn more about HDTV? Then turn to HDTVoice.com, the Web's premiere source for information about high-definition broadcasting. The site is just a collection of forums, but these are great forums peopled by lots of experts and knowledgeable hobbyists. Join the fun at www.hdtvoice.com.

May 13/14, 2006

SATURDAY/SUNDAY

THIS WEEK'S FOCUS
HDTV

SQUARE PROGRAMS ON A RECTANGULAR SCREEN

HDTV features a 16:9 aspect ratio picture, which is great for watching movies and other widescreen programming. But what do you do with non-widescreen programming—how do you display 4:3 programs on a 16:9 screen?

Most HDTV sets offer several options for displaying traditional 4:3 programming. My personal preference is to use what is called the windowpane mode. This is the opposite of letterboxing, in which 4:3 programming is displayed with black or gray bars on either side of the picture (called windowpanes or pillars), thus preserving the programming's original aspect ratio.

Many viewers don't like the black bars, however, which is why television manufacturers offer a few different ways to stretch the square picture to fill the entire width of the rectangular screen. The simplest form of stretching simply pulls the picture wider, which results in a somewhat distorted picture. I personally hate this mode, as it makes people look short and fat.

Some televisions feature a "smart" expand mode. This process maintains the original aspect ratio for the center portion of the picture, stretching only the edges of the picture to fill the entire widescreen display. The result appears somewhat less distorted than normal stretching.

Another option is the "zoom" mode, which enlarges the entire 4:3 picture to fill the entire width of the widescreen display. This process results in some cropping of the top and bottom of the original picture, however. Still, this method of enlarging the most important part of the picture might be the best choice if you want to fill your entire widescreen display.

ON THIS DAY: TABLE KNIFE INVENTED (1637)

On May 13, 1637, the table knife was created by Cardinal Richelieu in France. Until this time, daggers were used to cut meat, as well as to pick one's teeth. (Careful!) Richelieu had the points rounded off all the knives to be used at his table, thus creating the prototype for the modern table knife.

BLOG OF THE WEEK: HIGH-DEFINITION BLOG

Find out everything that's going on in the world of high-definition television—including all the latest HDTV broadcasts—at the HDTV Blog. Read it at www.highdefinitionblog.com.

May 15, 2006

MONDAY

THIS WEEK'S FOCUS
SPAM

I HATE SPAM!

Every Internet user with an email account knows what spam is—it's those unsolicited and unwanted marketing messages that are sent en masse to users across the Web. These messages seem to multiply over time, and quickly become a major annoyance—even if all you do is hit the Delete key when they pop up in your inbox.

In essence, email spam is like the junk mail you receive in your postal mailbox. It's easy enough to throw away, but you'd prefer not to get it at all—especially when more than half of your incoming messages are junk.

Of course, the easiest way to deal with a spam message is to simply delete it. But if you're receiving dozens—or, in my case, *hundreds*—of spam messages daily, even the task of deleting the messages can be time-consuming.

That's when you turn to aggressive methods to stop the flow of spam. There are lots of things you can do—which I'll discuss throughout this week—but probably the best approach is to get a new email address and try to keep it as private as possible. All the spam filters and blockers in the world aren't even necessary when the spammers don't have your address. As much as possible, you should do your best to hide your address online. *That will keep your inbox clean!*

LEARN MORE ABOUT SPAM

To learn more about spam, along with spyware, viruses, identity theft, and other Internet nasties, pick up a copy of Que's *Absolute Beginner's Guide to Security, Spam, Spyware, and Viruses*, by Andy Walker. Andy is my co-host on Call for Help, which airs in Canada on TechTV.

ON THIS DAY: LISTERINE TRADEMARKED (1923)

On May 15, 1923, Listerine was registered as a United States trademark. Today, that vile mouthwashing liquid is available in a half-dozen flavors, some of them actually pleasant-tasting, though even the makers of Listerine make light of its bad taste, but incredible results, in recent television ads.

FACT OF THE WEEK

FrontBridge (www.frontbridge.com) reports that spam accounted for 84.6% of all inbound email in April 2005. Offers for pharmaceuticals comprised a third of all spam messages, and stock tips a quarter. Surprisingly, porn spam was less than 6% of total spam volume.

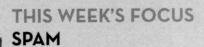

TUESDAY

THIS WEEK'S FOCUS
SPAM

HOW SPAMMERS GET YOUR ADDRESS

How do spammers know to send you their spam? Most spammers use a combination of techniques to assemble their spam lists, everything from sophisticated programming to brute force. Here's some of their techniques:

- Buy lists of email addresses from other websites.
- Pull email addresses from public directories of usernames and addresses.
- Use automated software (called *robots*) to troll newsgroups, message boards, and chat rooms and pull addresses from the postings and directory listings there.
- Use search programs to query Google and other search sites for email addresses from specific domains. (Try searching for @hotmail.com, for example; you'll be shocked at the results!)
- Use address generation software to automatically generate hundreds of thousands of email address combinations (abe@isp.com, abby@isp.com, abraham@isp.com, and so on).
- Use phishing techniques to drive you to a website that asks for your email address and other information, either to enter the site or to register for a contest of some sort.

As you can see, you can defeat some of these techniques by never posting your email address on the Internet. However, there's not much you can do to combat the automatic name generation programs—expect maybe change your address to 10p62xqauxZ@isp.com, which no program could ever guess!

ON THIS DAY: ROOT BEER INVENTED (1866)

On this day in 1866, Charles Elmer Hires invented root beer. In case you don't know, root beer is a sweetened, carbonated beverage originally made using the root of a sassafras plant. Unlike tea, coffee, and most soft drinks, root beer does not naturally contain caffeine—although some national root beer brands (such as Barq's) artificially add caffeine for an added buzz.

GADGET OF THE WEEK: OREGON SCIENTIFIC AWS888 WEATHER FORECASTER

I'm a fool for these weather station gizmos. The AWS888 from Oregon Scientific is especially neat in that it sports a 3.8-inch full-color display which can show current conditions and forecasts with photo-image animations; the time and forecasts come from a wireless connection to the U.S. atomic clock. Buy it for $249.95 from www.oregonscientific.com.

May 17, 2006

WEDNESDAY

THIS WEEK'S FOCUS
SPAM

PROTECTING YOUR EMAIL ADDRESS

One way to reduce the amount of spam you receive is to limit the public use of your email address. It's a simple fact: The more you expose your email address, the more likely it is that a spammer will find it—and use it.

To this end, you should avoid putting your email address on your web page or your company's web page. You should also avoid including your email address in any postings you make to web message boards, Usenet newsgroups, blogs, and the like. In addition, you should most definitely not include your email address in any of the conversations you have in chat rooms or via instant messaging.

Another strategy is to actually use *two* email addresses. Take your main email address (the one you get from your ISP) and hand it out only to a close circle of friends and family; do *not* use this address to post any public messages or to register at any websites. Then obtain a second email address (you can get a free one at Hotmail, Yahoo! Mail, or Google's Gmail) and use that one for all your public activity. When you post on a message board or newsgroup, use the second address. When you order something from an online merchant, use the second address. When you register for website access, use the second address. Over time, the second address will attract the spam; your first email address will remain private and relatively spam-free.

If you do have to leave your email address in a public forum, you can insert a *spamblock* into your address—an unexpected word or phrase that, although easily removed, will confuse the software spammers use to harvest addresses. For example, if your email address is johnjones@myisp.com, you might change the address to read johnSPAMBLOCKjones@myisp.com. Other users will know to remove the SPAMBLOCK from the address before emailing you, but the spam harvesting software will be foiled.

ON THIS DAY: JOHN DEERE DIES (1886)

American agricultural equipment inventor and manufacturer John Deere died on this date in 1886. (He was born on February 7, 1804.) Deere began by repairing and designing steel plows, and then expanded into the agricultural machine business. John Deere tractors mow countless lawns and help harvest millions of acres of farm crops to this day.

HARDWARE OF THE WEEK: GRIFFIN POWERMATE USB CONTROLLER

Griffin's PowerMate is easy to describe: It's a big knob for your PC. You can program the PowerMate to control just about anything you want on your computer—volume, window scrolling, video editing functions, and so on. Buy it for $45 at www.griffintechnology.com.

May 18, 2006

THURSDAY

THIS WEEK'S FOCUS
SPAM

BLOCKING SPAM IN MICROSOFT OUTLOOK

If you use Microsoft Outlook 2003 as your email client, you have a built-in spam blocking feature that can help you curb the junk in your inbox. This feature alone makes it worth upgrading from Outlook Express to Outlook.

You access Outlook's anti-spam tools by selecting Actions, Junk E-Mail, Junk E-Mail Options. This opens the Junk E-Mail Options dialog box. Select the Options tab to view the level of spam protection available. You can choose from the following options: No Automatic Filtering (turns off Outlook's spam filter), Low (blocks the most obvious spam messages), High (blocks the majority of spam messages—but might also block some non-spam email), and Safe Lists Only (blocks all email except messages from people on your Safe Senders List).

For most users, I recommend the High option; it will catch a high percentage of spam with minimal blocking of regular messages. You can minimize the downside by telling Outlook not to block messages from people in your Contacts list. Just select the Safe Senders tab in the Junk E-Mail Options dialog box and click the Also Trust E-Mail from My Contacts option.

You can also have Outlook block messages from specific senders. Just select a message from a sender you want to block, and then select Actions, Junk E-Mail, Add Sender to Blocked Senders List. All future messages from that sender will go right into the Junk E-Mail folder.

ON THIS DAY: MOUNT ST. HELENS ERUPTS (1980)

On May 18, 1980, the long-dormant Mount St. Helens volcano erupted in Washington state, hurling ash 15,000 feet into the air and setting off mudslides and avalanches. The explosion carried as much energy as 27,000 atomic bombs, and the resulting cloud of ash circled the globe.

SOFTWARE OF THE WEEK: EMAIL SENTINEL PRO

Email Sentinel Pro is an add-in utility for just about any email program that performs a number of different security operations—including spam blocking. It also includes a "whitelist" option that lets you accept emails only from approved senders. Download it for $14.95 from www.emailaddressmanager.com.

May 19, 2006

FRIDAY

THIS WEEK'S FOCUS
SPAM

ANTI-SPAM SOFTWARE

If the amount of spam in your inbox becomes particularly onerous, you might want to consider using an anti-spam software program. Most anti-spam software uses some combination of spam blocking or content filtering to keep spam messages from ever reaching your inbox; their effectiveness varies, but they will decrease the amount of spam you receive to some degree.

The most popular anti-spam software includes ANT 4 MailChecking (www.ant4.com), MailWasher (www.mailwasher.net), McAfee SpamKiller (www.mcafee.com), Norton AntiSpam (www.symantec.com), and RoadBlock (www.roadblock.net).

If this anti-spam software isn't powerful enough for you, you can subscribe to one of several online services that interactively block spam, using a variety of filtering and blocking techniques. Many of these services are also available for small business and large corporate networks; most are priced on a per-month subscription basis.

Some of the more popular spam-filtering services include Emailias (www.emailias.com), Mailshell (www.mailshell.com), SpamCop (www.spamcop.net), SpamMotel (www.spammotel.com), and Symantec Brightmail Antispam (enterprisesecurity.symantec.com).

ON THIS DAY: LAWRENCE OF ARABIA DIES (1935)

On May 19, 1935, T.E. Lawrence, a.k.a. "Lawrence of Arabia," passed away. (He was born on August 15, 1888.) Lawrence was an archaeological scholar who later became known as a military strategist in the Middle East during World War I.

WEBSITE OF THE WEEK: SPAMHAUS

The Spamhaus project tracks the Internet's worst spammers and offers a variety of spam-related information and services. Check it out at www.spamhaus.org.

May 20/21, 2006

SATURDAY/SUNDAY

THIS WEEK'S FOCUS
SPAM

TRACE AND REPORT

If you're aggressively anti-spam, you can take the battle to its source—by tracing the spam's sender and complaining to the spammer's Internet provider. This takes some work, however.

You might think that you have the spammer's email address—that's it listed in the From: field of the spam message, right? Wrong. The From: address is easily spoofed, and more often than not is totally bogus. To find out the spammer's *real* address, you have to display the message header, which your email program typically hides. (For example, in Outlook Express you'll need to open the message and select File, Properties.)

You'll need this header information to complain to the spammer's ISP. In fact, the best way to respond is to forward the spam message, with the header information copied into the body of the message, to each of the domains listed in From: field of the header. Send the message to the following addresses, which are typically used for handling spam complaints: abuse@*domain.name* and postmaster@*domain.name*.

If there's a URL in the spam message, you can trace down who owns the page (and how to contact them) by performing a WHOIS lookup at www.completewhois.org. The resulting page will list who owns the domain, which service hosts the domain, and how to contact the owners. You can then contact the spammer directly with your complaint.

ON THIS DAY: LEVIS PATENTED (1873)

On May 20, 1873, Jacob W. Davis of Reno, Nevada, received a patent for a rivet process for strengthening the pocket openings of canvas pants. He assigned the patent to himself and his business partner, Levi Strauss. (Yes, *that* Levi Strauss.) Today, Levis are nearly ubiquitous clothing items. Despite some changes to them over the years, today's Levis look very similar to the original Levis produced more than 130 years ago.

BLOG OF THE WEEK: SPAM KINGS BLOG

A terrific blog all about the world of spam—in particular, how to stop it. Host Brian McWilliams is the author of the book of the same name, and this is the best of several spam-related blogs on the Web. Check it out at spamkings.oreilly.com.

134

May 22, 2006

MONDAY

THIS WEEK'S FOCUS
GOOGLE

ADVANCED GOOGLE SEARCHING

Google is not only the largest search engine on the Web, most people think it's the easiest to use. (It certainly has the simplest home page!) However, there are some tricks you can employ to fine-tune your Google searching. Here are some of my favorites:

- Include synonyms in your search by adding a "~" before the keyword.
- Search for pages that include one or another keyword by using the OR operator between the two keywords.
- Include a common word (such as "and" or "the") in your search by entering the word, preceded by a "+" sign.
- Search for a range of numbers by entering the lowest and highest number, separated by two periods, like this: 200..240
- List all pages that are similar to a given page by using the related: operator before your query.
- Restrict your search to a given domain by using the site: operator in your query.

And if all that's too difficult for you, just click the Advanced Search link to use Google's Advanced Search page—which makes searching as easy as filling in the blanks.

ON THIS DAY: TOOTHPASTE TUBE INVENTED (1892)

On this day 1892, Connecticut dentist Dr. Washington Sheffield invented a collapsible metal toothpaste tube. Prior to this invention, toothpaste was sold in round pots.

FACT OF THE WEEK

Google claims to index more than 8 billion web pages. The site averages more than 2.3 billion searches per month—that's close to 900 searches every second! Some analysts claim that Google delivers more than 70% of the search traffic to the average website.

May 23, 2006

TUESDAY

THIS WEEK'S FOCUS
GOOGLE

GOOGLE IMAGES

Google isn't just for text searches. Most folks don't know that Google can also search for images on the Web, using the Google Images feature.

You get to Google Images by clicking the Images link on Google's home page. Conduct your search by entering the appropriate keywords, and Google will return a results page with thumbnails of all the graphics files that match your query. When you click a thumbnail, you're taken to the page that hosts that image.

To avoid unwelcome surprises, you might want to edit your Google Image preferences before you go searching. Click the Preferences link on Google's home page, then scroll down to the SafeSearch Filtering section and select the Use Moderate Filtering option. This will prevent dirty pictures from showing up in your search results.

You can restrict your searches to a particular type of graphics file by using the filetype: operator in your query. For example, to show only JPG files, include filetype:jpg in your query. You can then filter the results by file size by clicking the Large, Medium, or Small link at the top-right corner of the search results page.

ON THIS DAY: BIFOCALS INVENTED (1785)

On this date in 1785, our old pal Benjamin Franklin invented bifocal glasses, as a solution to carrying around two pairs of glasses to see objects at different distances. As Ben wrote to a friend, "I have only to move my eyes up and down as I want to see far or near."

GADGET OF THE WEEK: WALLFLOWER 2 MULTIMEDIA PICTURE FRAME

Show off your digital photos with this digital picture frame. The Wallflower 2 connects to your PC via WiFi, downloads the pictures you've selected, and then displays them in a rotating slideshow. The color screen is a big 14.1 inches, costs $899, and can be purchased at www.wallflower-systems.com.

May 24, 2006

WEDNESDAY

THIS WEEK'S FOCUS
GOOGLE

GOOGLE MAPS

Here's a new Google feature that I really like. Online mapping services have been around practically forever, but Google Maps (maps.google.com) takes the concept one step further.

For starters, the Google Maps interface works just like the regular Google search interface. Enter an address into the search box at the top of the page, click Search, and you get the map you want. You can also start directly from the default U.S. map, and use the Zoom control on the left to zoom into a particular region. (You can then use your mouse to drag the map in any direction—you don't have to use the normal directional buttons.)

You can also use Google Maps to find things near a given location. For example, to find hotels in the Chicagoland area, enter hotels near chicago. Google will display a map of the area, with all hotels pinpointed.

But the really neat thing about Google Maps is that it can display both traditional map and satellite photo views. Click the Satellite button at the top-right corner of any map and Google displays a high-resolution satellite map of the selected area. Zoom in to get a surprisingly close view. It's really cool!

ON THIS DAY: FIRST TELEGRAPH MESSAGE (1844)

On May 24, 1844, Samuel Morse used his newfangled telegraph machine to transmit a message from the U.S. Supreme Court room in Washington, D.C. to the Mount Clare station of the Baltimore and Ohio Railroad. The message? "What hath God wrought!"

HARDWARE OF THE WEEK: WESTERN DIGITAL MEDIA CENTER

Western Digital's Media Center is more than just an external hard disk; it also functions as an 8-in-1 digital media reader, two-port USB hub, and two-port FireWire hub. It connects to your PC via either USB 2.0 or FireWire and offers 250GB of capacity. Buy it for $259.95 from www.wdc.com.

May 25, 2006

THURSDAY

THIS WEEK'S FOCUS
GOOGLE

PERSONALIZE YOUR GOOGLE HOME PAGE

Tired of the Google's streamlined search page? Then personalize it!

You personalize your Google home page at www.google.com/ig/. This displays a modified version of the normal Google home page; click the Get Started link to begin the personalization process.

Google now displays a page that lets you select what content you want on your home page. Available options include Gmail, Google News, Stock Market, Weather, New York Times, and so on. Click the Save Personalization button when you're done.

Once you have the new home page created, you can edit any of the content you've chosen, by clicking the Edit link next to that content block. You can rearrange the content blocks by hovering over a block heading, and then using your mouse to drag the block into a new position on the page. It's probably the easiest start page I've seen to customize—and it puts that ol' Google search box front and center!

ON THIS DAY: FIRST INTERNATIONAL WWW CONFERENCE (1994)

CERN hosted the first international World Wide Web conference on this day in 1994. This conference led to the expansion of the Web from Tim Berners-Lee's original concept of a single information storage facility into the multi-faceted Web that we know today.

SOFTWARE OF THE WEEK: GOOGLE DESKTOP SEARCH

Google's great for searching the Web for information—what about searching your computer for files? That's where Google Desktop Search comes in. It lets you use the familiar Google Web interface to search email messages, document files, and the like stored on your computer's hard disk. Download it for free from desktop.google.com.

May 26, 2006

FRIDAY

THIS WEEK'S FOCUS
GOOGLE

GMAIL

Web-based email services have long been popular options for Internet users. The concept of webmail got expanded a bit, however, when Google introduced its Gmail service.

Like Hotmail and Yahoo! Mail, Gmail is a free webmail service. You can open a Gmail account at gmail.google.com. That's when the fun starts.

For openers, Gmail doesn't use the traditional folder metaphor for storing your email messages. Google is a search company, which explains why it uses a simple search function to help you find your stored messages. Forget the folders—just search for the stored emails you want.

And there's no need to bother with throwing away your old messages. Gmail gives you more than 2GB of free storage, which is more than enough to store all your email messages. Heck, that's enough storage space to use Gmail as a backup facility for all the valuable data files on your computer. Just email a file to yourself at your Gmail account, and keep it stored on Google's servers—just in case!

ON THIS DAY: FIRST MIDDLE EAST OIL STRIKE (1908)

Before 1908, the Middle East was just a bunch of sand. But on May 26 of that year, a gusher of black gold rose 50 feet above the top of a drilling rig at Masjid-i-Suleiman, Persia. A century of dependence on foreign oil followed.

WEBSITE OF THE WEEK: GOOGLE LANGUAGE TOOLS

Here's a hidden gem in the Google network of sites. The Google Language Tools page lets you translate blocks of text or complete web pages. Take a look for yourself at www.google.com/language_tools/.

May 27/28, 2006

SATURDAY/SUNDAY

THIS WEEK'S FOCUS
GOOGLE

OTHER COOL GOOGLE SITES

Google's engineers keep quite busy coming up with new tools for you to use. Most of these have something to do with searching, but not all; some of these sites are quite useful, and others just plain fun. Here are some of my favorites:

- **Google Answers** (answers.google.com)—Ask a question, get an answer from an expert.
- **Google Catalogs** (catalogs.google.com)— Search mail-order catalogs online.
- **Google Groups** (groups.google.com)— Search Usenet newsgroups and create your own topic-specific groups.
- **Google Local** (local.google.com)—Find local businesses and services.
- **Google News** (news.google.com)—The day's hottest headlines, from all over the Web.
- **Google Print** (print.google.com)—Search the full text of books.
- **Google Ride Finder** (labs.google.com/ridefinder/)—Use Google Maps to pinpoint the location of taxis and shuttles in selected cities.
- **Google Video** (video.google.com)—Search TV programs and videos.

ON THIS DAY: JELL-O INTRODUCED (1897)

On May 28, 1898, Pearl B. Wait, a manufacturer of cough medicine, produced the first varieties of Jell-O gelatin. (Those first flavors? Strawberry, raspberry, orange, and lemon.) Sales were poor, and Wait sold the Jell-O business for $450 to a neighbor. Today, almost any flavored gelatin dessert is referred to (often incorrectly) as Jell-O.

BLOG OF THE WEEK: THE UNOFFICIAL GOOGLE WEBLOG

As the title says, this is an independent blog that reports on all things related to the world's largest search engine. (I love the blog's subtitle: "Who watches the watchmen?") Read it at google.weblogsinc.com.

May 29, 2006

MONDAY

THIS WEEK'S FOCUS
DIGITAL PHOTOGRAPHY

CHOOSING A DIGITAL CAMERA

Whether you're looking for a $200 point-and-shoot camera or a $1,000 digital SLR, here are some key features you need to compare:

- **Picture resolution**—This is measured in megapixels—the bigger the number, the sharper your pictures will be. Most decent point-and-shoot cameras today have a resolution in the range of 3–5 megapixels; prosumer and D-SLR models go up to 8 megapixel resolution.
- **Zoom lens**—This helps you get up close to distant subjects. Ignore the *digital* zoom specs and focus on the *optical* zoom built into the camera's lens.
- **Size and weight**—Many point-and-shoot cameras are little bigger than a credit card and weigh less than half a pound. Prosumer models are bigger and heavier, and D-SLRs even more so. Choose a model that feels comfortable to you.
- **Operating speed**—That is, how long it takes to power up and shoot a picture after you've pressed the autofocus button. Many low-end cameras have a noticeable shutter lag that makes it difficult to take pictures of fast-moving subjects. D-SLRs are much faster, turning on almost instantly.
- **Storage capacity**—All digital cameras store their photos on a flash memory card of some sort. Most cameras come with a blank memory card, typically a low-capacity model—which means you'll probably want to spring for a higher-capacity card.

ON THIS DAY: HILARY CONQUERS MOUNT EVEREST (1953)

On this day in 1953, New Zealand explorer Sir Edmund Hilary (accompanied by his Sherpa, Tenzing Norgay) was the first man to reach the summit of Mount Everest, the world's highest mountain. Since then, nearly 2,000 climbers have reached the summit—while another 200 have lost their lives trying.

FACT OF THE WEEK

Men and women are different, even when it comes to digital cameras. Lyra Research (www.lyra.com) says that the most popular digital camera brand among women is Kodak, followed by Canon, Sony, and HP. Men rank Canon as their favorite brand, followed by Sony, Olympus, and then Kodak. Overall, Kodak holds the number-one sales spot in the U.S., shipping almost 4 million cameras in 2005.

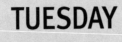

May 30, 2006
TUESDAY

THIS WEEK'S FOCUS
DIGITAL PHOTOGRAPHY

PROSUMER CAMERAS

Pocket-sized snapshot digital cameras are fine, but when you want really good pictures, you want a camera with a little beef between the bun—what the industry calls *prosumer* digital cameras. Most of these cameras have similar specs and performance, and look and feel similar, too.

One thing all prosumer cameras have in common is that they offer a lot of professional-level features in a consumer-friendly package. These cameras let you shoot high-resolution pictures, in the 6–8 megapixel range, that can be blown up to 8 × 10 or larger print sizes. You'll also get a better-quality lens than you would with a point-and-shoot model, typically with a longer zoom (6X–10X), as well as the capability to bypass the automatic shooting mode and shoot with either aperture- or shutter-priority modes.

Most prosumer cameras also feature metal bodies (instead of the plastic bodies found in most point-and-shoot cameras), electronic viewfinders (in addition to the standard large LCD monitor on the back), and intelligent hot shoes to which you can attach external flashes, strobe lights, and the like. Many of these cameras also let you save photos in the RAW file format, which is more versatile than JPEG or TIFF when it comes to post-photo processing in Adobe Photoshop or some similar program. Of course, you pay extra for all these features; a typical prosumer camera will cost you close to a thousand bucks, give or take.

What are some of my favorite prosumer digicams? A short list includes the Sony Cyber-shot DSC-828, Nikon Coolpix 8800, Konica Minolta DiMage A2, Canon Powershot Pro 1, and Olympus Camedia C-8080 Wide Zoom. All sell for around $1,000, although bargains can sometimes be had, depending on where you shop.

ON THIS DAY: KRYPTON DISCOVERED (1898)

No, not the planet—the element. On this date in 1898, English chemist Morris William Travers discovered the elusive element later named krypton. The name, by the way, derives from the Greek word for "hidden."

GADGET OF THE WEEK: SANDISK PHOTO ALBUM

SanDisk's Photo Album lets you easily view your digital photos on any TV. Just connect the Photo Album to your TV, insert a memory card or USB flash drive, and settle back for a digital slideshow—no computer necessary. Press the "store" button and the Photo Album automatically creates an onscreen "photo album," complete with background music. Buy it for $49.99 from www.sandisk.com.

May 31, 2006

WEDNESDAY

THIS WEEK'S FOCUS
DIGITAL PHOTOGRAPHY

DIGITAL SLRS

A digital single lens reflex (D-SLR) camera is the digital equivalent of a high-end SLR film camera. This is the same type of camera that the pros use, and it takes higher-quality pictures than even the prosumer digital models.

What makes a D-SLR so great? First, a D-SLR camera uses a reflex mirror apparatus so that, when you look through the optical viewfinder, you're actually viewing through the lens itself. That helps you take better pictures, and it also helps battery life—because you're not powering the LCD display every time you take a picture. (In fact, you can't use the LCD display as a viewfinder; it's only there when you need to make a selection from the camera's menu system.)

D-SLRs use larger image sensors than traditional digital cameras, often as big as 35mm film, which provide better results in low-light conditions. You also get the ability to use interchangeable lenses, which gives you lots of flexibility. In addition, D-SLRs are usually designed by the companies' 35mm camera divisions (instead of their consumer electronics divisions) and provide all the operating flexibility you need to make great photos, fast.

Unfortunately, D-SLRs aren't cheap, although they're not that much more expensive than a good prosumer model. Prices start at around a grand, and go up from there.

ON THIS DAY: JOHN KEMENY BORN (1926)

John Kemeny co-developed (with Thomas Kurtz) the BASIC programming language. Prior to that milestone, he worked on the Manhattan Project during World War II, no small accomplishment in itself. He was born on this day in 1926 in Budapest, Hungary.

HARDWARE OF THE WEEK: EPSON PERFECTION 4180 PHOTO SCANNER

A digital camera isn't the only way to create digital photos; you can also scan in photo prints and slides via a photo scanner. One of the best mid-price photo scanners out there is the Epson 4180, which offers 4800 × 9600 resolution as well as a built-in adapter for scanning 35mm negatives, slides, and medium-format transparencies. Buy it for $199.99 from www.epson.com.

June 2006

June 2006

SUNDAY	MONDAY	TUESDAY	WEDNESDAY	THURSDAY	FRIDAY	SATURDAY
				1 1961 FM stereo broadcasting is authorized by the FCC	**2** 1928 Kraft invents Velveeta Cheese	**3** 1856 Machine for making pointed screws is patented; prior to this invention, screws had blunt ends
4 1984 Scientists in Berkeley, CA, clone DNA sequences from an extinct animal, the quagga	**5** 1977 Apple II released	**6** 1949 George Orwell's *1984* published	**7** 1887 Monotype typesetting machine is patented in Washington, D.C.	**8** 1957 Scott Adams, creator of "Dilbert," is born	**9** 1934 Donald Duck makes his first film appearance in *The Wise Little Hen*	**10** 1936 First cable television broadcast made from Radio City to the Empire State Building in New York City
11 Father's Day 1962 Three prisoners escape (and are never found) from Alcatraz	**12** 1897 Swiss army knife is patented	**13** 1983 Space probe Pioneer 10 crosses the orbit of Neptune and becomes the first man-made object to leave our solar system	**14** 1834 Sandpaper is patented in Springfield, VT	**15** 1752 Benjamin Franklin conducts his famous kite-flying experiment	**16** 1903 Pepsi-Cola is trademarked in New Bern, N.C.	**17** 1994 O.J. Simpson is arrested after a nationally televised attempt to flee from police
18 1983 Sally K. Ride becomes the first U.S. woman in space	**19** 1941 Cheerios cereal is invented	**20** 1977 Trans-Alaska pipeline opens	**21** First day of summer 1948 First long-playing records, more commonly known as LPs, are introduced	**22** 1946 Jet airplanes are first used to transport mail	**23** 1964 The hula-hoop is patented	**24** 1963 The first television video recorder is demonstrated in London
25 1876 General George Custer is killed at the Battle of Little Big Horn in Montana	**26** 1974 Bar codes first used in supermarkets; the first item scanned is a pack of Wrigley's chewing gum	**27** 1978 First ink pen with erasable ink is introduced	**28** 1965 First satellite phone call is made	**29** 1995 U.S. space shuttle Atlantis docks with Soviet space station Mir and orbits the Earth	**30** 1971 Three Russian astronauts become the first humans to die in space aboard Soyuz II	

June 1, 2006

THURSDAY

THIS WEEK'S FOCUS
DIGITAL PHOTOGRAPHY

BETTER LIGHTING = BETTER PICTURES

One of the worst photographic offenses is to shoot under standard indoor room light. The result is not only a dark picture, but one with poor color balance—your photo will probably have a reddish or yellowish cast due to the color temperature of the fluorescent or incandescent light.

A simple solution is to take your photos outdoors, where there is a lot of natural lighting. Shooting on an overcast or hazy day is easiest because you don't have to worry about direct sunlight washing out your pictures. On a sunny day, it's good to shoot in open shade, like that on the side of a building. Shooting in direct noon-time sunlight can result in washed-out details.

If you really want to shoot indoors, you need some sort of auxiliary light. Your camera's built-in flash might do the job, but it can make your subjects look two-dimensional and creates a glare from shiny surfaces. (You also have the red-eye effect to deal with.) Try it first to see if you like the results.

A better approach for indoor shooting is to use an external lighting kit. I like the low-priced kits from Smith-Victor, but you can also make your own lights with regular tungsten bulbs in clamp-on reflectors, which you can get from Home Depot or any similar store. Two 150-watt lights are enough for most digital cameras. Position the lights at 45-degree angles to the subject (on either side of your camera), and you're good to go.

ON THIS DAY: FM STEREO BROADCASTING BEGINS (1961)

On the first of June, 1961, the Federal Communications Commission received the first notification of FM stereo broadcasting. The first two stereo FM stations were WEFM in Chicago and WGFM in Schenectady.

SOFTWARE OF THE WEEK: PAINT SHOP PRO

Although Photoshop is the photo-editing software of choice for professional photographers, it's pricy and not always easy to use. For many casual photographers, a better choice is Paint Shop Pro, which can do just about everything Photoshop does for a lot less money and effort. Buy it for $129 from www.corel.com.

June 2, 2006

FRIDAY

THIS WEEK'S FOCUS
DIGITAL PHOTOGRAPHY

HOW TO TAKE GREAT DIGITAL PICTURES

Want to take great-looking photos with your digital camera? Here are some tips that can make the rankest amateur look like a seasoned pro:

- **Use a tripod**—You need good solid support to take shake-free pictures—which means investing a few bucks in a tripod.
- **Use a neutral background**—Don't let the background compete with the subject. One solution is to take your photo against a plain white wall. If you want to get fancier, you can use seamless photography paper, curved up behind your subject.
- **Shoot at an angle**—Don't shoot your subject head-on. Most people and objects look best when placed at an angle to the camera. Try positioning your light at a 45-degree angle, and then facing the subject into the light for a flattering three-quarter view.
- **Fill the frame**—To take effective photographs, you must learn proper composition. That means centering the subject in the center of the photo and getting close enough to the object so that it fills up the entire picture.
- **Add "hair lighting"**—Many photographers use what they call a "hair light" *behind* the subject to add highlights around the edges and to better separate the subject from the background. This is a small light placed high on the back side of the subject, sometimes a little to the side. Try it—you'll like the results!

ON THIS DAY: VELVEETA CHEESE INVENTED (1928)

Kraft's Velveeta cheese was invented on this date in 1928. What makes Velveeta special is that when melted, it's as smooth as velvet (hence the name) and it won't curdle when heated. Snackers everywhere rejoice!

WEBSITE OF THE WEEK: DIGITAL PHOTOGRAPHY REVIEW

Want the latest digital camera news—or just to read detailed product reviews? Then check out Digital Photography Review (www.dpreview.com), the Web's premiere source for digital camera news, reviews, and information.

June 3/4, 2006

SATURDAY/SUNDAY

THIS WEEK'S FOCUS
DIGITAL PHOTOGRAPHY

DIGITAL PHOTO VAULTS

As any digital photographer knows, taking a lot of high-resolution photos takes up a lot of storage space. Although you could stock up on multiple flash memory cards, a better alternative is to offload your photos to a digital photo vault.

A digital photo vault is a portable device that combines hard disk storage with a color LCD viewing screen. You transfer your photos from your camera to the vault either via memory card or USB connection, and then view the photos when you're on the go. You get the advantage of freeing up storage space on your camera to take more pictures, as well as easily sharing your photos with others via the photo vault's LCD screen.

Actually, most photo vaults double for portable video players; the technology is the same, in any case. For all practical purposes, just about any portable video player can be used as a photo vault, as can some portable audio players (those with bigger screens, anyway, such as the color iPods). But the best photo vaults are those dedicated to the task.

Some of my favorite photo vaults include Epson's P-2000, the Archos AV420, the SmartDisk FlashTrax XT, and the Jobo GIGA Vu Pro. Prices range from $400 to $600, depending on storage capacity and other features.

ON THIS DAY: ROBERT NOYCE DIES (1990)

Robert Noyce was the co-inventor (with Jack Kilby) of the semiconductor, and a co-founder (with Gordon Moore and Andy Grove) of Intel Corporation. He died on June 2, 1990, at the age of 62.

BLOG OF THE WEEK: PHOTOGRAPHYBLOG

This UK-based blog will keep you up to date on developments in the world of digital photography, particularly with new camera and accessory releases. Check it out at www.photographyblog.com.

June 5, 2006

MONDAY

THIS WEEK'S FOCUS
DIGITAL MOVIE MAKING

DIGITAL VIDEO FORMATS

Digital video recording not only records audio/video information digitally, but it also lets you use your PC as a movie editing studio to create sophisticated home movies you can distribute on DVDs. The whole digital editing chain starts with a digital camcorder, in one of several different recording formats:

- **MiniDV**—This is the most popular and most common digital camcorder format today. It records broadcast-quality video (500+ lines of resolution) on small, low-priced cassettes, about 1/12 the size of a standard VHS tape. When in doubt, choose a MiniDV camcorder.

- **HDV**—This is a high-definition version of the MiniDV format, found on only a few high-priced high-definition camcorders. HDV uses standard MiniDV cassettes but produces either 720p or 1080i resolution, along with Dolby Digital surround sound.

- **DVD**—DVD camcorders don't use tape at all; they record directly to DVDs, in either DVD-RAM or DVD-R/RW format. You can get up to 120 minutes on a blank DVD.

- **Digital8**—This is an older, larger, and generally lower-priced, digital format. For compatibility with older analog recorders, Digital8 camcorders can view 8mm or Hi-8 tapes.

- **MicroMV**—This was a smaller digital format, somewhat proprietary to Sony camcorders, that never quite caught on with the public. Sony has since abandoned the format.

ON THIS DAY: APPLE II INTRODUCED (1977)

On June 5, 1977, The Apple II personal computer first went on sale. The Apple II was the brainchild of Steve Wozniak and Steve Jobs. It ran a 1MHz 6502 microprocessor, had the capability to display color graphics, and used a cassette tape for input/output. It soon had competition from the Commodore Pet and the Radio Shack TRS-80.

FACT OF THE WEEK

According to the NPD Group (www.npdtechworld.com), the top five camcorder brands, in terms of dollar sales, are Sony (1), Canon (2), Panasonic (3), JVC (4), and Samsung (5). These top five brands accounted for 96.4% of total category sales in 2004.

June 6, 2006

TUESDAY

THIS WEEK'S FOCUS
DIGITAL MOVIE MAKING

CHOOSING THE RIGHT VIDEO CAMERA

Even the lowest-priced MiniDV recorders will take surprisingly good pictures; most of the picture quality is in the format itself rather than in additional features, which means many people can get by with a simple $500 camcorder, no problem. But higher-priced models are available and worthy of your consideration.

In general, a bigger budget buys you one or more of the following features: smallness, ease of use, special features (such as transition effects, night-vision shooting, and so on), or pro-level performance. And when I say pro-level performance, I mean pro-level performance; the very best consumer camcorders (those costing more than $1,000) deliver digital pictures good enough for television or film use.

Even if you don't spring for one of these uber-expensive pro-level camcorders, you should expect near-professional performance and features when you spend more than $600 or so on a camcorder. Any camcorder selling in this range should have a good-quality zoom lens, an image stabilization system (to keep your pictures steady even if your hands aren't), a variety of automatic exposure modes, and some sort of video editing built in. This last feature lets you perform in-camera edits between scenes, including audio dubbing, fade in and out, and other special effects.

You should also pay particular attention to the camcorder's image-sensing system. Most lower-priced camcorders use a single CCD to capture the video image; higher-priced models use a 3-CCD system that splits the image optically and feeds color-filtered versions of the scene to three CCD sensors, one for each color—red, green, and blue. Naturally, a 3-CCD camera will deliver better color than a single-CCD model.

ON THIS DAY: FIRST HOUSEHOLD DETERGENT INTRODUCED (1907)

On this day in 1907, Persil, the first household detergent, was introduced. It was marketed by Henkel & Cie of Dusseldorf as the first "self-acting" washing powder; it combined both washing and bleaching agents in one powder. The brand name derived from the beginning syllables of its two most important chemical components: perborate (a bleaching agent) and silicate. Persil remains a very popular detergent, particularly in the UK.

GADGET OF THE WEEK: SUNPAK READYLIGHT 20

The quickest way to improve the picture quality of your home movies is to better light the scenes you shoot. To that end, check out the Sunpak Readylight 20, a cordless (battery operated) and compact video light that can attach to any camcorder. Buy it for $39.99 from www.sunpak.com.

June 7, 2006
WEDNESDAY

THIS WEEK'S FOCUS
DIGITAL MOVIE MAKING

HIGH-DEFINITION CAMCORDERS

The best camcorders today move beyond the traditional standard definition format to record movies in true high-definition video (HDV). Today, HDV camcorders are few and far between (and priced like small cars), but expect more (and lower-priced) models to hit the market over the next year or two.

An HDV camcorder offers all the features of a pro-level standard definition camcorder, but with the capability of recording high-definition signals onto a MiniDV tape. Depending on the camera, you're looking at recording in either the 720p or 1080i format, both of which should be playable on any HDTV-capable television.

Naturally, an HDV camcorder will shoot in the 16:9 aspect ratio, which is part of the high-definition format. You can also record Dolby Digital surround sound, although you'll probably need an external surround sound microphone for this.

Should you spring for a HDV camcorder? Well, the high-definition picture is definitely nice—better than nice, actually. But, for now anyway, you're limited in how you can distribute your high-definition videos. That's because we don't yet have a high-definition DVD format, so you can't burn your HDV video to DVD. You're pretty much limited to connecting your HDV camcorder to your HDTV set and playing back right from the camcorder. HDV models are available from Sony and JVC; expect to pay anywhere from $1,700 to $3,700.

ON THIS DAY: FIRST SOLAR POWER PLANT (1980)

On June 7, 1980, the first solar power plant in the U.S. was dedicated. The plant was built near the Natural Bridge National Monument in Utah and featured more than 250,000 solar cells arrayed in 12 long rows.

HARDWARE OF THE WEEK: HP DC5000 MOVIE WRITER

HP's dc5000 is an external DVD burner with built-in analog video capture. Just connect your VCR or camcorder to the dc5000, insert a blank DVD, and activate the Video Transfer Wizard for automated and unattended conversion of your home movies to DVD. Buy it for $249.95 from www.hp.com.

June 8, 2006

THURSDAY

THIS WEEK'S FOCUS
DIGITAL MOVIE MAKING

PROSUMER CAMCORDERS

Some consumer-level camcorders are small enough to hold in the palm of your hand. Prosumer camcorders, in contrast, are big enough to require a shoulder rest; these camcorders look, feel, and perform just like the type of camcorder you see TV news crews or independent filmmakers lugging around.

Prosumer camcorders are big and bulky, yes, but for a reason. Many prosumer camcorders let you use interchangeable lenses (for more shooting versatility) and shoot in the 16:9 widescreen format. More important, they come with a bevy of automatic recording modes and manual adjustments that let you custom-tailor your movies to a variety of shooting styles and situations. Plus, picture quality is second to none, especially under difficult lighting conditions. Lots of technospeak, I know, but it all translates into lots of flexibility to deliver eye-popping widescreen pictures.

The technology behind the better picture is the use of a 3-CCD image sensing system. This system splits the image optically and feeds color-filtered versions of the scene to three CCD sensors, one for each color—red, green, and blue. (All TV and film production is done with 3-CCD cameras, BTW.) The bigger the CCDs, the better; 1/3" CCDs are better than 1/6" ones. And, for even better picture quality, most prosumer models feature progressive scan technology and true 16:9 framing for film-like results.

The uber-camera of choice for many independent filmmakers is the Canon XL2. What the pros especially like is the interchangeable lens system that lets you use any of Canon's XL and EF 35mm camera lenses. Even though it costs a mind-boggling $6,500 (and doesn't even shoot in high-def!), it's a true pro-level camera—used by real pros.

ON THIS DAY: ALAN TURING FOUND DEAD (1954)

Alan Turing was a mathematical genius, a successful code-breaker during WWI, and the developer of a "universal computing machine." He committed suicide on June 7 and was found on June 8, 1954; he was just 42.

SOFTWARE OF THE WEEK: ADOBE PREMIERE ELEMENTS

If you're video editing on a Mac, you can use Final Cut Pro, which is one of my favorite programs. But you're probably using a Windows machine, which means your video editing program of choice should be Adobe Premiere Elements. Buy it for just $99.99 at www.adobe.com.

June 9, 2006

FRIDAY

THIS WEEK'S FOCUS
DIGITAL MOVIE MAKING

RECORDING BETTER SOUND WITH YOUR MOVIES

The built-in microphone in most camcorders is functional at best. If you want to pick up more and better sound when you're shooting, you need an auxiliary microphone. Fortunately, most camcorders let you easily connect an external mic, and you have lots of different types of mics to choose from. Here's a brief selection:

- **Lavaliere**—For even better recording of dialogue, consider a "newscaster-style" lavaliere microphone that clips onto the subject's tie or shirt. One good model to consider is the Audio Technica ATR35s.
- **Shotgun**—When you want to record someone from across the room, you need a shotgun microphone—essentially, a unidirectional condenser mic that works for close-, medium-, and long-distance pickup. My favorite shotgun mic is the Audio Technica ATR55.
- **Wireless**—Another solution to the across-the-room problem is to use a wireless microphone system. Most systems use a lavaliere mic that clips onto the subject, a transmitter pack that tucks into the subject's back pocket, and a receiver that connects to your camcorder. Check out the Audio Technica PRO 88W to see what I mean.

- **Stereo**—Most built-in microphones are mono only. For stereo recording, go with a dual-capsule external mic, such as the Sony ECM-S930C.
- **Surround-sound**—If you're using an HDV camcorder, you can record in Dolby Digital Surround Sound. For this you need to use either multiple mics or a single mic capable of surround recording, such as Sony's ECM-HQP1.

ON THIS DAY: FIRST AUTOMAT OPENS (1902)

June 9, 1902, saw the opening of the first restaurant with vending machine service. The Hron & Hardart Automat Restaurant was located at 818 Chestnut Street in Philadelphia, serving sandwiches and other fast foot for a nickel apiece.

WEBSITE OF THE WEEK: CAMCORDERINFO.COM

CamcorderInfo.com offers the most comprehensive collection of camcorder news, reviews, and information on the Web. See for yourself at www.camcorderinfo.com.

June 10/11, 2006

SATURDAY/SUNDAY

THIS WEEK'S FOCUS
DIGITAL MOVIE MAKING

STOP SHAKY PICTURES!

Only amateur movies—and professional movies trying for an artsy "shaky cam" effect—bounce around like a monkey on caffeine. If watching your home movies makes you nauseas (and it's not because of the content), then it's time to get serious about going steady.

What you need is a way to steady your camera when you shoot. This can be as simple as using a monopod or tripod or as fancy as using a shoulder mount with some sort of motion-stabilization rig. If the pros can shoot steady pictures, then you can, too.

Assuming you already know about monopods and tripods (just count the number of legs), let's move quickly to the concept of a camera brace. At its most basic, this is a brace that attaches to a strap that hangs around your neck; by letting the camcorder rest against the bottom of the neckstrap, you get better support and less-shaky pictures. A good low-priced camera brace is SIMA's VideoProp (www.simacorp.com), which costs about $30.

For even better stability, prepare to spend more money. VariZoom's VZ-LSP (www.varizoom.com) is a lightweight system that provides three-point support for your camcorder via a shoulder brace, abdomen brace, and hand grip; it costs about $430. VariZoom also offers the DV Sportster system, which consists of an articulating arm connected to a support vest; the whole rig helps provide stability and eliminate arm fatigue during long shoots. It's the same type of stabilization that the pros use, and it costs a whopping $1,200. (You get what you pay for!)

ON THIS DAY: SPEAK & SPELL INTRODUCED (1978)

And the cow goes "moo." June 11, 1978, saw the debut of Texas Instruments' Speak & Spell, a talking learning aid/toy for kids aged 7 and up. It was the first product to offer electronic speech on a single silicon chip. Literally millions of children have been entertained ad nauseam by the Speak & Spell—and its many incarnations—since this date.

BLOG OF THE WEEK: PVRBLOG

PVRblog is devoted exclusively to digital video recorders—also known as personal video recorders, or PVRs. (That means TiVo and its ilk.) Check it out at www.pvrblog.com.

June 12, 2006

MONDAY

THIS WEEK'S FOCUS
HIGH-TECH OUTDOORS

FUN WITH GPS

If you're new to the technology, GPS stands for *global positioning system*, which is a way to determine location based on signals beamed from a network of 24 satellites positioned in six geosynchronized orbital paths around the Earth. A GPS unit receives signals from several of these satellites simultaneously, measures the speed of each satellite, and compares it relative to the unit's location—thus determining your latitude, longitude, and altitude. That is, it tells you where you're at. Precisely.

It's rather complicated technology, originally developed for military use, which has now filtered down to the consumer level. You punch in a location and the GPS software maps out a route and displays it on an onscreen map; your current position is marked on the map, and the map scrolls as you move. Alternatively, you can use the GPS unit to show you where you are, which is great if you ever get lost while hiking or camping.

A handheld GPS unit is a must for any serious outdoorsman; decent units are available for less than $200. Of course, you can also spend a lot more than $200, especially if you like lots of gee-whiz features. Spend more money and you get a more compact unit (it costs big bucks to make things smaller); a color display; and more creature comforts, in the form of automatic route generation, more programmed points of interest, more storage capacity, and so on.

While many adventurers are happy with general handheld GPS units they can use across several different activities, you also have the choice of activity-specific units. That is, there are GPS units targeted specifically at campers and hikers, at hunters, at runners, at boaters, and so on. For example, a GPS unit for hunters might include a built-in compass and onscreen icons to track hunting stands and such; a unit for runners might include a chronometer and lap counter. You get the picture.

ON THIS DAY: 3COM AND U.S. ROBOTICS MERGE (1997)

On June 12, 1997, networking companies 3Com and U.S. Robotics completed their merger. But let's call it what it actually was—3Com acquiring U.S. Robotics. Remember, the company that keeps its own name in a "merger" is the one doing the acquiring.

FACT OF THE WEEK

According to Jupiter Research (www.jupiterresearch.com), online travel sales hit $54 billion in 2004, accounting for 23% of total U.S. travel revenue. That number is expected to rise to $77 billion by 2007, when online travel will account for 30% of all travel revenues.

June 13, 2006

TUESDAY

THIS WEEK'S FOCUS
HIGH-TECH OUTDOORS

HIGH-TECH RUNNING

Like all sports, running has become particularly high-tech in recent years. From the simple pedometer to the complete performance monitoring systems, computers are being put to good use to help you track and monitor your activities and performance.

At the high end, performance monitoring systems incorporate GPS receivers to help track your location, distance traveled, pace, speed, and so on. You'll also get separate heart monitors, which are typically worn around your chest; signals are transmitted from the heart monitor to the main unit via wireless technology. You can set a target heart rate and get an alert if you go beyond it.

Most high-end performance monitoring systems also let you transfer the collected data to your PC, either via USB, infrared, or RF signals. You can then upload the data to performance-tracking software and generate all manner of graphs and reports. It's a cool way to track and analyze your performance over time.

When shopping for a performance monitoring system of any sort, probably the most important factor is size. You don't want to be weighted down by a bulky monitor and display; you want something small and sleek and lightweight. Some systems are incorporated into sports watches, which is nice. Others come in separate units that you can clip to your belt or wear on an armband. Whichever type of system you choose, be sure it's something you'll be comfortable with during a long run.

ON THIS DAY: FIRST V1 BOMB ATTACK (1944)

On the evening of June 13, 1944, the first German V1 flying bomb hit London, England. The V1 resembled a 25-foot long aircraft, with a wingspan of 17 feet; it carried a 1,870-pound warhead and had a maximum speed of about 400 MPH. It wasn't very accurate, but it could hit a target the size of a city—which was all that was required.

GADGET OF THE WEEK: SOUNDSAK SONICBOOM

The SonicBoom is a backpack with built-in speakers and power amplifier—ideal for jamming on the go. The whole thing is powered by six AA batteries, and the 4-inch speakers are powered by a 5 watt/channel amplifier. You can even get it with an optional hydration reservoir, in case you get dry mouth while grooving on the trail! Buy it for $149.99 from www.soundkase.com.

June 14, 2006

WEDNESDAY

THIS WEEK'S FOCUS
HIGH-TECH OUTDOORS

HIGH-TECH FISHING

Fishing used to be a relaxing sport. You bait your hook, drop your line, and then settle back with a cold beer and a hat over your eyes. (Come to think of it, the fishing part of the activity wasn't nearly as important as the beer and the hat.) Today, however, there are all sorts of high-tech devices that make fishing more like hunting, an active activity rather than a passive one.

First off, consider the concept of the fishfinder. This is a device that uses sonar technology to beam back electronic images of underwater life, so you can see onscreen whether there are any fish nearby. Extremely high-tech and extremely practical, a good fishfinder will keep you from wasting time fishing where the fishes aren't.

If sonar is good, why not utilize an underwater camera to actually see any fish that are nearby? It's doable, thanks to today's miniaturized and ruggedized video cameras. Just drop a special fishcam off the side of your boat, and watch the show on an attached LCD monitor. Low-light technology ensures that you can actually see through the murky depths—although not as far as you can with sonar. (That's why the best bet is to use a combination of video and sonar technology.)

Finally, we have the option of trying to attract the fish, not just look around and see if they're nearby. To that end, consider a submersible fish light, which uses a special colored bulb to make fish think there's a full moon out. Baitfish and insects are attracted to the light, which in turn attracts the bigger fishies. Once the bigger fishies arrive, it's up to you to catch 'em. Once again, technology delivers!

ON THIS DAY: UNIVAC 1 DEDICATED (1951)

On this date 1951, the Univac 1 was dedicated in Washington, DC as the world's first commercial computer. The Univac 1 was manufactured for the U.S. Census Bureau by Remington Rand Corp.; it was 8 feet high, 7 1/2 feet wide, and 14 1/2 feet long, could retain a maximum of 1,000 numbers in memory, and was capable of adding, subtracting, multiplying, dividing, sorting, collating, and calculating square and cube roots.

HARDWARE OF THE WEEK: ACT LABS LIGHT GUN

When you want to do your shooting inside, on your PC, instead of outdoors, check out the Act Labs PC USB Light Gun. This is a highly accurate light gun, good for use with all manner of PC shooting games. Buy it for $79.99 from www.act-labs.com.

June 15, 2006

THURSDAY

THIS WEEK'S FOCUS
HIGH-TECH OUTDOORS

HIGH-TECH BOATING

Boating has really gone high-tech, with the express goal of keeping you on course and away from dangerous conditions—on, under, or above the water. There are so many fancy gadgets available today, you wonder how Christopher Columbus ever made it across the ocean with just a sextant and a compass.

Chris would have especially liked the gizmo that high-tech boaters call a chartplotter. A chartplotter is like a traditional GPS device but with detailed maps of waterways rather than roadways. If you're navigating any large expanse of water—from the Great Lakes to the Atlantic ocean—one of these devices is essential for determining exactly where you are at any given point in time. As with any GPS device, you want to look for highly detailed maps and lots of storage for different routes and trackpoints.

What gets fun is combining a chartplotter with other technology. Here you have two different ways to go. The first combines the chartplotter with a sonar device, so you can track your progress underwater as well as on the waves. (It's also great for fishing!) The other option combines the chartplotter with a radar transmitter, so you can track oncoming weather while you're on the water. The best of these units function much like the professional units used by your local TV weatherperson, with color displays for various levels of precipitation and such.

If you go with one of these combo devices, pay particular attention to the display screen. The more data you display, the bigger the screen you want. Some of these displays go to 10 inches or more and are designed for use in both shade and direct sunlight. A bigger screen lets you display two types of info side-by-side—GPS chart and sonar map, for example, or GPS map and radar screen. Even better, some units let you superimpose one type of map on top of the other. Trust me, it's extremely useful to see your normal map with radar conditions overlaid on top.

ON THIS DAY: BEN FRANKLIN FLIES A KITE (1752)

Founding father Benjamin Franklin conducted an important experiment on this date in 1752. While flying a kite with a key attached during a lightning storm, he proved that lighting and electricity were related. What a shock!

SOFTWARE OF THE WEEK: FISH-N-LOG PROFESSIONAL SUITE

Here's a neat piece of software for the serious fisherman. The Fish-N-Log suite lets you log all the details of your fishing trips, detail your catch results and fishing patterns, manage your fishing equipment, and track fishing tournament schedules and results. Buy it for $49.95 at www.taysys.com/thpsuite.htm.

June 16, 2006

FRIDAY

THIS WEEK'S FOCUS
HIGH-TECH OUTDOORS

HIGH-TECH SCUBA DIVING

I'm not a scuba diver myself, but I know some guys who are, and I really think it's a fun sport—not just for what you do while underwater, but for all the cool gear you get to buy. Scuba diving has long been a high-tech sport, and it keeps getting higher tech as time goes on.

The first high-tech device for any serious diver is the dive computer. This is a gizmo that measures depth, time, temperature, ascent/descent rate, and the like, and calculates how much time you have left on your oxygen tank, when you need to ascend (and how quickly), and so on. Newer technology is helping to miniaturize the traditional dive computer, to the point that you can now get a dive computer on a wristwatch. This tech is so cool it tempts me to take scuba lessons myself!

There's also the issue of how you get around while you're underwater. If you remember the James Bond movie *Thunderball*, you recall how 007 jetted from place to place using an underwater jetpack of sorts, which was extremely cool. Well, underwater jetpacks haven't quite come into the mainstream, but the concept of an underwater propulsion device has become viable. I'm talking about a small propeller-driven gadget that you hold onto with both hands and let it drag you behind it. You only go a couple of miles an hour, but that's a couple of miles an hour faster than you can manage with a pair of flippers. It's actually quite cool, and a lot of fun.

But these two types of gadgets just scratch the surface. There's a lot more out there for the high-tech diver, as well as for high-tech adventurers of all sorts. Check with your local dive store to see all that's available!

ON THIS DAY: CTR INCORPORATED (1911)

On June 16, 1911, the Computing-Tabulating-Recording Company (CTR) was incorporated. CTR was the result of the merger of the Computing Scale Company (C), the Tabulating Company (T), and the International Time Recording Company (R). The company later mutated once again, this time into the beast we now know as International Business Machines (IBM).

WEBSITE OF THE WEEK: GORP

No, GORP isn't some sort of berries-and-nuts outdoor snack, it's the premiere website for adventure travel and recreation. Turn to the GORP site for news and reviews related to hiking, camping, fishing, climbing, birding, and the like. It's all at gorp.away.com.

June 17/18, 2006

SATURDAY/SUNDAY

THIS WEEK'S FOCUS
HIGH-TECH OUTDOORS

HIGH-TECH CAMPING

Used to be, going camping for the weekend meant ditching all the creature comforts and getting away from it all. That's not the case today: With all the current miniaturized digital gadgets, you can take it all with you—and it will all fit on your belt.

That's not to say that there aren't useful camping gadgets, because there are. Obviously, one of the most essential gadgets for campers is a handheld GPS device. You don't need anything fancy, just something to tell you where you are when all you can see are trees and rocks. Some extra features are worth considering, though; I particularly like those units with a built-in electronic compass, altimeter, and the like. The more info the better!

Speaking of info, for those long hikes it's nice to know what to expect of the weather. To that end, consider a portable weather tracker. The best of these provide dozens (if not hundreds) of measurements, from temperature and humidity to wind chill and dew point. Most units come with built-in thermometers and barometers, as well as little fan-like thingies to measure wind speed. Be sure the unit is small enough to carry comfortably, of course.

Other useful gadgets for campers tend to fall into the creature comforts category. By this I mean backpacks, coolers, water carriers, and the like. There are even portable ice cream and espresso makers designed especially for use out in the wild. Heck, take along your cell phone, PDA, and portable video player, and it's like you've never left home!

ON THIS DAY: AMELIA EARHART CROSSES THE ATLANTIC (1928)

On June 18, 1928, Amelia Earhart became the first woman to fly across the Atlantic Ocean. The crossing from Newfoundland to Wales took about 21 hours. Earhart went on to establish herself as a noted aviator and a respected role model for women everywhere.

BLOG OF THE WEEK: DIE IS CAST

Die is Cast (www.dieiscast.com) is a blog by and for gamblers. It's all Vegas, all the time, baby!

June 19, 2006

MONDAY

THIS WEEK'S FOCUS
APPLE

APPLE HISTORY (PART I)

The Bible says that apples were present in the Garden of Eden, and an apple was certainly instrumental in the whole Adam and Eve fall of man thing. Similar stories and traditions involving a garden of paradise filled with fruit trees exist in other cultures, as well.

For example, in Greek mythology Gaia presented a tree with golden apples to Zeus and his bride Hera on their wedding day. The apple tree was planted in the garden of the Hesperides and guarded by a serpent named Ladon. (Sound familiar?) These Greek apples were central to many tales of love, bribery, and temptation, ranging from the abduction of Helen of Troy to the defeat and marriage of Atlanta.

Back in the real world, apple-based agriculture spread from the Tien Shan mountains of eastern Kazakstan throughout the civilized world sometime around 8,000 B.C. During this time previously isolated gene pools from 25 distinctly different species of apples are brought into contact with each other, and gene transfer among these apple species occurs.

Meanwhile, a tablet found in northern Mesopotamia, dating to 1,500 B.C., recorded the sale of an apple orchard by Tupktilla, an Assyrian from Nuzi, for the sum of three prized breeder sheep. Hittite Law Codes of the time specified a three-shekel penalty for anyone allowing a fire to destroy an apple orchard.

Oh, wait a minute—you wanted history about Apple *computers*? My mistake; I'll get to that tomorrow.

ON THIS DAY: BLAISE PASCAL BORN (1623)

French mathematician Blaise Pascal was born on this date in 1623. He was responsible for producing many important theorems and treatises on geometry, physics, theology, and other subjects. Pascal also invented the Pascaline, a calculating machine; the Pascal programming language was named after him.

FACT OF THE WEEK

As much as I like my Mac, it's worth reminding myself that Apple remains a niche player in the PC market. At the end of 2004, IDC (www.idc.com) reported that Apple was the number five U.S. computer manufacturer, with just 2.9% of total market sales. Contrast that with Dell, which held a 34.7% market share, or HP, with 20.8% share, and you get the picture.

June 20, 2006

TUESDAY

THIS WEEK'S FOCUS
APPLE

APPLE HISTORY (PART I-A)

Okay, now to the serious Apple history.

High-school friends Steve Wozniak and Steve Jobs (the two Steves) were both interested in electronics, and both ended up working for Silicon Valley companies (Woz for HP, Jobs for Atari). Together, in 1976, they designed a kit computer that they called the Apple I.

Things got serious when Woz and Jobs created their second computer, the Apple II, which debuted in 1977 at a local computer trade show. Users were impressed with the integrated design and color graphics, and sales took off.

In 1980, when the Apple III was released, the company founded in a garage had grown to several thousand employees. Apple was becoming a real company, complete with a board of directors and experienced mid-level managers. The next year was a difficult one, however, when slowing sales forced the company to lay off 40 employees and Woz was injured in a plane crash. In Woz' absence, Jobs became chairman of Apple.

In 1983 Apple released the Lisa, the first computer to include a built-in monitor and mouse. Lisa was a groundbreaking computer but priced too high to succeed in a now-crowded marketplace. It spawned the much more successful Macintosh in 1984.

Apple truly became a big-time company in 1983, when John Sculley, then president of Pepsi-Cola, became president and CEO. The professional Sculley and the techie Jobs eventually clashed, leading Jobs to resign from Apple in 1985. (More to come....)

ON THIS DAY: TRANS-ALASKA PIPELINE OPENS (1977)

This day in 1977 saw the opening of the trans-Alaskan oil pipeline. The pipeline cost $7.7 billion to build and connected the oil fields in Prudhoe Bay to the shipping port of Valdez. The pipe itself is a tube of 1/2-inch-thick steel with a diameter of 48 inches, wrapped with four inches of fiberglass insulation and a cover of aluminum sheet metal.

GADGET OF THE WEEK: GARAGEBAND GUITAR CABLE

Griffin's GarageBand Guitar Cable lets you connect any electric guitar directly to your Mac, for use in GarageBand and similar music applications. The 10-foot cable sports a 1/4-inch guitar plug on one end and a stereo mini-jack on the other. Buy it for $24.99 at www.griffintechnology.com.

June 21, 2006

WEDNESDAY

THIS WEEK'S FOCUS
APPLE

APPLE HISTORY (PART II)

When Steve Jobs left Apple in 1985, he founded a new company named NeXT. The NeXT computer, which shipped in 1989, was a workstation-like machine designed specifically for the university market. The NeXTcube workstation ran a Motorola 6830 processor and incorporated a magneto-optical disk drive; the NeXTstep operating system was highly graphical and very Mac-like. Because of the high price, NeXT never really took off in the marketplace. By 1993, the company had changed its focus to licensing the NeXTstep operating system to other platforms, including SPARC and Sun.

Back in Cupertino, John Sculley was forced to pit Apple against the more successful Microsoft Corporation. The Mac fought a losing battle with Windows and lower-priced PC clones, in spite of the 1987 release of the Mac II and Mac SE, the 1988 release of the Mac IIx, and the 1989 release of the Mac IIcx and Mac SE/30. The following year, in an attempt to regain some of the company's lost magic, Apple released the Mac Classic, a return to the original 1984 Mac form factor.

In 1991, Apple released its first successful portable computer, the PowerBook. In 1993 Apple created the PDA market with the Newton; unfortunately, it was too far ahead of its time and failed in the marketplace.

That same year saw John Sculley leave the company. He was replaced as CEO by Michael Spindler, who himself was replaced by Gil Amelio in 1996. Throughout this period Apple continued to lose presence in the marketplace, in spite of the 1994 release of the PowerMac family, the first computers to use the PowerPC chip jointly developed with IBM and Motorola. The simple fact was that Microsoft was making Windows better and better (and more Mac-like), which made it harder for Apple to compete. (More to come....)

ON THIS DAY: FIRST FERRIS WHEEL OPENS (1893)

On June 21, 1893, the first Ferris wheel premiered at the world's fair in Chicago. It was invented by George Washington Ferris, a bridge builder from Pittsburgh. Each of the 36 cars carried 60 passengers; the whole thing stood 264 feet tall and weighed 2,100 tons. Today, a county fair isn't a county fair without at least one Ferris wheel.

HARDWARE OF THE WEEK: APPLE 30-INCH HD CINEMA DISPLAY

Apple's 30-inch Cinema Display is a widescreen (16:10 aspect ratio) LCD display, perfect for playing or editing movies. It's a high-resolution display, with more than 4 million pixels in a 2560 × 1600 pixel configuration. Do the math and you see that this display offers better-than-HDTV specs, which is saying something. Buy it for a wallet-numbing $2,999 from www.apple.com.

June 22, 2006

THURSDAY

THIS WEEK'S FOCUS
APPLE

APPLE HISTORY (PART III)

Apple's salvation came in late 1996, when Apple acquired NeXT and brought Steve Jobs back into the fold. Jobs took over as "interim" CEO in 1997—a post he continues to hold to this day.

Jobs shook up the company's board of directors and management and refocused Apple on its core products and markets. The company also began selling computers direct, rather than through a network of traditional computer dealers, a decision that ultimately resulted in a chain of Apple stores and a high-profit retail website.

By 1998 Jobs had also revitalized the Apple product line. The flagship product was the trendsetting iMac, an all-in-one machine that found tremendous consumer acceptance. The iMac was followed in 1999 by the iBook, a similarly trendsetting notebook computer, which met with similar market success.

Apple expanded beyond the computer market in 2001, with the extremely popular iPod portable music player. This was followed (in 2003) by the launch of the iTunes Music Store, the Internet's first successful online music download service.

Today, Apple continues to be a market trendsetter, in both the computer and consumer electronics markets. The iPod has become one of the most successful consumer electronics products in history, and Apple computers (including the low-priced Mac Mini, introduced in 2005) continues to hold its own in the marketplace. Apple might be a niche player, but it's very successful in its niche—and the company is definitely here to stay.

ON THIS DAY: KONRAD ZUSE BORN

Konrad Zuse developed and built the world's first binary digital computing device, the Z1. He also created the Z3, the first computer controlled by an external program—fed by punched paper tape. Zuse was born on this date in 1941, in Germany.

SOFTWARE OF THE WEEK: SUPERDUPER

SuperDuper is a hard disk copying/cloning program for Mac OS X. It lets you clone your entire hard disk, which you can then use to restore your system in the event of a major system error. Buy it for $19.95 from www.shirt-pocket.com/SuperDuper/.

June 23, 2006

FRIDAY

THIS WEEK'S FOCUS
APPLE

APPLE PORTABLES

Apple currently sells two different lines of notebook computers: PowerBooks and iBooks. They're both very popular and are targeted at different audiences.

The iBook line is designed for casual and bargain computer users, and for mobile professionals who desire a lightweight portable solution. The iBook is the lightest notebook computer on the market (4.9 pounds for the smallest model) and extremely thin (just 1.35 inches), which makes it easy to carry around. It's available in two screen sizes (12.1-inch and 14.1-inch), both with built-in WiFi wireless networking. Prices start at $999.

Apple's PowerBooks are aimed at power users—or any user who wants a more full-featured experience. PowerBook processors are more powerful (the top-end G4 runs at 1.67GHz), and all models come with more memory and bigger hard disks. Three screen sizes are available: 12-inch, 15-inch, or 17-inch (the last two with a widescreen aspect ratio). You pay a little more for this performance, of course; the base model starts at $1,499, and prices go all the way up to $2,699 (for a fully loaded 17-inch unit).

Which model should you buy? If you're just an average user, not doing anything too heavy duty, or if you want something small and light, go with one of the iBooks. On the other hand, if you're doing high-end operations (video editing, photo editing, and so on) or just want a bigger screen, go with one of the larger PowerBooks. In either instance, you get that great Apple performance and design—which you can't get in the Windows world.

ON THIS DAY: SAXOPHONE PATENTED (1848)

On this date in 1848, a patent was awarded for the saxophone. Who got the patent, you ask? A man named Adolphe Sax, of course. (And now you know....)

WEBSITE OF THE WEEK: THINK SECRET

Think Secret is *the* site for Apple news and info—and rumors. This is the place to go when you want the low-down on what's cooking in Cupertino; these guys know what's coming before Steve Jobs makes the official announcement. Take a peek at www.thinksecret.com.

June 24/25, 2006

SATURDAY/SUNDAY

THIS WEEK'S FOCUS
APPLE

TROUBLESHOOTING MAC OS X

Unlike Windows, Mac OS X is a very forgiving operating system. Not that it never crashes, but they're fewer and further between than what you get with Windows, and easier to recover from.

When you encounter an OS X problem, the first thing to do is restart your Mac with the Shift key held down, which restarts your machine in Safe Boot mode and launches the First Aid file system check utility. Let your Mac start up completely in Safe Boot mode, and then restart it again, normally. This fixes a lot of oddball problems.

If that doesn't do the job, go to your Utilities folder and open Disk Utility. On the left side of the Disk Utility window, select the hard drive you want to repair, click the First Aid tab, and then click the Repair Permissions button. Often, repairing permissions like this will fix a memory-related problem.

Next, you should run Apple's latest Combo update. You can find this update on the Apple website (www.apple.com).

If worse comes to worst, you might need to back up your documents (or, even better, clone your entire system with SuperDuper), reinitialize your hard disk, and then reinstall OS X (or restore your system from the clone files).

ON THIS DAY: BARBED WIRE PATENTED (1867)

On June 25, 1867, Lucien B. Smith of Kent, Ohio, received a patent for barbed wire. His idea was for an artificial "thorn hedge," consisting of wire with short metal spikes twisted on by hand at regular intervals. Since then, barbed wire has been used for everything from preventing livestock from wandering away to keeping prisoners imprisoned.

BLOG OF THE WEEK: THE UNOFFICIAL APPLE WEBLOG

If you're looking for Apple news of all stripes and colors, head on over to The Unofficial Apple Weblog (www.tuaw.com). These folks have all the latest poop from Cupertino!

June 26, 2006

MONDAY

THIS WEEK'S FOCUS
SPYWARE

I SPY...SPYWARE!

As bad as computer viruses and worms are, an even bigger nuisance is the proliferation of spyware programs. These are programs that install themselves on your computer and then surreptitiously send information about the way you use your PC to some interested third party. In this aspect, a spyware program is kind of like a Trojan horse; once it's inside the gates of your PC, it does its dirty work.

Spyware typically gets installed in the background when you're installing another program. One of the biggest sources of spyware are KaZaA and other peer-to-peer music-trading networks; when you install the file-trading software, the spyware is also installed.

Having spyware on your system is nasty, almost as bad as being infected with a computer virus. Some spyware programs will even hijack your computer and launch pop-up windows and advertisements when you visit certain Web pages. If there's spyware on your computer, you definitely want to get rid of it.

Unfortunately, most antivirus programs won't catch spyware because spyware isn't a virus. To track down and uninstall these programs, you need to run a dedicated anti-spyware utility—or two or three of them. That's because a single anti-spyware program won't catch all the spyware that's out there. (There's that much of it!) You'll probably need to run two or three overlapping programs to catch all the possible spyware programs.

ON THIS DAY: BAR CODE DEBUTS (1974)

On June 26, 1974, at 8:01 a.m., a package of Wrigley's chewing gum with a bar code printed on it passed over a scanner at the Marsh Supermarket in Troy, Ohio. This marked the first product sold using IBM's new Universal Product Code (UPC) computerized recognition system. Today, it's difficult to imagine a time without barcodes. Ah, history!

FACT OF THE WEEK

According to Webroot (www.webroot.com), two-thirds of all personal computers were infected with spyware in the first quarter of 2005. The average PC has 25 different instances of spyware surreptitiously installed.

June 27, 2006

TUESDAY

THIS WEEK'S FOCUS
SPYWARE

WHAT DAMAGE CAN SPYWARE DO?

Unlike a computer virus, spyware typically doesn't harm your system. Instead, it runs in the background, hidden from view, and monitors your computer and Internet usage. That probably means performing some or all of the following operations:

- Recording the addresses of each Web page you visit
- Recording the addresses of each email you send and receive—and, perhaps, the contents of those messages
- Recording the contents of all the instant messages you send or receive—along with the usernames and addresses of your IM partners
- Recording every keystroke you type with your computer keyboard
- Recording all your Windows-related activities, including the movement and operation of your mouse

The information recorded by the software is saved to a log file. That log file, at a predetermined time, is transmitted (via the Internet) to a central source. That source can then aggregate your information for marketing purposes, use the information to target personalized communications or advertisements, or steal any confidential data for illegal purposes. Nasty, eh?

ON THIS DAY: CHLOROPHYLL SYNTHESIZED (1960)

On this date in 1960, so-called chlorophyll "a" was first synthesized by Harvard scientist Robert Burns Woodward. This molecule consists of 55 carbon atoms linked with 72 hydrogen atoms, 5 atoms of oxygen, and 1 atom of magnesium. In case you were wondering, Chlorophyll is a molecule that absorbs sunlight and uses the sunlight's energy to synthesize carbohydrates from CO_2 and water. This process is known as photosynthesis and is what allows plant life to survive. Because animals and humans obtain a portion of their food supply by eating plants, photosynthesis obviously is an important part of the life cycle.

GADGET OF THE WEEK: JB1 JAMES BOND 007 SPY CAMERA

The JB1 is the kind of gadget you'd expect Agent 007 to carry—a digital camera built in to a cigarette lighter case. It takes pictures at either 640 × 480 or 320 × 240 resolution; the internal memory holds up to 150 of the higher-resolution pictures. Buy it for $99.99 at www.jbcamera.com.

June 28, 2006

WEDNESDAY

THIS WEEK'S FOCUS
SPYWARE

HOW TO DEFEAT SPYWARE

To defeat spyware, you first have to be aware of its presence. Naturally, you should try to defeat spyware at the source—by not installing it in the first place. The installation of many spyware programs is actually optional when you install the host program; if you look close, you're given the option *not* to install these so-called "companion programs." Check (or uncheck) the proper box on the installation screen, and you avoid installing the spyware (which is sometimes referred to as *adware*).

Other spyware programs are *not* optional components; they install automatically when you install the host program. If you know that a particular program includes piggyback spyware, and you don't have the option not to install the spyware, then you can always opt not to install the main software itself.

You might think that you could remove a spyware program by removing the host program—the KaZaA client, for example. This isn't the case; simply removing the host software seldom (if ever) removes tag-along spyware programs. You have to remove the spyware program separately from the host.

One way to cleanse your system of spyware programs is to manually search your computer for such programs, and then use Windows' Add/Remove Programs utility to do the removal. You can find a list of known spyware programs at SpywareGuide (www.spywareguide.com).

Yet another way to detect and disable spyware is to click Start, choose Run, type msconfig and press Enter. The System Configuration Utility will appear. Click the Startup tab and have a look at all the items that Windows loads every time you start up your PC. Deciding what's important, what isn't, and what's purely malicious is the hard part. For some help with the heavy lifting, fire up your web browser and go to www.sysinfo.org/startuplist.php. Here, you'll find a detailed list of common startup programs. Search through the alphabetical listing to determine what you can disable and what to leave alone.

ON THIS DAY: MACKINAC BRIDGE DEDICATED (1958)

On this date in 1958, the world's longest suspension bridge was dedicated. The Mackinac Bridge connected Michigan's upper and lower peninsulas, reducing the crossing time from several hours to just 10 minutes.

HARDWARE OF THE WEEK: PANASONIC BL-C30A NETWORK CAMERA

Here's a network webcam you can control when you're away from home, via a simple Internet connection. This Panasonic camera operates over an 802.11b or g WiFi connection, so you can place it anywhere within range of your wireless network. Use your web browser, cell phone, or PDA to control panning and tilting, or go directly to one of eight preset shooting positions. Buy it for $299.95 at www.panasonic.com.

June 29, 2006

THURSDAY

THIS WEEK'S FOCUS
SPYWARE

ANTI-SPYWARE UTILITIES

Fortunately, spyware can be easily sniffed out and removed from your system when you use an anti-spyware utility. These programs are designed to identify any and all spyware programs lurking on your computer, and will also uninstall the offending programs and remove their entries from the Windows Registry.

Some of the most popular of these anti-spy and spyware-removal programs include

- Ad-Aware (www.lavasoftusa.com)
- McAfee AntiSpyware (www.mcafee.com)
- Microsoft AntiSpyware (www.microsoft.com/athome/security/spyware/software/)
- Spy Sweeper (www.webroot.com)
- Spybot Search & Destroy (www.safer-networking.org)

I particularly like Microsoft AntiSpyware, which not only does a good job finding all manner of spyware programs, but also runs in the background to prevent spyware from being installed on your system. This program also automatically updates itself over the Internet and can be scheduled to scan your system at predetermined times. (I have my scan scheduled to run every night at two in the morning, while I'm asleep.)

ON THIS DAY: PYGMY MAMMOTH DISCOVERED (1994)

Yeah, I know it sounds like an oxymoron (kind of like jumbo shrimp), but there actually was a beast called a pygmy mammoth. On this date in 1994, the near-complete fossil of just such a beast was found in the sea cliffs on Santa Rosa Island off the coast of California. Scientists estimate that the specimen stood 5 1/2 feet tall and weighed about a ton.

SOFTWARE OF THE WEEK: ACTIVE@ ERASER

This neat little utility lets you completely sanitize your hard disk, erasing and shredding files so that no one can recover them. It even cleans all the tracks of your Internet activities. Download it for $29.95 from www.active-eraser.com.

June 30, 2006

FRIDAY

THIS WEEK'S FOCUS
SPYWARE

LEGITIMATE SPYWARE PROGRAMS

It's an unfortunate fact (for privacy advocates, anyway) that there are numerous spyware programs legitimately available on today's market. Many of these programs are targeted toward the corporate market and are used to spy on a company's employees. Other programs are targeted at the home market and are used to monitor children's Internet-related activities. Still other programs are designed for covert remote operation—which means they can be used by crackers and other third parties to track computer usage over the Internet.

What programs am I talking about? Here's just a sampling:

- ActMon (www.actmon.com)
- Pearl Echo (www.pearlecho.com)
- RemoteSpy (www.remotespy.com)
- SpyAgent (www.spytech-web.com)
- SpyBuddy (www.exploreanywhere.com)
- Stealth Activity Reporter (www.stealthactivityreporter.com)
- Win-Spy (www.win-spy.com)

I find it a bit disconcerting that there are so many computer surveillance programs currently on the market. Some are remote monitoring programs, some alert monitoring personnel (in real time) if banned websites are accessed, some alert the corporate overlords if you dare to type objectionable text. (No offense if you're one of the corporate overlords—I just happen to value my privacy, even when at work.)

ON THIS DAY: LEAP SECOND ADDED (1972)

June 30, 1972, was the first Leap Second Day. No, it's not one of those stupid Hallmark holidays, like Senior Executive Assistant Day. It's simply the day that one second was added to the world's time in order to keep the super-accurate atomic clocks in step with the Earth's actual rotation. Since then, more than 20 seconds have been added to the Atomic Clock because the Earth's rotation keeps slowing down.

WEBSITE OF THE WEEK: COUNTEREXPLOITATION

A good source of information about finding and removing spyware is the Counterexploitation website, located at www.cexx.org. This site provides much valuable information about all sorts of computer nuisances, including spyware, spam, viruses, and the like.

July 2006

July 2006

SUNDAY	MONDAY	TUESDAY	WEDNESDAY	THURSDAY	FRIDAY	SATURDAY
						1 Canada Day / 1858 Darwin's *Theory of Evolution* is first published in London
2 1937 Amelia Earhart's plane disappears somewhere in the South Pacific	**3** 1929 Foam rubber is invented	**4** Independence Day / 1997 Pathfinder probe enters Mars's atmosphere sending back the first-ever photos	**5** 1946 The bikini is introduced to the U.S. by French designers	**6** 1957 Paul McCartney and John Lennon meet at a church picnic near Liverpool	**7** 1936 Phillips-head screwdriver patented	**8** 1776 First public reading of the Declaration of Independence in Philadelphia, PA
9 1815 The first natural U.S. gas well is discovered accidentally in Charleston, WV	**10** 1933 First police radio system begins operation in Eastchester Township, NY	**11** 1979 Skylab, the first U.S. space station, crashes back into Earth	**12** 1962 The Rolling Stones give their first public performance in London	**13** 1881 Billy the Kid shot dead by Pat Garrett in Lincoln County, NM	**14** 1868 Spring tape measure is patented	**15** 1965 Mariner 4 satellite transmits the first close-up photos of Mars back to Earth
16 1945 First atomic bomb tested at the Alamogordo Air Base near Albuquerque, NM	**17** 1989 First flight of the B-2 Stealth bomber over the California desert	**18** 1939 MGM screens a sneak preview of *The Wizard of Oz*	**19** 1799 Rosetta Stone is discovered in Egypt	**20** 1969 Neil Armstrong walks on the moon	**21** 1946 First jet is launched from a ship deck	**22** 1933 First around-the-world solo flight
23 1996 First commercial HDTV broadcast made by WRAL in Raleigh, NC	**24** 1897 Amelia Earhart born	**25** 1997 Stem cells are lab cultured for the first time from aborted human embryos	**26** 1908 FBI is founded	**27** 1940 Bugs Bunny makes his debut in "A Wild Hare"	**28** 1858 Fingerprints first used for identification purposes in Jungipoor, India	**29** 1954 J.R.R. Tolkien's *The Lord of the Rings: The Fellowship of the Ring* published
30 1928 Charles Francis Jenkins starts the first television station broadcast	**31** 1975 Controversial labor leader Jimmy Hoffa disappears and is presumed dead					

July 1/2, 2006

SATURDAY/SUNDAY

THIS WEEK'S FOCUS
SPYWARE

OFFICIAL U.S. GOVERNMENT SPYWARE

Spyware isn't the exclusive province of unscrupulous Internet marketers and even less scrupulous crackers. It's also part of your government's strategy for combating terrorism. Really.

The U.S. government's exercise in high-tech spying comes courtesy of Carnivore, a spyware program developed and used exclusively by the Federal Bureau of Investigation. The FBI uses Carnivore to track down potential criminals and terrorists, by tracking their online activities.

Carnivore is a packet sniffing program designed to "sniff" all the information flowing across a single Internet service provider, and to filter that data based on user. In theory, the FBI obtains a court order to tap the Internet usage of a suspected criminal or terrorist, installs Carnivore at that user's ISP, and then uses the packet sniffer to record all the data sent to or from the targeted individual. It's kind of like a telephone wiretap, but for the Internet.

Post-9/11, the FBI is using Carnivore to track a wide variety of potential lawbreakers. The list of targeted perpetrators extends beyond terrorists to include individuals engaged in child pornography, espionage, fraud, and information warfare. As might be expected, privacy advocates don't like this, while security hounds think it's great. Like it or not, Carnivore is perfectly legal, so it's something we'll have to learn to live with.

Although Carnivore supposedly can be used only with a court order (as with wiretapping), and only to sniff email from a specific user, it's easy to see how this powerful software could be abused—and used to sniff all traffic routed through a given ISP. Do you really want the government reading your email? If not, you need to get more involved in the current privacy debate!

ON THIS DAY: IBM 650 ANNOUNCED (1953)

On July 2, 1953, IBM announced the 650 series of mainframe computers. The IBM 650 received information and instructions on programmed punch cards and stored that information on a rotating magnetic drum.

BLOG OF THE WEEK: SPYWARE WARRIOR

Get smarter about spyware at the Spyware Warrior blog. There's lots of news and advice here; I particularly like the weekly "spyware week in review" posts. Read it at www.netrn.net/spywareblog/.

July 3, 2006

MONDAY

THIS WEEK'S FOCUS
MICROSOFT WORD

CREATING NUMBERED LISTS

The fastest way to create a numbered list in Word is to highlight the paragraphs in your list, and then click the Numbering button on the Formatting toolbar. This indents the selected paragraphs and applies the default numbering style.

To change the style of your numbered list (from 1., 2., 3. to I, II, III or A, B, C), select the numbered text then select Format, Bullets and Numbering. When the Bullets and Numbering dialog box appears, select the Numbered tab, choose a new number type, and then click OK.

To continue the numbering from one numbered list to another (separated by one or more paragraphs of normal text), start by positioning your cursor in the first numbered item in the second list; then open the Bullets and Numbering dialog box and select the Numbered tab. From here you can choose to continue the numbering from the previous numbered list (by checking the Continue Previous List option) or to start the numbering of this new list at the number "1" (by checking the Restart Numbering option).

You can also change the style of your numbered list, from 1., 2., 3. to I, II, III or A, B, C. Just select the numbered text you want to reformat, open the Bullets and Numbering dialog box, and select the Numbered tab. From here you can select a new number type from the list. And if you don't like any of the numbered lists presented, feel free to create a custom list style by clicking the Customize button.

By the way, don't confuse numbered lists with line numbering, which are simply numbers displayed for each line of your printed document. By numbering each line, it's easy to go to a specific phrase or item. To turn on line numbering for a document, select File, Page Setup, then select the Layout tab. Click the Line Numbers button to display the Line Numbers dialog box, and then check the Add Line Numbering option. Select any other options that apply, and then click OK.

ON THIS DAY: FOAM RUBBER DEVELOPED (1929)

On July 3, 1929, the Dunlop Latex Development Laboratories in Birmingham, England, developed foam rubber. The first batch was whipped up by scientist E.A. Murphy using an ordinary kitchen mixer.

FACT OF THE WEEK

The Pew Internet and American Life Project (www.pewinternet.org) reports that, as of the first quarter of 2005, 25% of Internet users have read blogs, while 9% of all users have created their own blogs.

July 4, 2006
TUESDAY

THIS WEEK'S FOCUS
MICROSOFT WORD

USING STYLES

If you have a preferred paragraph formatting that you use over and over and over, you don't have to format each paragraph individually—you can assign all your formatting to a paragraph *style*, and then apply that style to multiple paragraphs across your entire document.

Styles include formatting for fonts, font style, font size, font color, paragraph alignment, tabs, borders, language, frames, and numbering. When you apply a style to a paragraph, all these formatting options are applied. As an example, you might have a style you use for section headings. This style would include formatting for the heading text (font, style, size, color), the paragraph options (line spacing, paragraph spacing, and margins), and perhaps a border or section number. When you apply this style to a paragraph, that paragraph is automatically reformatted with all the properties of the style—which is both quicker and easier than applying each of those individual formatting options individually.

By the way, Word's predesigned templates come with a selection of styles specific to that type of document. For example, the Contemporary Letter template includes more than 100 different styles (although only about three dozen are actively used), ranging from Attention Line to Caption to CC List to Mailing Instructions to Return Address to Title. You apply the appropriate style to the appropriate paragraphs.

Applying a style to a paragraph is easy. Just position the cursor anywhere in the paragraph, pull down Style list on Word's Formatting toolbar (or press Ctrl+Shift+S), and then select the desired style. That's it.

To modify an existing style, select Format, Styles and Formatting to display the Styles and Formatting pane. The current style is noted in the top box; click this box, click the down arrow, and select Modify. This displays the Modify Style dialog box; from here you can change basic style properties, or click the Format button to change more detailed settings.

ON THIS DAY: FIRST DIRECT KEYBOARD INPUT (1956)

On Independence Day, 1956, direct keyboard input debuted on MIT's Whirlwind computer. This method of input was revolutionary at a time when programmers issued instructions by either inserting punch cards or twiddling dials and switches.

GADGET OF THE WEEK: VIOLIGHT TOOTHBRUSH SANITIZER

The VIOlight is a toothbrush sanitizing system that uses a germicidal UV light bulb to kill 99.9% of the nasty bacteria found between your bristles; the entire process takes less than 10 minutes. It doubles as an ordinary mild-mannered toothbrush holder, so no extra counter space is necessary. Buy it for $49.95 from www.violight.com.

July 5, 2006

WEDNESDAY

THIS WEEK'S FOCUS
MICROSOFT WORD

ADDING BACKGROUND COLORS AND GRAPHICS

Black text on white paper is fine, but there will come a time when you want something a little less conventional. When you need to make your document stand out from the crowd, use Word to add backgrounds and borders to your pages—with just a few clicks of the mouse!

To change from a white page to a different-color background, select Format, Background and choose a format from the color submenu. To choose from a wider palette of colors, select More Colors to display the More Colors dialog box. To use a background gradation or texture, select Fill Effects to display the Fill Effects dialog box.

Although you can't use a picture in the background of your printed documents, Word does let you insert a watermark to appear behind your text. Watermarks are actually linked to headers and footers but can be positioned anywhere on your page; they appear on all pages that include the linked-to header or footer.

To create a watermark, select View, Header and Footer. Position the insertion point in either the header or footer, and then select Insert, Picture *item*, where *item* is the type of object (picture, clip art, Word Art, and so on) you want to insert. Now double-click the picture to display the Format Picture dialog box, select the Layout tab, select Behind Text, and then click OK. (You should use your mouse to reposition and resize the object as appropriate.)

By the way, if you're changing to a dark background, check to see whether your normal black text is still readable. If not, consider changing your font color to white throughout.

ON THIS DAY: INTEL PRICE CUTS (1994)

On July 5, 1994, Intel announced the first of two major price cuts on its Pentium line of microprocessors. The goal was to keep rival companies from taking a larger share of the market. Today, Intel and AMD continue to slug it out in the PC microprocessor ring.

HARDWARE OF THE WEEK: LOGITECH LASER CORDLESS MOUSE

An optical mouse is better than a rollerball mouse, and Logitech's MX1000 is even better than that. That's because the MX1000 uses a tiny laser beam to "read" the desktop, providing 20 times the tracking resolution of a typical optical mouse. It's also cordless and sells for $79.95 from www.logitech.com.

July 6, 2006

THURSDAY

THIS WEEK'S FOCUS
MICROSOFT WORD

USING SECTION BREAKS

If you have a short document, you might be able to use the same page formatting throughout the entire document. However, if your document is longer, you might need to vary certain formatting or layout options—you might need one part of a single-column document to utilize a two-column layout, or you might want to number various sections of your document differently.

When you need to make major layout changes to parts of your documents, you have need of Word's *sections* feature. Word lets you break documents up into multiple sections, with each section having its own formatting and layout.

When you need to start a new section in your document, position your cursor where you want to start a new section, and then select Insert, Break. When the Break dialog box appears, select the type of break you want: Next Page, Continuous, Even Page, or Odd Page. Once you've inserted the break, you can format the sections before and after the break separately.

Note that the new section you create maintains all the formatting of the previous section—until you change it. At that point, any layout or page formatting changes you make apply only to the current section, not to other sections (either earlier or later) in your document.

To delete a section break—and thus "merge" the first section with the second section—simply delete the section break marker. Note that when you merge two sections, the second section assumes the formatting (including header/footer contents and formatting) of the first section.

ON THIS DAY: AOL SETTLES LAWSUITS

On this day in 1996, America Online settled several California lawsuits that accused the company of misleading subscribers about how it computed monthly service charges. As part of the settlement, customers received $22 million in cash rebates and free online time.

SOFTWARE OF THE WEEK: WORDWARE

Wordware is a suite of more than 45 add-in utilities and templates for Microsoft Word. Tools include a personal information manager, daily journal, Web link extractor, fax composer, document backup tool, and more. Buy it for $39.95 from www.amfsoftware.com/word/.

July 7, 2006

FRIDAY

THIS WEEK'S FOCUS
MICROSOFT WORD

USING HEADERS AND FOOTERS

Headers and footers are repeating sections at the top (header) or bottom (footer) of your document. You can use headers and footers to display information about your document, such as title, subject, author, date, chapter number, or page number.

Know, however, that headers and footers do not appear as part of your normal document text; in fact, they aren't visible at all unless you're in Word's Print Layout mode. You can't even use the arrows on your keyboard to move the insertion point to a header or a footer.

To work within a header or footer, select View, Header and Footer. This switches Word to Print Layout view (if you weren't in that view already), opens the header for editing, and displays the Header and Footer toolbar. You use the commands on this toolbar to edit and format your header and footer.

To add text to a header or footer, select View, Headers and Footers to open the header editing area. You can then enter text as normal within the header. To switch to the footer editing area, click the Switch Between Header and Footer button on the toolbar. Click the Close button to close the header/footer editing areas.

Note that you can format text within a header or footer as you'd format any normal text elsewhere in your document. You can apply individual text/font formatting, or select Format, Paragraph to format the header or footer "paragraph" attributes. You can even select Format, Style to edit either the Header or Footer styles (individually). And don't forget to use Word's ruler to reposition the tabs within the header or footer, if necessary.

ON THIS DAY: PHILLIPS-HEAD SCREW PATENTED (1936)

A new, improved type of screw was patented on this date in 1936. The so-called Phillips-head screw was invented by Henry F. Phillips. The patent described a fastening system that involved a shallow cruciform recess and a matching driver with a tapering tip that self-centered in the screw head.

WEBSITE OF THE WEEK: WOODY'S OFFICE PORTAL

Woody Leonhard is a Microsoft Office guru extraordinaire (and a fellow author), and the WOPR site is chock-full of tips, advice, downloads, and other information about all the Microsoft Office applications—Word included. Check it out for yourself at www.wopr.com. If you like what you see there, note that Woody Leonhard has teamed with Ed Bott to write *Special Edition Using Microsoft Office*, a great book published by Que. Check it out your favorite bookstore or online retailer.

July 8/9, 2006

SATURDAY/SUNDAY

THIS WEEK'S FOCUS
MICROSOFT WORD

CREATING A MULTIPLE-COLUMN LAYOUT

Certain types of documents—newsletters, especially—lend themselves to multiple-column layouts. Word lets you quickly and easily apply multiple-column formatting to your entire document, or just to selected sections. The result is a very professional look, when done right.

To change the layout of the current section of your document (or, if your document isn't divided into sections, your entire document) to a multiple-column layout, start by positioning your cursor where you want the multiple-column layout to begin. Next, select Format, Columns to display the Columns dialog box, and choose one of the preset column types (one column, two column, three column, and so on). If you want a line between your columns, check the Line Between option; then pull down the Apply To list and select whether you want to apply this column layout to your Whole Document, or just This Point Forward to the end of your document. Click OK, and you have your multiple-column document.

If you don't like any of Word's preset column layouts, you can create your own by specifying custom column widths and spacing. Just open the Columns dialog box and select how many columns you want from the Number of Columns list. If you want your columns to be of equal width, check the Equal Column Width option; if you want columns of different widths, *uncheck* this option. If you selected unequal column widths, you can select the width and spacing for each column in the Width and Spacing area. Do the Apply To thing, click OK, and you're ready to go.

By default, Word aligns all your column text to the top of the column. This sometimes results in ragged column bottoms. If you prefer even column bottoms, you can set Word's vertical alignment to justify your text vertically. Position your cursor where you want the new vertical alignment to begin—typically the same place you started the multiple-column layout. Select File, Page Setup to display the Page Setup dialog box, and then select the Layout tab. Pull down the Vertical Alignment list and select Justified, and then pull down the Apply To list and select This Point Forward. Click OK when you are done.

ON THIS DAY: *TRON* RELEASED (1982)

On July 9, 1982, computer geeks flocked to the movie theaters for the release of Disney's *Tron*. Starring Jeff Bridges, *Tron* was the first mainstream film to use extensive computer-generated graphics and special effects. Better yet, the Hollywood version of *Tron* was Al Gore's favorite movie.

BLOG OF THE WEEK: THE OFFICE WEBLOG

This is a blog for users of Microsoft Office and its sundry applications. Lots of good news and advice here; the URL is office.weblogsinc.com.

July 10, 2006

MONDAY

THIS WEEK'S FOCUS
WEB SEARCHING

WHY IT'S SO HARD TO FIND WHAT YOU WANT ONLINE

You search and you search and you search the Web, yet you never seem to find exactly what you're looking for. There's a simple reason for this: The Internet isn't nearly as orderly as you might like to believe.

You see, the Internet is like a bunch of file cabinets in a great big office. You can walk from file cabinet to file cabinet (surf from website to website), rifle through the file folders in any particular file drawer (browse through the pages on any site), and scan the papers within any individual folder (read the contents of any individual web page). After all that work, you still might not be able to find what you were looking for.

Think about it: How many times have you tried to find a specific piece of information in your office—and failed? Did you always go to the right file cabinet? Did you always find the right folder? Were the folders always organized the way you thought they'd be organized? Did the folders you looked in always contain the papers you thought would be there? And did the papers you read always contain the information you wanted—worded in precisely the manner you expected?

Of course not. Papers and folders and files are all created and organized by human beings, and human beings (1) are not perfect, (2) seldom think perfectly logically, and (3) rarely think alike. And it's the same way on the Web: Things aren't always as organized as they should be. There's simply no single organized repository of online information, which makes searching for information that much more difficult.

Which is why we have search engines—which will learn more about tomorrow.

ON THIS DAY: TELSTAR LAUNCHED (1962)

On this day in 1962, the world's first geosynchronous—not to mention privately funded—communications satellite was launched from Cape Canaveral. Developed by AT&T, Telstar 1 was designed to relay television and telephone signals between the United States and Europe. The first live transatlantic telecast included a press conference by U.S. President John F. Kennedy, a performance by French singer Yves Montand, and the changing of the guard at England's Buckingham Palace. It ceased operations in 1963 after being damaged by radiation and was replaced by Telstar 2, which was launched in May 1963.

FACT OF THE WEEK

Nielsen-Netratings (www.nielsen-netratings.com), which monitors these things, says that as of January 2005, Google was far and away the most popular search engine, with 47% of total searches. Yahoo! is number two with a 21% share; MSN Search is a distant third with just 13%. Nobody else is big enough to matter.

July 11, 2006

TUESDAY

THIS WEEK'S FOCUS
WEB SEARCHING

SEARCH ENGINES AND DIRECTORIES

As the number of individual web pages grew from tens of thousands to hundreds of thousands to millions to tens of millions to hundreds of millions, it became imperative for people to quickly and easily find their way around all those pages. With the explosion of the Web, then, came a new industry of cataloging and indexing the Web—and the two main ways of organizing the Web were *search engines* and *directories*.

The first approach to organizing the Web was to physically look at each web page and stick them into a hand-picked category. Once you got enough web pages collected, you had something called a *directory*. A directory doesn't search the Web—in fact, a directory only catalogs a very small part of the Web. But a directory is very organized and very easy to use, and lots and lots of people use Web directories every day. The big directories on the Web today are LookSmart (search.looksmart.com) and the Open Directory (www.dmoz.org).

A different and more effective approach was to let technology do the searching for you—via a Web search engine. Unlike a directory, a search engine isn't powered by human hands; instead, a search engine uses a special type of software program (called a spider or crawler) to roam the Web automatically, feeding what it finds back to a massive bank of computers. These computers hold indexes of the Web. When you use a search engine, you're actually searching the index stored on the site, not the Web itself. The biggest search engine today is Google (www.google.com), hands down.

So, which is better, a directory or a search engine? It all depends. If you want the most (and the most current) results, you should use a search engine. On the other hand, if you want the highest-quality, most organized results, you should use a directory.

Although it's tempting to say that search engines deliver quantity and directories deliver quality, that isn't always the case. Some of the best and most powerful search engines can deliver quality of results matching or beating those from the top directories. And, to complicate matters even further, most search engine sites include Web directories as part of their services—and the major directories often include search engine add-ons.

ON THIS DAY: SURGICAL ZIPPERS ANNOUNCED (1985)

On July 11, 1985, zippers for stitches were announced by Dr. H. Harlan Stone. The zippers were used for patients on which the surgeon thought he might have to re-operate (because of internal bleeding), and lasted between 5 and 14 days.

GADGET OF THE WEEK: OAKLEY THUMP

The Oakley THUMP grafts a flash MP3 player and pair of integrated earphones onto some tres cool Oakley shades. You get your choice of 128MB ($400) or 256MB ($500) models; check them out at www.oakley.com/thump/.

July 12, 2006

WEDNESDAY

THIS WEEK'S FOCUS
WEB SEARCHING

TOP SEARCH SITES

When you think search, you probably think either Google or Yahoo! But these aren't the only search sites on the Web; here's an alphabetical list of some of the other search engines:

- AllTheWeb (www.alltheweb.com)
- AltaVista (www.altavista.com)
- AOL Search (search.aol.com)
- Ask Jeeves! (www.ask.com)
- HotBot (www.hotbot.com)
- MSN Search (search.msn.com)
- Teoma (www.teoma.com)

Let's not forget directories, the most popular of which are About (www.about.com), Google Directory (directory.google.com), LookSmart (search.looksmart.com), and Open Directory (www.dmoz.org). And remember: Yahoo! offers both a directory and a search engine, which gives you the best of both worlds!

ON THIS DAY: SMOKING (OFFICIALLY) CAUSES CANCER (1957)

On July 12, 1957, U.S. Surgeon General Leroy Burney issued a report on a connection between smoking and lung cancer. The report stated, "It is clear that there is an increasing and consistent body of evidence that excessive cigarette smoking is one of the causative factors in lung cancer."

HARDWARE OF THE WEEK: DECK KEYBOARD

This is one cool-looking keyboard. The entire Deck keyboard—including the individual keys—is backlit in either Gold, Fire (red), or Ice (blue). The lighting is via LEDs under a tough polycarbonate housing, great for heavy-handed typists. Buy it for $99 from www.deckkeyboards.com.

July 13, 2006

THURSDAY

THIS WEEK'S FOCUS
WEB SEARCHING

METASEARCH ENGINES

In addition to traditional search sites, there are also a number of search engines that let you search multiple search engines and directories from a single page. This is called a *metasearch*; results can be displayed in a consolidated fashion (all the results from multiple sites displayed in the same list) or in separate lists (for each site searched).

Why would you use a metasearch site, as opposed to just going to Google and doing your thing there? As good as today's search sites are, they all tend to scour the Web in slightly different ways. This means that you can enter the same query at different sites and get slightly—if not drastically—different results. If Google doesn't satisfy, Yahoo! might, or maybe Teoma will do the job. Instead of searching each site separately, a metasearch site lets you enter one query and get results from all these sites, and more.

The top metasearch sites today include

- Beaucoup (www.beaucoup.com)
- CNET's Search.com (www.search.com)
- Dogpile (www.dogpile.com)
- Mamma (www.mamma.com)
- MetaCrawler (www.metacrawler.com)
- OneSeek (www.oneseek.com)
- Search Spaniel (www.searchspaniel.com)
- WebTaxi (www.webtaxi.com)

ON THIS DAY: ERNO RUBICK BORN (1944)

Hungarian mathematician and educator Erno Rubick was born on this date in 1944. Rubick is more famously known as the inventor of Rubick's Cube, one of the most popular puzzle toys of the 1980s. (In case you were wondering, there are 43 quintillion possible configurations of the cube; only one is correct.)

SOFTWARE OF THE WEEK: WEBFERRET

WebFerret is a metasearch utility you run from the Windows desktop. Launch the program, enter your query, and get results from multiple search engines. Download it for free from www.ferretsoft.com.

July 14, 2006

FRIDAY

THIS WEEK'S FOCUS
WEB SEARCHING

THE CORRECT WAY TO SEARCH

Is there one single correct way to perform an online search? No, of course not; every searcher has a preferred style and approach. That said, there is a basic search procedure that works on just about every search site on the Web. It's a seven-step process that looks like this:

1. Start by thinking about what you want to find. What words best describe the information or concept you're looking for? What alternative words might some use instead? Are there any words that can be excluded from your search to better define your query?
2. Determine where you should perform your search. Do you need the power of Google or the better-qualified results of a directory? Are there any topic-specific sites available you should use instead of these general sites?
3. Construct your query. If at all possible, try to use Boolean expressions (they're more flexible). Use as many keywords as you need—the more the better. If appropriate (and available), use the site's advanced search page or mode.
4. Click the Search button to perform the search.
5. Evaluate the matches on the search results page. If the initial results are not to your liking, refine your query and search again—or switch to a more appropriate search site.
6. Select the matching pages that you wish to view and begin clicking through to those pages.
7. Save those pages that best meet your needs.

ON THIS DAY: DYNAMITE DEMONSTRATED (1867)

On this day in 1867, Alfred Nobel demonstrated dynamite for the first time at a quarry in Redhill, Surrey, in England. Dynamite is a relatively safe and manageable form of nitroglycerin. (Nobel is also known as the man behind the prize that bears his name.)

WEBSITE OF THE WEEK: SEARCH ENGINE WATCH

Get all the latest news about search engines, directories, and related technology at Danny Sullivan's Search Engine Watch. Check it out at www.searchenginewatch.com.

July 15/16, 2006

SATURDAY/SUNDAY

THIS WEEK'S FOCUS
WEB SEARCHING

BOOLEAN SEARCHING

Just about every search site on the Web (Google excluded) lets you fine-tune your search with Boolean operators. These are expressions, based on algebraic Boolean logic, that result in a value of either true or false. The most common Boolean operators include

- **AND** (a match must contain *both* words to be true)—Searching for **monty AND python** will return Monty Python pages or pages about pythons owned by guys named Monty, but not pages that include only one of the two words.
- **OR** (a match must contain either of the words to be true)—Searching for **monty OR python** will return pages about guys named Monty or pythons or Monty Python.
- **NOT** (a match must exclude the next word to be true)—Searching for **monty NOT python** will return pages about guys named Monty but will not return pages about Monty Python—because you're *excluding* "python" pages from your results.

In addition, Boolean searching lets you use parentheses, much as you would in a mathematical equation, to group portions of queries together to create more complicated searches. For example, if you wanted to search for all pages about balls that were red or blue but not large, the query would look like this: **balls AND (red OR blue) NOT large**.

A handful of other Boolean operators are also available, such as ADJ or NEAR or FAR, that have to do with *adjacency*—how close words are to each other. However, very few search engines incorporate these adjacency operators, so you probably won't have much of an opportunity to use them.

ON THIS DAY: ENIGMA ENCODES FIRST MESSAGE (1928)

On July 15, 1928, the ENIGMA machine encoded its first message. The ENIGMA was a machine invented by the Germans, about the size of a portable typewriter, that converted text messages into encrypted code. It was used extensively in World War II to transmit battle plans and other secret information.

BLOG OF THE WEEK: SEARCH ENGINE BLOG

Searching for a blog that's all about search engines? Look no further than the aptly named Search Engine Blog, located at the imaginative URL of www.searchengineblog.com.

July 17, 2006

MONDAY

THIS WEEK'S FOCUS
CREATING YOUR OWN WEB PAGES

USING A HOME PAGE COMMUNITY

It seems like everybody and his brother has his own personal web pages these days. If you want to keep up with the Joneses (and the Smiths and the Berkowitzes), you need to create a personal web page of your own.

If you want a one-stop solution to creating and hosting your own web pages, without having to learn HTML, turn to one of the major home page communities on the Web. These sites not only help you create your own web pages, they even host your pages on the Web.

When you join a home page community, you're provided with a specified amount of space on their servers, typically in the 15MB–50MB range. (In case you're wondering, 50MB should be *more* than enough to host your personal pages—unless you're attempting to re-create the *Encyclopedia Britannica* online!) You can then use the tools on the site to create your web pages, or upload previously created pages to their site.

The most popular of these home page communities include Angelfire (angelfire.lycos.com), Tripod (www.tripod.lycos.com), and Yahoo! GeoCities (geocities.yahoo.com). All these sites offer a variety of hosting plans, the most basic of which are free. (Free is good.) Both Angelfire and Tripod offer tools for building and hosting your own personal blog.

Many Internet service providers also offer free personal home pages to their subscribers; check with your ISP to see what services are available. In addition, if you're an America Online member, you can avail yourself of the AOL Hometown home page community. (AOL Hometown is accessible from within the AOL service, or on the Web at hometown.aol.com.)

ON THIS DAY: MAJOR EMAIL DISRUPTION (1997)

On July 17, 1997, email went screwy. At 2:30 a.m. Eastern Daylight Time, a computer operator in Virginia ignored alarms on a computer that updated Internet address information, leading to problems at several other computers with similar responsibilities. This led to a problem with accessing addresses across the entire Internet, resulting in millions of unsent email messages.

FACT OF THE WEEK

Okay, this is more of an opinion than a fact, but design guru Jakob Nielsen (www.useit.com) lists the top 10 mistakes in web page design as (1) Bad search, (2) PDF files for online reading, (3) Not changing the color of visited links, (4) Non-scannable text, (5) Fixed font size, (6) Page titles with low search visibility, (7) Anything that looks like an advertisement, (8) Violating design conventions, (9) Opening new browser windows, and (10) Not answering users' questions.

July 18, 2006

TUESDAY

THIS WEEK'S FOCUS
CREATING YOUR OWN WEB PAGES

FINDING A WEB HOST

A professional website is more than just a collection of web pages. It's a complete, thoughtfully envisioned, well laid-out experience, where the home page leads naturally to subsidiary pages, and where all the pages share a similar design and navigation system. No matter which page users ultimately gravitate to, they should know that that page is part of a greater site, and be able to navigate back to a home page that truly serves as a portal to all that lies within.

If you want to create a professional website—or simply a collection of more sophisticated personal pages—you need to use a more sophisticated HTML editing program, such as Adobe GoLive CS (www.adobe.com/products/golive/), Dreamweaver (www.macromedia.com), or Microsoft FrontPage (www.microsoft.com/frontpage/). Then you have to find a site on the Web to host your pages.

One place to look is at Yahoo! GeoCities and the other home page communities, most of which offer separate page hosting services (usually for a fee) geared toward personal web pages. Alternatively, if you need a host for a complete website (or for small business purposes), you should probably examine a service that specializes in more sophisticated website hosting.

A website hosting service will manage all aspects of your website. For a monthly fee, you'll receive a fixed amount of space on their servers, your own website address (and your own personal domain, if you want to pay for it), and a variety of site management tools. Pricing for these services typically start at $10 or so a month, and goes up from there.

The best way to look for a website host is to access a directory of hosting services. Most of these directories let you search for hosts by various parameters, including monthly cost, disk space provided, programming and platforms supported, and extra features offered. Among the best of these host search sites are HostIndex.com (www.hostindex.com), HostSearch (www.hostsearch.com), and TopHosts.com (www.tophosts.com).

ON THIS DAY: INTEL FOUNDED (1968)

On July 18, 1968, three very wise men named Robert Noyce, Andy Grove, and Gordon Moore founded a company to produce microprocessors. They named their company Intel, and it was very successful. You've probably heard of it.

GADGET OF THE WEEK: PROAIM GOLFING GOGGLES

Better golfing through technology—thanks to ProAim's "virtual alignment trainer." This gizmo is actually a set of goggles that use night-vision technology to project a virtual alignment grid into your field of vision. Line up your shot with the grid, and your shots will be perfectly aligned with the hole. Buy a pair for $59.95 at www.proaim.com.

July 19, 2006

WEDNESDAY

THIS WEEK'S FOCUS
CREATING YOUR OWN WEB PAGES

UNDERSTANDING HTML

HTML stands for *hypertext markup language*, and though the concept of HTML coding might sound difficult, it's really pretty easy—something you can do yourself. (It's not nearly as complicated as a fancy computer programming language, such as BASIC or C++—trust me.) HTML is really nothing more than a series of hidden codes. These codes tell a Web browser how to display different types of text and graphics. The codes are embedded in a document, so you can't see them; they're visible only to your web browser.

The first thing you need to know is that HTML is nothing more than text surrounded by instructions, in the form of simple codes. Codes are distinguished from normal text by the fact that they're enclosed within angle brackets. Each particular code turns on or off a particular attribute, such as boldface or italic text. Most codes are in sets of "on/off" pairs, called tags; you turn "on" the code before the text you want to affect and then turn "off" the code after the text. A typical tag looks like this:

<tag>text</tag>

There is always a tag name, it is always a single word, and it always comes first in the tag, at the very left, right after the opening bracket. Some HTML tags only have the tag name part. Others have one or more attributes; still others let you define one or more parameters. Just remember that almost all tags work in pairs, with an opening and closing tag. The affect of the tag pairs is applied to the area or text between them. (Read tomorrow's article to learn specific HTML tags.)

When you view the HTML code for a document, you're actually viewing a plain text document. Because of this, you can use any text editor—such as Windows Notepad or Wordpad—to edit HTML documents. Although this makes for easy editing, entering raw code is not always the most intuitive way to build a web page.

ON THIS DAY: SAMUEL COLT BORN (1814)

Samuel Colt was born on this day in 1814; he died on January 10, 1862. Colt was an American firearms manufacturer who popularized the Colt 45 revolver and other guns. The Colt factory was one of the most innovative in its use of mass-production techniques and still manufactures some of the finest American firearms available, including the now legendary Colt .45-caliber handgun.

HARDWARE OF THE WEEK: GRIFFIN AIRCLICK USB REMOTE

Want to operate your PC's media functions (audio playback, and so on) via remote control? Then check out Griffin's AirClick USB. The receiver connects to your PC or Mac via USB; the hand unit has five buttons for play/pause, next track, previous track, volume up, and volume down. Buy it for $39.99 from www.griffintechnology.com.

July 20, 2006

THURSDAY

THIS WEEK'S FOCUS
CREATING YOUR OWN WEB PAGES

COMMON HTML TAGS

Okay, now that you know how HTML works, what are some of the more useful HTML tags? Here's a short list:

- **<!>text</!>**—Adds invisible comments to your document.
- **<b>text</b>**—Boldfaces text.
- **
**—Forces a line break.
- **<center>text</center>**—Centers content.
- **<em>text</em>**—Designates emphasis in text.
- **<h1>text</h1>**—Heading level one.
- **<h2>text</h2>**—Heading level two.
- **<h3>text</h3>**—Heading level three.
- **<hr>**—Inserts a horizontal rule or line.
- **<i>text</i>**—Italicizes text.
- **<p>text</p>**—Designates a paragraph.
- **<strong>text</strong>**—Designates stronger emphasis than <em>.
- **<u>text</u>**—Underlines text.

Remember, with very few exceptions (such as the <br> tag), your HTML tags must be applied in pairs. Insert the first tag to turn on the attribute, and the second tag (typically preceded with a slash mark) to turn off the attribute. If you forget to insert the "off" tag, all the rest of the text in your document will retain the formatting specified with the "on" tag. (This is what's wrong when you see boldface or italic text starting in the middle of a paragraph and then continuing to the end of the document!)

ON THIS DAY: MAN WALKS ON MOON (1969)

On July 20, 1969, Apollo 11 astronaut Neil Armstrong became the first human being to set foot on the surface of the moon. His first words? "That's one small step for a man, one giant leap for mankind." Without a doubt, his is one of the most famous quotes ever uttered.

SOFTWARE OF THE WEEK: HOTDOG PAGEWIZ

HotDog PageWiz is an easy-to-use HTML editor you can use to create your personal web pages. It lets you create great-looking (and quite sophisticated) pages without imposing a steep learning curve. Buy it for $69.95 at www.sausage.com/products.html.

July 21, 2006

FRIDAY

THIS WEEK'S FOCUS
CREATING YOUR OWN WEB PAGES

INSERTING HYPERLINKS

One of the most common uses of HTML is to create hyperlinks to other web pages. This is accomplished with a single piece of code that looks like this: `<a href="`*URL*`">This is the link</a>`.

The text between the on and off codes will appear onscreen as a typical underlined hyperlink; when users click that text, they'll be linked to the URL you specified in the code. Note that the URL is enclosed in quotation marks and that you have to include the `http://` part of the address.

Hyperlinks aren't limited to just text. You can also hyperlink a graphic by surrounding the `<img>` tag with the `<a>` and `</a>` tags, like this: `<a href="`*URL*`"><img src="`*IMAGEURL*`"></a>`. When users click the hyperlinked graphic, they jump to the linked page.

You can also add links that visitors can click to send you email messages. Just enter the following piece of HTML code: `<a href="mailto:`*yourname@domain.com*`">Click here to email me</a>`. Replace *yourname@domain.com* with your own email address, of course. When the visitor clicks on this link, they'll open a new blank email message in their email program, with your specified address already inserted into the "To" field.

Another use of hyperlinks is to create (and jump to) bookmarks on really long web pages. You insert the code `<a name="#`*jumphere*`">` to create a bookmark. (Replace *jumphere* with the name for the bookmark, of course.) Then you create a hyperlink to the bookmark, with this code: `<a href="#`*jumphere*`">Click to jump</a>`. (Again, replace *jumphere* with the name of the bookmark.) When visitors click on this link, their browser will automatically scroll down to the bookmarked section of your page.

ON THIS DAY: XEROX WITHDRAWS FROM COMPUTER MARKET (1975)

Xerox used to be a big player in the computer market; that all changed in the 1970s. The first five years of the decade saw Xerox lose $264 million on its computer business, which convinced the company to get out of the computer market. Xerox announced the decision on July 21, 1975. Of course, its name is virtually synonymous with "copy machine" today, as in "can I make a Xerox of that?"

WEBSITE OF THE WEEK: WEB PAGES THAT SUCK

The website that inspired the book that publicized the website, this site by author/designer Vincent Flanders highlights the worst-designed sites on the Web—for learning purposes only. As you might expect, it's an extremely well-designed site, which you can see for yourself at www.webpagesthatsuck.com.

July 22/23, 2006

SATURDAY/SUNDAY

THIS WEEK'S FOCUS
CREATING YOUR OWN WEB PAGES

CHANGING FONT TYPE AND SIZE

One common use of HTML is to change the font type and size on your web pages. You can change the font of all the text on a page (by placing the start tag at the very beginning and the end tag at the very end), or just for selected text.

To change the font type or size, you use the <font> code. Put the <font> opening code just before the text you want to change, and the </font> closing code just after the selected text.

To specify a font type for selected text, use the <font> code with the face attribute, like this: <font face="xxxx">text</font>. Replace the xxxx with the specific font, such as Arial or Times Roman—in quotation marks. It's best to use one of the "safe" fonts that most people will have installed on their PCs: Arial, Comic Sans MS, Roman, Times New Roman, and Helvetica.

To change the size of selected text, use the size attribute in the <font> tag. The code looks like this: <font size="xx">text</font>. Replace the xx with the size you want, from 1 to 7, with 1 being the smallest and 7 being the biggest. You get "normal" sized text by leaving this attribute blank.

Adding color to your text works much the same as changing the font face or size. The code you use looks like this: <font color="#xxxxxx">text</font>. Replace the six x's with the code for a specific color. For example, the code for red is #FF0000.

Of course, you can change font face, size, and color within a single <font> selection. Just "gang" the attributes together, one after another. For example, here's the code that specifies extra-large red Arial text: <font face="Arial" size="7" color="#FF0000">text</font>.

ON THIS DAY: TYPEWRITER PATENTED (1829)

On July 23, 1829, surveyor William Austin Burt received a patent for a device he called the typographer. This was a forerunner of the typewriter, with type mounted on a metal wheel with a rotating semicircular frame. When you turned the crank you moved the wheel until it came to the letter you wanted.

BLOG OF THE WEEK: THOMAS HAWK'S DIGITAL CONNECTION

Thomas Hawk is a member of the digerati, which doesn't mean all that much except he has a lot to say about the current digital revolution. It's one of the most literate tech blogs on the Web; find out for yourself at www.thomashawk.com.

July 24, 2006

MONDAY

THIS WEEK'S FOCUS
GAMES

CHOOSING A VIDEO GAME SYSTEM

Serious gamers don't limit themselves to a single gaming system. Great games are available for all the major video game systems, and because it's the games that drive things, you need to be prepared to play any new game that comes along. This means investing in all the current game consoles—Sony's market-leading PlayStation 2, Microsoft's Xbox, and Nintendo's GameCube. Yeah, it's a pain to have all these systems hooked up to your TV, but that's the price you pay.

If you're on a tight budget, you might have to limit your gaming to just one or two of these console game systems. If this is the case, know that Sony's PS2 is far and away the market leader, with the largest number of games in all categories; if you can buy only one game console, this is the one to have. Microsoft's Xbox is well-liked by online and network gamers, even though the number of really good games for the Xbox is rather small. Nintendo's GameCube is getting a little long in the tooth, but it's still a good system for younger gamers.

Of course, you don't want to limit your game playing to your living room, which means you'll also need to invest in a couple of portable game systems—specifically, Sony's PSP and Nintendo's DS. The nice thing about one of these gadgets—well, about the PSP, anyway—is that you can use it for more than just playing games. The PSP is an eminently hackable little gizmo, capable of being used as a portable music player, video player, and who knows what all else.

All that said, we're on the cusp of a new generation of video gaming. These things go in cycles, starting with the original PONG game through the Atari 2600 and the first Nintendo Entertainment System, on through today's three major video game systems. Well, the big three game systems are about ready to be upgraded, which means that if you're a serious gamer, you should start saving your money now for the next-generation video game systems—Microsoft's Xbox 360, Sony's PlayStation 3, and Nintendo's Revolution. More about these systems tomorrow.

ON THIS DAY: INSTANT COFFEE INTRODUCED (1938)

No more beans! On this day in 1938, Nescafé instant coffee was commercially introduced in Switzerland by the Nestlé company. It wasn't a new concept; Dr. Satori Kato of Japan presented the first instant coffee during the Pan-American World Fair in 1881.

FACT OF THE WEEK

The Entertainment Software Association (www.theesa.com) reports that in 2004, U.S. computer and video game software sales grew 4% from prior-year levels, to $7.3 billion—more than double the sales level in 1996. More than 248 million computer and video games were sold, representing almost two games for every household in America.

July 25, 2006

TUESDAY

THIS WEEK'S FOCUS
GAMES

VIDEO GAMES: THE NEXT GENERATION

The upcoming next-generation video game systems will be much more powerful than the current-generation systems (and also more powerful than your average desktop PC, believe it or not), and offer a lot more features, especially in the digital entertainment realm. All will be able to play music CDs and DVD movies and connect to your home network (wirelessly) and the Internet. Sit one of these puppies next to your living room TV and you have a full-fledged digital entertainment hub. In other words, you'll be able to use your video game system to replace your VCR, CD player, and DVD player—and to play games, as well.

So what do we know about the next-generation video game consoles? Here's a quick peek:

- **Xbox 360**—Due November 2005. Wireless controllers; HDTV output; detachable 20GB hard drive; three 3.2GHz IBM PowerPC-based processors; Dolby Digital 5.1-channel surround sound; built-in Ethernet networking, optional WiFi connectivity; also plays CDs and DVDs. The price for the base unit is $299.
- **PlayStation 3**—Due spring 2006. Up to seven wireless Bluetooth controllers; HDTV output; detachable hard drive; seven (!) 3.2GHz PowerPC-based Cell processors; Dolby Digital 5.1-channel surround sound; network connectivity via Ethernet and WiFi; built-in web browser; also plays CDs, DVDs, and the new Blu-ray HD DVDs. The price isn't set yet, but the Sony folks keep warning us that the PS3 will be "expensive," whatever that means.
- **Nintendo Revolution**—Due "mid-2006." Uses four 2.5Ghz IBM G5 processors; dual-core ATI RN520 graphics chipset; 512MB RAM; all other specs TBA. The console is said to be the size of three standard DVD cases and uses a proprietary 12-centimeter optical disc format for games.

Of course, anyone who's been around the video game industry for more than, say, 20 minutes, knows the release dates are purely speculative. The dates included here are the ones available at the time this book went to print.

ON THIS DAY: MICROSOFT REVENUES EXCEED $1 BILLION (1990)

Boy, Microsoft got real big real fast. It was founded in 1975, and just 15 years later it was a billion-dollar company. More specifically, July 25, 1990, Microsoft reported revenues of more than $1 billion for its 1990 fiscal year. Wowzers!

GADGET OF THE WEEK: SPHEREX RX2 GAME CHAIR

Gaming becomes an immersive experience with this unique surround-sound game chair. The RX features its own 5.1 speaker system and 300 watt power amplifier installed into the frame of the chair. Just plug the RX2 into the audio out jacks of your PC or game consoler. Buy it for a meager $1,600 at www.spherexinc.com.

July 26, 2006

WEDNESDAY

THIS WEEK'S FOCUS
GAMES

BUILDING THE PERFECT GAMING PC

The serious gamer plays more than just video games, of course; he also plays a lot of PC-based games. And the thing with the latest PC games is that they require a lot of computing horsepower, in terms of both basic processing and graphics display. In fact, playing games is probably the most demanding thing you can do with your PC, more demanding even than editing digital movies and photographs. Here's the bare minimum you need:

- First and most important, you need a fairly powerful processor to handle today's complex game play. Think Pentium 4 or AMD Athlon64, running at 3GHz or more.
- You'll need at least 1GB of memory, with 2GB or more being ideal. Faster memory is better, as well.
- You'll also need a lot of hard disk storage. Go for at least a 200GB hard disk, and consider the two-disk option—one hard drive for your operating system and basic programs, with a second hard drive dedicated to games.
- Today's hottest games require a 256MB video card with 3D graphics accelerator and DirectX 9 compatibility, but that's just the minimum. The best gaming PCs feature 512MB NVIDIA or ATI graphics cards; some even feature dual cards, for running two monitors.
- For the ultimate gaming experience, go with a high-quality 3D sound card with built-in surround sound, and be sure you have a quality 5.1- (or 6.1 or 7.1) speaker system, complete with subwoofer.

And if you plan on doing any online gaming, you'd better plan on having high-speed Internet access via cable, DSL, or satellite. Don't even think about trying to game online using dial-up. Ick!

ON THIS DAY: CARL JUNG BORN (1875)

Carl (Gustav) Jung is a Swiss psychologist considered the father of modern psychoanalysis. He met and collaborated with Freud in Vienna, but then developed his own theories, which he called "analytical psychology." Jung was born on July 26, 1875, and died on June 6, 1961.

HARDWARE OF THE WEEK: ATARI FLASHBACK

The Atari Flashback is a new millennium version of the classic 1980s Atari 7800 game console, complete with 20 classic games built-in. The console looks and feels like a genuine Atari 7800 (the successor to the popular 2600) and features two classic 7800-style controllers. Buy it for just $19.95 from www.atari.com.

July 27, 2006

THURSDAY

THIS WEEK'S FOCUS
GAMES

CHOOSING A GAME CONTROLLER

Game controllers are a lot more sophisticated today than they were when I was a youngster—and there are more different types available. If you're a serious gamer, get used to disconnecting one controller and connecting another when you switch from first-person shooters to flight games to racing games; each type of game demands its own type of controller.

When you're shopping for a controller, here are the major styles to choose from:

- **Gamepad**—The default controller for most video game consoles, complete with a variety of buttons and the directional D-pad. It's versatile enough for just about any type of game.
- **Flight/combat stick**—A type of joystick with 360° movement and firing buttons; it's ideal for flight games.
- **Racing wheel**—Combines a full-function steering wheel, gear shift, and gas and brake pedals for playing racing games. (Some wheels implement the gas and brake functions as buttons on the wheel.)
- **Light gun**—Lets you shoot at onscreen objects; it's ideal for all types of shooting games.

And, of course, some games actually use the computer keyboard and mouse as input devices, so you have that to take into consideration, as well. It also helps to have an ultra-fast, ultra-precise mouse (corded is probably better than cordless) for those instances where mousing is the way to play.

ON THIS DAY: GRASSHOPPER PLAGUE (1931)

July 27, 1931, was the day of the locust in Iowa, Nebraska, and South Dakota. This was when a swarm of grasshoppers descended and destroyed thousands of acres of crops. Old-timers who were around back then said that the buggies were so thick you could scoop them up in a shovel.

SOFTWARE OF THE WEEK: ROLLERCOASTER TYCOON 3

It's tough to recommend a single game that all my readers will like, but RollerCoaster Tycoon 3 just might be that game. You get to design your own theme park and rollercoasters, and then view your park (and ride your rides) from just about any angle. It's a fun and rewarding game; buy it for $39.95 from www.atari.com.

July 28, 2006

FRIDAY

THIS WEEK'S FOCUS
GAMES

POPULAR GAME CONTROLLERS (PART I)

Without further ado, here are five of my favorite PC game controllers:

- **Logitech Cordless Rumblepad 2** (www.logitech.com, $39.99). This award-winning game controller is one of the best general gamepads on the market—and it's wireless, to boot!

- **Gravis Xterminator Force Feedback Gamepad** (www.gravis.com, $69.99). The Xterminator Force is a general-purpose gamepad that offers one of the best force feedback systems today, complete with proportional-control D-pad and flippers.

- **Saitek X52 Flight Control System** (www.saitekusa.com, $129.95). This is a two-piece control system for flying games, like Microsoft Flight Simulator. Piece one is a precision joystick with four fire buttons, while piece two is a progressive throttle with clutch button.

- **CH FighterStick USB** (www.chproducts.com, $149.95). This is simply the best joystick/flightstick controller available today, modeled after the Air Force's F-16 control column.

- **Saitek Cyborg evo Wireless** (www.saitekusa.com, $59.99). This is the world's first fully adjustable wireless stick; you can personalize it for either left- or right-hand play.

As with all game-related peripherals, it's always good to try before you buy. A controller that feels good to me might be one that you totally hate to use!

ON THIS DAY: TRICYCLE CROSSES ENGLISH CHANNEL (1883)

I'm not making this up. On July 28, 1883, a special "water tricycle" was pedaled from Dover to Calais by a Mr. Ferry. It was constructed from a normal land tricycle, with bulky paddlewheels replacing the front wheel.

WEBSITE OF THE WEEK: IGN.COM

IGN.com is *the* site for PC and video game news, reviews, and cheats; if it's any way at all game-related, it's covered here. Bookmark it at www.ign.com.

July 29/30, 2006

SATURDAY/SUNDAY

THIS WEEK'S FOCUS
GAMES

POPULAR GAME CONTROLLERS (PART II)

For your further gaming pleasure, here are five more first-rate PC game controllers:

- **X-Arcade Solo** (www.x-arcade.com, $99.95). The X-Arcade Solo is an accurate reproduction of a classic arcade controller, complete with joystick and nine operation buttons.
- **MonsterGecko PistolMouse FPS** (www.monstergecko.com, $39.94). This is a stylish, high-resolution pistol controller for first-person shooting games.
- **Zboard Modular Keyset** (www.zboard.com, $49.99). This is a modular keyboard that's kind of like a high-tech Lego game; just move this module here and that one there for different PC games.
- **Logitech MOMO Racing Wheel** (www.logitech.com, $99.95). This is a great-feeling force-feedback racing wheel, complete with paddle shifters and foot pedals.
- **ThrustMaster Enzo Ferrari Force Feedback Wheel** (www.thrustmaster.com, $79.99). This is an exact replica of the steering wheel on the latest Enzo Ferrari, complete with force-feedback effects.

ON THIS DAY: FIRST ASPHALT PAVEMENT (1870)

On July 29, 1870, America's first asphalt pavement was laid in Newark, New Jersey. The pavement was laid in front of the City Hall by chemist Edmund J. DeSmedt.

BLOG OF THE WEEK: GAME*BLOGS

If you're a serious PC gamer, check out Game*Blogs. This blog aggregates the latest news about PC games, games research and theory, and game industry issues. Check it out at www.gamesblogs.org.

July 31, 2006

MONDAY

THIS WEEK'S FOCUS
PORTABLE MEDIA PLAYERS

HARD DRIVE PLAYERS

There are three types of digital music players available today, defined by the type of storage they use: hard drive, MicroDrive, and flash memory. Hard drive players are definitely the most popular because they have much more available storage (up to 60GB) and can hold much more music. These players, like the Apple iPod, store data on 1.8-inch hard disks that provide up to 40GB of storage; they're also prone to skipping, especially if you listen while jogging.

Aside from the iPod, which I've already talked about enough in this book, my other favorite hard drive players include

- **Archos Gmini XS200** (www.archos.com). This is a tiny hard-drive player with a big 2-inch screen and 10 hours of battery life. Capacity is 20GB; price is $249.95.
- **Creative Zen Touch** (www.creative.com). The Zen Touch uses a vertical touchpad, has a huge capacity, and offers 24 hours of battery life. What's not to like? Capacity is 40GB; price is $329.99.
- **Olympus m:robe 500i** (www.olympusgroove.com). The m:robe 500i has a huge 3.7-inch high-resolution color screen that helps the player do double duty as a photo vault. Capacity is 20GB; price is $399.99.
- **iRiver H340** (www.iriveramerica.com). This puppy offers huge storage capacity and a really nice 2-inch color LCD display—plus a built-in FM tuner and voice recorder. Capacity is 40GB; price is $399.99.

ON THIS DAY: FIRST CLOSE-UP MOON PICTURES (1964)

On the last day of July in 1964, the American space probe Ranger 7 transmitted the first close-up images of the moon's surface. The probe carried six video cameras and took a total of 4,308 pictures before it crashed into the Mare Cognitum (Sea of Clouds).

FACT OF THE WEEK

The Consumer Electronics Association (www.ce.org) says that sales of portable audio players in 2004 hit 6.9 million units, up from 3 million units in 2003. Sales are expected to hit 10 million units in 2005, with an average selling price of $164.

August 2006

SUNDAY	MONDAY	TUESDAY	WEDNESDAY	THURSDAY	FRIDAY	SATURDAY
		1 1994 NASDAQ trading halts for 34 minutes when a squirrel chews a power line	**2** 1876 "Wild Bill" Hickok is murdered in Deadwood, SD	**3** 1958 First undersea voyage to the North Pole is completed	**4** 1922 Every telephone in the U.S. and Canada shuts down for one minute in memory of Alexander Graham Bell	**5** 1962 Marilyn Monroe is found dead
6 1945 U.S. drops atomic bomb on Hiroshima, Japan	**7** 1959 First pictures of Earth taken by Explorer VI	**8** 1974 President Richard Nixon resigns amid the Watergate scandal	**9** 1945 U.S. drops atomic bomb on Nagasaki, Japan, killing 150,000 people	**10** 1846 Smithsonian Institution founded in Washington, D.C.	**11** 1966 Chevrolet introduces the Camaro	**12** 1981 IBM announces its first PC, powered by a 4.77MHz Intel 8088 CPU
13 1889 Pay phone patented in Hartford, CT	**14** 1953 The Wiffle Ball is patented	**15** 1961 East Germany begins construction of the Berlin Wall	**16** 1977 Elvis Presley dies in Memphis, TN	**17** 1835 The wrench is patented in Springfield, MA	**18** 1823 First parachute jumper dies	**19** 1848 California's gold rush officially begins
20 1862 National Labor Union officially creates the eight-hour workday	**21** 1888 Adding machine patented in St. Louis, Missouri	**22** 1865 Liquid soap is patented	**23** 1904 Snow tire chains are patented in New York	**24** 1853 Potato chips are first created in Saratoga Springs, NY	**25** 1925 Television tube patented in New Jersey	**26** 1920 19th Amendment to the Constitution is adopted, securing the rights of women to vote
27 1910 First radio broadcast made from a plane	**28** 1859 First U.S. oil well discovered in Titusville, PA	**29** 1885 First motorcycle is patented in Germany	**30** 1881 First stereophonic sound system patented in Germany	**31** 1897 Movie camera patented by Thomas Edison		

August 1, 2006

TUESDAY

THIS WEEK'S FOCUS
PORTABLE MEDIA PLAYERS

MICRODRIVE PLAYERS

Occupying the middle ground between hard drive and flash players are MicroDrive players, such as the Apple iPod Mini and Creative Zen Micro. These units use smaller hard drives—called MicroDrives—that offer a decent compromise between size, storage, and price. You'll get storage in the 1.5GB–4GB range, in a fairly compact form factor. Price is typically in the $200–$250 range, and they suffer from the same skipping tendency as the larger hard drive players. My favorite MicroDrive players include

- **Creative Zen Micro** (www.creative.com). The Zen Micro offers an easy-to-use control pad and comes in 10 cool colors. I like it better than the iPod Mini. It comes in 4GB, 5GB, and 6GB versions; prices range from $179.99 to $229.99.
- **Rio Carbon Player** (www.digitalnetworksna.com/rioaudio/). The Rio Carbon is a nice little (and I mean little) player with a big LCD display, 20-hour battery life, and a built-in microphone and voice recorder. It comes in 5GB and 6GB versions, priced at $199.99 and $229.99.
- **iRiver H10** (www.iriveramerica.com). What makes iRiver's H10 unique is its 1.5-inch color LCD display; all the other mini players are just black and white. You also get a built-in FM radio and voice recorder, along with 12-hour battery life. It comes in 5GB and 6GB versions, priced at $249.99 and $279.99.

ON THIS DAY: ATOMIC ENERGY COMMISSION ESTABLISHED (1946)

On July 1, 1946, Congress established the Atomic Energy Commission. The AEC's charter was to foster and control the peacetime development of atomic science and technology.

GADGET OF THE WEEK: AUDIOTRONIC ICOOL SCENTED MP3 PLAYERS

What's cool about these iCool mini-players is that they smell—in a good way. All iCool players have changeable embossed faceplates, and selected faceplates come with an aroma smell. Sniff them out for $65 at www.audiotronic.com.

August 2, 2006

WEDNESDAY

THIS WEEK'S FOCUS
PORTABLE MEDIA PLAYERS

FLASH MEMORY PLAYERS

Flash players are the smallest (and lowest priced) digital players available today. These devices, such as the iPod Shuffle, store their digital music in flash memory and offer anywhere from 64MB to 1GB of storage capacity. They're small enough to fit in any pocket, are relatively inexpensive (as low as $100 or so), and won't skip if you're jogging—which makes them great for active lifestyles. My favorite flash players include

- **iPod Shuffle** (www.apple.com). This is a player without a readout; playback is strictly random, which Apple pushes as a positive. It is a cool device, however, available in 512MB and 1GB versions, priced at $99 or $149, respectively.
- **Creative Zen Nano Plus** (www.creative.com). The Zen Nano is a flash player with reversible LCD display (for right- or left-handed operation), 18-hour battery life, and built-in FM radio and recorder. It comes in 512MB and 1GB versions (priced at $129 and $169) and in 10 different colors.
- **Rio Forge Sport Player** (www.digitalnetworksna.com/rioaudio/). This is a neat little player designed for on-the-go action that is extremely compact and rugged, with 20-hour battery life and built-in FM radio. It comes in 128MB, 256MB, and 512MB versions priced between $129.99 and $219.99.

- **iRiver iFP-900 Series** (www.iriveramerica.com). These are compact, squarish units that fit well in any pocket. All models offer 40 hours of battery life, a cool color display, and built-in FM tuner and voice recorder. Choose from 256MB, 512MB, or 1GB versions; prices range from $199.99 to $299.99.

ON THIS DAY: GREENWICH MEAN TIME ADOPTED (1880)

On this date in 1880, Greenwich Mean Time (GMT) was officially adopted by the British Parliament. The United States soon followed, adopting GMT on November 18, 1883; the International Meridian Conference in 1884 tackled the details and divided the world into 24 separate time zones.

HARDWARE OF THE WEEK: RAZER DIAMONDBACK MOUSE

Razer's Diamondback is a wired mouse with the ultra precision necessary for playing today's PC games, provided by a high-quality infrared optical sensor. It also features seven programmable buttons optimized for gaming response. Buy it for $59.99 from www.razerzone.com.

August 3, 2006

THURSDAY

THIS WEEK'S FOCUS
PORTABLE MEDIA PLAYERS

PORTABLE VIDEO PLAYERS

Now that you know all about portable audio players, how about portable *video* players? Well, there's nothing surprising or complex here: A portable video player is a portable device that plays back video (as well as audio) files on a built-in LCD screen.

That said, this is a relatively new consumer electronics category, and there aren't a lot of devices on the market yet. That's beginning to change, of course, as more and more consumers discover the benefits of watching movies on a handheld device that looks like a portable game machine.

Most portable video players have at least a 20GB hard disk. Movies are recorded in MPEG-4 format, which is the visual equivalent of audio MP3 files; you get four hours of programming per gigabyte. Picture quality is typically around 300 × 225 pixels, which is okay for watching on a small screen. Most units play back movies using Microsoft's Windows Media Center software.

All portable video players also function as portable music players, so you don't have to carry two devices around. Some also let you view digital photographs on the built-in screen, thus doing additional duty as digital photo vaults. There are even a few units that function as pure hard disk storage, for any type of PC file. The more options the merrier, is what I say.

For what it's worth, my favorite portable video players are the Archos PMA400 ($799.95) and Gmini 400 ($349.95), the Creative Zen Portable Media Player ($499.99), and the RCA Lyra 2780 ($399.99).

ON THIS DAY: RADIO SHACK ANNOUNCES TRS-80 (1977)

On August 3, 1977, Radio Shack announced the company's first personal computer, the TRS-80 Model 1. Dubbed the "Trash 80" by critics, it came with 4KB RAM and cassette tape storage. It was a successful launch; Radio Shack ended up selling 200,000 of the things.

SOFTWARE OF THE WEEK: BLAZE MEDIA PRO

Want to convert your WMA files to MP3 format, or vice versa? Then use Blaze Media Pro, a full-featured digital audio conversion/editing utility. Buy it for $50 from www.blazemp.com.

August 4, 2006

FRIDAY

THIS WEEK'S FOCUS
PORTABLE MEDIA PLAYERS

BETTER EARPHONES = BETTER SOUND

You'd be surprised at the difference a good set of earphones can make. Most portable music players deliver pretty good sound, but you need a good set of phones to hear it. Even the standard iPod earphones, which are better than some, still kind of suck in my personal opinion. The very first thing I do when I buy a new portable player is to throw away the stock earphones and replace them with something that sounds a lot better. Plus, you need really good headphones to appreciate the compact discs you play on your home audio system.

Headphones come in two types: open-air and closed-ear. Open-air headphones sit lightly on top of your ears; this type of phone is lightweight and very comfortable to wear, even for extended periods. Closed-ear phones provide even better sound quality because of their sealed design, but they're not really suited for portable use.

What you typically get with a portable player are earbuds, which are like headphones without the phones. Instead of bulky foam cups that enclose your entire ear, earbuds are tiny earphones that fit inside your ear. The advantage of in-ear devices is that they do a better job of blocking background noise than standard over-the-ear headphones. Plus, they're small and light, perfect for portable use.

You might think that earbuds, because of their small size, wouldn't sound as good as full-size headphones. You'd be wrong. Most earbuds sound very good, and the best deliver even better sound than similarly priced headphones. That's because the tiny transducers fit right inside your ear, with nothing to interfere with the sound—the better the fit, the better the sound quality. Earbuds also deliver good isolation from external sounds—and the better the isolation, the better the sound you hear.

ON THIS DAY: CHAMPAGNE INVENTED (1693)

On this day in 1693, champagne was invented by—wait for it—Dom Pierre Perignon. (And you thought it might be Bob Bubbly?) Dom and his pal Frére Jean Oudart were the cellar masters of the Saint-Pierre d'Hautvillers and Saint-Pierre aux Monts de Châlons abbeys, and they jointly developed naturally sparkling wine in its purest and most perfect form.

WEBSITE OF THE WEEK: MP3.COM

If you want to learn more about MP3 technology and portable audio players, turn to MP3.com. It's also a great site for free MP3 music streams and downloads; check it out for yourself at www.mp3.com.

August 5/6, 2006

SATURDAY/SUNDAY

THIS WEEK'S FOCUS
PORTABLE MEDIA PLAYERS

CELEBRITY PLAYLISTS

Apple has been assembling iPod playlists from various celebrities; here are some of the most notable tunes being listened to by various celebrities:

- Singer Mariah Carey recommends Aretha Franklin's "Ain't No Way," Prince's "I Wanna Be Your Lover," and The Gap Band's "Outstanding"; she also has the audacity to put three of her own songs on her playlist.
- NASCAR driver Dale Earnhardt, Jr., listens to more than just country, including America's "A Horse with No Name," The Black Crowes' "Seeing Things," Fleetwood Mac's "Sara," and Ray Charles and Willie Nelson singing "Seven Spanish Angels."
- Comic and political radio host Al Franken likes the oldies; his playlist includes CCR's "Fortunate Son," Dire Straits' "Sultans of Swing," James Taylor's "Sweet Baby James," and contemporary folksinger Dar Williams doing a cover of the Kinks' "Better Things."
- Actress Nicole Kidman is listening to "Walking on the Moon" by the Police, "Lips Like Sugar" by Echo and the Bunnymen, David Bowie's "Heroes," and U2's "Where the Streets Have No Name."
- Musician/funnyman "Weird Al" Yankovic has a rather weird playlist, as you might expect; his faves include "Surfin' Bird" by the Trashmen, "Der Fuehrer's Face" by Spike Jones, the Devo version of "(I Can't Get No) Satisfaction," and Brian Wilson's goofy-but-kinda-cool "Vega-Tables," from the *SMiLE* album.

So what's on *your* iPod playlist?

ON THIS DAY: FIRST ELECTRIC TRAFFIC LIGHTS (1914)

On August 5, 1914, Cleveland, Ohio, activated the first electric traffic signal designed to control the flow of different streams of traffic. The traffic signal (with only red and green lights) was installed at the intersection of Euclid Avenue and East 105th Street.

BLOG OF THE WEEK: MP3 PLAYER BLOG

As the name implies, this blog is devoted exclusively to news and reviews of MP3 players. It's surprisingly comprehensive; find out for yourself at www.playerblog.com.

August 7, 2006

MONDAY

THIS WEEK'S FOCUS
WIFI

CITYWIDE WIFI

Many tech-savvy consumers have availed themselves of WiFi hotspots, like those found in many coffeehouses across the country. The only problem with a normal WiFi hotspot is that it's relatively small; venture out the door and down the street, and the signal dissipates.

Some cities think they have a solution to this problem—and a way to offer free Internet access to all their citizens. The age of city-wide WiFi is coming, thanks to the development of wireless mesh networks. It's not quite as simple as setting up a string of wireless access points; instead, it involves various technologies designed to eliminate overlap and interference. Still, once you get it up and running, you have free public wireless access virtually everywhere.

While some smaller cities, such as Chaska, Minnesota, have deployed smaller mesh networks, Philadelphia, Pennsylvania plans to install the first big-city wireless network. If all goes well, Philly's citywide network will be up and running by the summer of 2006.

There's no guarantee that offering citywide WiFi is viable in the long-run, however. First of all, it's costly; at $20 to $25 per household, Philadelphia will spend $10 to $15 million on their network. Second, offering free service ticks off the traditional for-profit ISPs, which today include the phone (DSL) and cable companies—and you don't want to tick off the cable company. Third, there's really no guarantee that it will work exactly as promised, at least without a lot of technical and maintenance issues. Still, if Philly and other cities can pull it off, getting wireless Internet access for free isn't a bad thing. Am I right?

ON THIS DAY: HARVARD GETS A GIANT BRAIN (1944)

On August 7, 1944, IBM president Thomas J. Watson, Sr., formerly presented the Automatic Sequence Controlled Calculator, otherwise known as the Mark I or "Giant Brain," to Harvard University. The Mark I was 51 feet long, 8 feet high, and weighed 5 tons.

FACT OF THE WEEK

According to Intel's "Most Unwired Cities" survey, the top 10 best WiFi cities in 2005 are (1) Seattle/Bellevue/Tacoma, WA; (2) San Francisco/San Jose/Oakland, CA; (3) Austin/San Marcos, TX; (4) Portland, OR; (5) Toledo, OH; (6) Atlanta, GA; (7) Denver, CO; (8) Raleigh/Durham/Chapel Hill, NC; (9) Minneapolis/St. Paul, MN; and (10) Orange County, CA.

August 8, 2006

TUESDAY

THIS WEEK'S FOCUS
WIFI

FINDING A WIFI HOTSPOT

When you're out and about and need to connect to the Internet, you want to find the nearest WiFi hotspot. But if you're new to an area, how do you know where all the WiFi is?

If you're in a do-it-yourself mode, you can drive around from likely place to likely place, testing for wireless signals with a WiFi hotspotter. This is a device, like the Canary Digital Hotspotter discussed below, that senses WiFi signals. When your hotspotter lights up, you know you can connect.

Perhaps a better option (if you plan ahead of time) is to use a Web-based hotspot directory. There are several of these directories on the Internet, including JiWire (www.jiwire.com), Wi-Fi-FreeSpot Directory (www.wififreespot.com), WiFi411 (www.wifi411.com), and WiFinder (www.wifinder.com).

Each of these directories offers various degrees of completeness, so you might want to try more than one. Just enter your city name or Zip code and see what's nearby!

Of course, if you're near a Starbucks, you can be fairly sure that you have WiFi service. That's because most Starbucks locations are also WiFi hotspots, with service offered by T-Mobile. The only bad thing about Starbucks WiFi is that you have to pay for it; it's not free. So if you don't mind paying $30–$40 per month for T-Mobile WiFi service, get out your credit card and start surfing!

ON THIS DAY: NETSCAPE GOES PUBLIC (1995)

On this date in 1995, Netscape Communications Corp. went public. Trading opened at $28 per share, jumped to $75 midday, then closed at $58. Not a bad day's profit, if you were one of the lucky early traders.

GADGET OF THE WEEK: CANARY WIRELESS DIGITAL HOTSPOTTER

The Digital Hotspotter is a second-generation device that not only senses WiFi hotspots, but also tells you (via its LCD display) the name (SSID) of the available network, signal strength, and encryption status. Buy it for $59.95 at www.canarywireless.com.

August 9, 2006

WEDNESDAY

THIS WEEK'S FOCUS
WIFI

TROUBLESHOOTING A BAD CONNECTION

Not all wireless connections go smoothly. Here are some tricks for maximizing your WiFi connections:

- **Check the network**—Right-click the Wireless Network Connection icon (the little TV set) in the Windows system tray, and then select View Available Wireless Networks. Be sure you're connected to the correct network in the list and, perhaps, connect to a stronger network if available. If you're having problems, do a disconnect then reconnect; this sometimes cleans up the connection.

- **Repair the connection**—Wireless connections sometimes get bumfoozled. A quick solution to many problems is to right-click the Wireless Network Connection icon and select Repair. This will disconnect you from the network, redo all the settings, and then reconnect you. It works more often than not.

- **Disable security measures**—If your wireless connection is fine but you can't access any websites, then try disabling your firewall, WEP, and any other security measures that might be blocking your way. Be sure to reenable the security once you've identified and fixed the problem.

- **Avoid interference**—WiFi signals operate in the 2.4GHz band and can receive interference from cordless phones, microwave ovens, and the like. Try moving to a different spot to reduce the interference—or provide a more direct signal from the access point.

If worse comes to worst, you might need to reboot your PC to reconnect to a bad WiFi connection. Yeah, it's a pain, but sometimes that's the only thing that works. (And if even *that* doesn't work, then you'll probably have to avail yourself of the WiFi service's technical support department.)

ON THIS DAY: MARVIN MINKSY BORN (1927)

Marvin Minksy is a biochemist and the founder of the MIT Artificial Intelligence Project. He has made many contributions to the fields of AI, cognitive psychology, mathematics, computational linguistics, robotics, and the like. He was born on this date in 1927.

HARDWARE OF THE WEEK: LINKSYS WIRELESS ADAPTER

Add WiFi to any home computer with the Linksys WUSB54GS wireless adapter. It connects to your PC via USB and provides access to 802.11b/g WiFi networks, with SpeedBoost technology. Buy it for $89.95 from www.linksys.com.

August 10, 2006

THURSDAY

THIS WEEK'S FOCUS
WIFI

SETTING UP A WIRELESS HOME NETWORK

When you're setting up a wireless network in your home, here are the basic steps to take.

First, if your main PC doesn't have a network interface card or Ethernet connection, add one. You'll need to run an Ethernet cable from this computer to your wireless router, and another Ethernet cable from the router to your broadband modem.

When you have the basic cables connected, connect the router to a power source and run the Windows Network Setup Wizard (or the router's included software) on your main PC. This will set up the basic core of your network.

All the other computers on your network will connect wirelessly. This means installing a wireless networking adapter on each computer—either internally, via an expansion card, or externally via USB. (Or, if it is a laptop, you can use a wireless networking PC card.) Run the Windows Network Setup Wizard or other installation software on each PC to properly configure the wireless adapter.

After you've connected all the computers on your network, you can proceed to configure any devices (such as printers) you want to share over the network. For example, if you want to share a single printer over the network, it connects to one of the network PCs (*not* directly to the router), and then is shared through that PC.

At this point, you're good to go. If you find that one or another computer on your network is having difficulty connecting (or is connecting at a slower speed than you expected), then you might have reception problems. Remember, WiFi signals can only broadcast so far, and they can be subject to interference from other wireless devices and large metal objects. You might need to reposition your router or wireless adapters to ensure a stronger, more direct transmission.

ON THIS DAY: LEO FENDER BORN (1909)

Leo Fender was the guy who invented the first mass-produced solid-body electric guitar. This was the Fender Broadcast (later renamed the Fender Telecaster), which came into existence in 1948. Leo himself came into existence on August 10, 1909, and played his final chord on March 21, 1991. Today, Fender guitars remain one of the most popular ever made and are used by the likes of Eric Clapton, John Mayer, and Buddy Guy.

SOFTWARE OF THE WEEK: LUCIDLINK WIRELESS CLIENT

This freeware program helps you access and connect to wireless networks and hotspots; it automatically detects network security settings, alerts you to incompatible settings, and provides instructions for resolving them. Download it for free at www.lucidlink.com.

August 11, 2006

FRIDAY

THIS WEEK'S FOCUS
WIFI

WIFI SECURITY

One of the problems with a wireless network is that all your data is just out there, broadcast over the air via radio waves, for anyone to grab. Fortunately, to keep outsiders from tapping into your wireless network, you can add wireless security to your network. This is done by assigning a fairly complex encryption code to your network; to tap into the network, a computer must know the code.

You can assign security codes to all the PCs in your network in one of several ways. First, most wireless hubs, routers, and adapters come with configuration utilities that let you activate this type of wireless security. If your wireless equipment has this type of configuration utility, use it.

Otherwise, you can use Windows XP's Wireless Network Setup Wizard, which is new with XP Service Pack 2. Despite the name, you don't use this wizard to actually set up your wireless network; instead, its purpose is to add security to your already installed wireless network.

You run the Wireless Network Setup Wizard (accessible from the Windows Control Panel) on each of the PCs connected to your wireless network. You can give your network an SSID name (different from the name assigned in the Network Setup Wizard) and then either automatically assign or manually enter a network key.

The type and length of the key you choose depends on the type of encryption you choose; the strongest encryption comes from a 26-character WEP key.

I'm somewhat amazed how many home users don't enable this wireless security. Without wireless security, any yahoo in the neighborhood (or driving by in a car) can access your wireless network and steal your Internet bandwidth. This might not seem like a big deal, until you realize that with your home network wide open you're one step closer to getting hacked. It might seem like the neighborly thing to do to share your Internet connection in this manner, but not if it puts your valuable files at risk!

ON THIS DAY: STEVE WOZNIAK BORN (1950)

Apple co-founder Steve Wozniak was born on this date in 1950. The Woz designed the single-board personal computer that he and Steve Jobs marketed as the Apple I. When he's not off spending his large fortune, he teaches computer science to school children in his home town of Los Gatos, California.

WEBSITE OF THE WEEK: WI-FI ALLIANCE

This is the home site of the Wi-Fi Alliance, the industry group formed to advance and promote the WiFi standard. Take a look at www.wi-fi.org.

August 12/13, 2006

SATURDAY/SUNDAY

THIS WEEK'S FOCUS
WIFI

WHAT'S WITH ALL THOSE LETTERS?

WiFi's technical name is the IEEE 802.11 wireless networking standard. The thing is, there are lots of different 802.11 standards, each designated by a single-letter suffix. Here's what they all mean:

- **802.11a**—The earliest WiFi standard, operating in the 5GHz band at a rate of 54Mbps. Not really designed for consumer networks.

- **802.11b**—The first form of WiFi intended for general consumers, operating in the 2.4GHz band at a rate of 11Mbps.

- **802.11g**—The latest extension of the WiFi standard, still operating in the 2.4GHz but at a rate of 54Mbps.

- **802.11n**—An upcoming extension of the WiFi standard, operating in the 2.4GHz range at speeds of up to 100Mbps. This standard is not yet finalized, and in fact there are two competing approaches to 802.11n being considered—so you can't buy any 802.11n products as of yet.

If you have older WiFi networking equipment, chances are it's of the slower 802.11b variety—which is good, but not great, in terms of speed. When you go shopping for new equipment, you want to go with the newer, faster current standard, which is 802.11g. When 802.11n is finalized, however, that'll be the way to go—it will be at least twice as fast as current equipment.

ON THIS DAY: IBM PC INTRODUCED (1981)

On August 12, 1981, IBM introduced the Model 5150 personal computer. IBM's first PC used a 4.77MHz Intel 8088 microprocessor and ran Microsoft's MS-DOS operating system. It met with some degree of success in the marketplace.

BLOG OF THE WEEK: WIFI NETWORKING NEWS

This is a blog devoted primarily (but not exclusively) to public WiFi hotspots. Lots of good information here; check it out at 80211b.weblogger.com.

August 14, 2006

MONDAY

THIS WEEK'S FOCUS
SURROUND SOUND

MATRIX SURROUND SOUND

Surround sound was first heard in the home via matrixed technology, where engineers tried to stuff four channels of information into the space normally used by two. This technology combined three or four streams of information into the traditional left and right tracks for distribution, and then retrieved the original three or four channels on playback. Channel separation wasn't great, as there was a lot of leakage during the combining and extraction processes, but it did enable surround sound movies to be played in the home environment.

The original surround sound format for home video was Dolby Surround. This was actually a three-channel format, with left, right, and surround channels—no center channel. The surround channel was matrixed into the left and right channels; the Dolby Surround decoder separated the surround channel from the mix and ran it into one or two surround speakers. (When you used two surround speakers, they both played the same surround channel.) Dolby Surround was first used in some prerecorded videotapes and in some augmented stereo television broadcasts.

Next came Dolby Pro Logic, which was a more sophisticated version of Dolby Surround. Pro Logic added a center channel to the mix; it was a four-channel system with left, center, right, and surround channels. The center and surround channels were matrixed into the left and right channel information, then separated out with a decoder. As with Dolby Surround, the single Dolby Pro Logic surround channel was often fed to two surround speakers, which both played the same track. Dolby Pro Logic was (and still is) used in many television, cable, and satellite broadcasts, as well as in most prerecorded videotapes.

The latest matrix technology is Dolby Pro Logic IIx, which is a version of Dolby Pro Logic with a different purpose. This surround format is designed to simulate a surround effect from a two-channel source. You use Dolby Pro Logic IIx to play stereo soundtracks and CDs on a surround sound system. A/V receivers with Dolby Pro Logic IIx decoders also decode standard Dolby Pro Logic soundtracks.

ON THIS DAY: SOCIAL SECURITY ACT SIGNED (1935)

On this date in 1935, President Franklin D. Roosevelt signed into law the Social Security Act. This established an economic safety net for seniors and dependants and remains the cornerstone of America's social insurance. (Until the politicians decide to dismantle it, that is.)

FACT OF THE WEEK

Two British professors did a little research on online dating and discovered that 94% of the people using online dating sites had more than one date with the "most significant" person they hooked up with. The average relationship lasted about seven months; 18% lasted more than a year.

August 15, 2006

TUESDAY

THIS WEEK'S FOCUS
SURROUND SOUND

DISCRETE SURROUND SOUND

Discrete surround sound keeps each channel separate from start to finish. This improved channel separation, whichever format is used, results in a more realistic surround effect. In the home, all discrete surround formats have always been digital in nature, which also contributes to improved sound quality.

The most popular discrete surround format today is Dolby Digital. As the name implies, it's a completely digital process, as opposed to the analog matrix technology used in the older Dolby Pro Logic format. There are actually several variations of Dolby Digital, depending on how many channels are used; Dolby Digital can include everything from 1.0 (one center channel, no subwoofer) to 5.1 (front left, front center, front right, surround left, surround right, and low-frequency effects) soundtracks. Dolby Digital is used in most commercial DVDs and some satellite and cable programming, and is the format specified for all HDTV broadcasts.

The Digital Theater Systems company markets a discrete surround format, called DTS, that competes head-to-head with Dolby Digital. Like Dolby Digital 5.1, the DTS format includes five main channels (front left, front center, front right, surround left, and surround right) plus a separate low-frequency effects channel. DTS works similar to Dolby Digital but with higher data rates, which results in slightly better sound quality. Unfortunately, few commercial DVDs come with DTS encoding.

If you want more than 5.1 channels, you can move up to the Dolby Digital EX format, which adds one or two rear channels in addition to the normal surround channels. In this 6.1/7.1 system, the rear channels use matrix technology; they're not discrete, like the other channels. If you want a discrete rear channel, you have to move to DTS ES, which is a 6.1-channel system with a single discrete rear channel. It offers the best sound of all.

ON THIS DAY: PANAMA CANAL OPENED (1914)

On August 15, 1914, the Panama Canal was officially opened. Before the Panama Canal was built, sea trade had to travel all the way around South America's Cape Horn. The canal offers a shorter, more direct route between the two oceans, using a series of huge locks.

GADGET OF THE WEEK: SONY SURROUND SOUND HEADPHONES

Sony's MDR-DS8000 replicates six-channel surround sound experience in a single set of wireless headphones. Not only do you get surround sound without the speakers, you eliminate the cord from the phones to your receiver via infrared technology. You can buy these phones for a paltry $799.99 from www.sonystyle.com.

August 16, 2006

WEDNESDAY

THIS WEEK'S FOCUS
SURROUND SOUND

COMPARING SURROUND SOUND FORMATS

With all the different surround sound formats available, which should you listen to? Well, if you have a choice, you should always choose the Dolby Digital or DTS soundtrack on a DVD and skip the Dolby Pro Logic soundtrack. That's because either of the discrete formats have much better channel separation than you find with matrix surround. The surround channel in Dolby Surround and Dolby Pro Logic soundtracks is often indistinct and always nondirectional; it sounds mushy and has a fairly narrow dynamic range.

In comparison, all the discrete surround formats have very clear surround channels, with distinct channel separation. Listen to *The Matrix* in Dolby Digital and you'll clearly hear bullets shoot from right to left behind you, something that just isn't possible with Dolby Pro Logic. An even better test is the subtle surround effects on the *Days of Heaven* DVD; in Dolby Digital, you can hear the soft rustling of the wheat fields and the constant droning of the crickets envelope your listening position. The discrete formats are also better at reproducing the reverberation of a concert hall, as witnessed on the *James Taylor: Live at the Beacon Theater* DVD. It really feels like you're in the middle of the audience.

That said, if you have the choice, I recommend listening to the DTS soundtrack over the Dolby Digital version. DTS simply has a greater dynamic range than does Dolby Digital, which makes it the preferred surround sound format—when you can get it. The DTS version of *Saving Private Ryan*, for example, features much more realistic rear-channel effects than the Dolby Digital DVD; the whizzing bullets are more clearly positioned in the sound field and the explosions have a greater depth.

For the ultimate surround experience, however, you have to add another speaker and use a DTS ES decoder. I find Dolby Digital EX a tad gimmicky, the additional rear speakers more useful for ambient noise than anything else. The discrete rear channel in DTS ES Discrete, however, delivers the goods—on those rare occasions where its available.

ON THIS DAY: QUEEN TELEGRAPHS PRESIDENT (1858)

No, not Freddie Mercury, although that would have been cool. I'm talking about England's Queen Victoria, who sent the first official telegraph message across the Atlantic Ocean to U.S. President Buchanan on this date in 1858.

HARDWARE OF THE WEEK: USB ALARM CLOCK

The USB Alarm Clock is a small analog clock that you can set as a desktop alarm. The neat part is that the alarm can be programmed from any sound file stored on your PC's hard disk—including MP3 and WMA music files! Buy it for $48 from www.jbox.com.

August 17, 2006

THURSDAY

THIS WEEK'S FOCUS
SURROUND SOUND

FANTASIA—THE FIRST SURROUND-SOUND MOVIE

In many ways, Walt Disney can be considered the inventor of surround sound. When the groundbreaking movie *Fantasia* was still in development, Disney met with conductor Leopold Stokowski to discuss the film's classical music score. Stokowski suggested that Disney contact the engineers at Bell Labs, who were working on a nascent multiple-microphone stereo recording technology. Disney, intrigued by the technology, thought it would be wonderful if, during the movie's "Flight of the Bumblebee" segment, the musical sound of the bumblebee could be heard flying all around the audience, not just in front of them.

Disney put his engineers to work on the challenge, with the result that *Fantasia*, released in 1941, became the very first film to incorporate surround sound. Disney employed the proprietary Fantasound technology to create a surround sound field with left front, center front, right front, left rear, and right rear channels. The main soundtrack incorporated only the front three channels; the two rear channels were recorded on a separate reel of film, and "steered in" separately when needed.

Unfortunately, the additional equipment necessary to reproduce Fantasound was too costly to roll out on a widespread basis. Only two Fantasound systems were sold to theaters: New York's Broadway Theater and the Carthay Circle Theater in Los Angeles. These installations cost $85,000 apiece and included 54 speakers placed throughout the auditorium. Disney also produced two scaled-back road show versions of the Fantasound system, at $45,000 each, although these traveling versions didn't include the surround speakers.

The Fantasound system ultimately proved too complex and too costly for other studios to adopt. It took another decade for Hollywood to embrace a more affordable multiple-channel sound technology—which it did in conjunction with the birth of the new widescreen film formats, in particular Cinerama and CinemaScope. These multiple-channel technologies were succeeded by Dolby Stereo in the 1970s and the more advanced Dolby Digital Surround in the 1990s.

ON THIS DAY: PIERRE DE FERMAT BORN (1601)

On this date in 1601, French mathematician Pierre de Fermat was borne. Often called the founder of the modern theory of numbers, Fermat anticipated differential calculus with his method of finding the greatest and least ordinates of curved lines.

SOFTWARE OF THE WEEK: CUBASE SX

If you're serious about home recording, Cubase SX is the software to get. More affordable than the competing Pro Tools, Cubase SX turns any personal computer into a full-featured recording studio, complete with surround sound mixing. Buy it for $799.99 from www.steinberg.net.

August 18, 2006

FRIDAY

THIS WEEK'S FOCUS
SURROUND SOUND

THE HISTORY OF SURROUND SOUND (PART I)

Multiple-channel sound in the home had its origins with the reproduction of music in stereo. Initial industry experiments with stereo recording took place in the early 1930s, employing an "infinite" number of microphones deployed in a type of curtain in front of the recording musicians, and an equal number of speakers used to reproduce the sound. Practical multiple-channel recording came into being in the 1950s, with the introduction of the two-channel stereophonic system. The first commercial two-channel stereo LP was released in 1958, and by the early 1960s stereo recordings became the norm. Stereo reproduction over the airwaves began in 1961, when the first stereo FM radio broadcast took place.

The desire to reproduce a 360-degree sound field led to the development of four-channel, or quadraphonic, recording. Quadraphonic sound officially debuted in 1969, with the release of the first consumer-level four-channel reel-to-reel tape deck. Soon the quadraphonic process was being applied to both eight-track tapes and vinyl records.

By the early 1970s there were multiple quadraphonic technologies competing in the marketplace. JVC's CD-4 system, introduced in 1971 for vinyl records, employed four discrete channels of audio information: front left, front right, rear left, and rear right. The SQ and QS systems, introduced in 1972 by CBS and Sansui, respectively, were both matrix technologies for vinyl records, in which the rear channel information was matrixed into the two front channels and then separated out by a surround decoder. And RCA's Quad-8 format, introduced in 1970, was designed specifically for eight-track tape players.

Unfortunately, the confusion generated by these competing technologies, along with the high cost of four-channel amplifiers and additional rear speakers, led to the abandonment of quadraphonic sound by the end of the decade. (If you still have any old quad equipment or records lying around, they're worth real money to collectors—which means it's time to do the eBay thing!)

ON THIS DAY: HEWLETT-PACKARD INCORPORATED (1947)

One of the tech industry's oldest companies was incorporated on this date in 1947. Hewlett-Packard was founded nine years previously by William Hewlett and David Packard. The first products they sold were oscillators; their first customer was Walt Disney, who used them to produce the movie *Fantasia*.

WEBSITE OF THE WEEK: THE SURROUND SOUND DISCOGRAPHY

This fan site offers a comprehensive database of surround sound recordings in all historical formats—including quadraphonic sound, Dolby Surround, and DTS. You can browse for hours at members.cox.net/surround/.

August 19/20, 2006

SATURDAY/SUNDAY

THIS WEEK'S FOCUS
SURROUND SOUND

THE HISTORY OF SURROUND SOUND (PART II)

True surround sound came to the home in 1982, with the introduction of Dolby Surround technology. Dolby Surround was designed to reproduce the cinema's Dolby Stereo surround soundtracks in a home environment. The Dolby Surround technology consisted of three channels: left, right, and a single surround channel, which was matrixed in with the front two channels. (This surround channel was frequently sent to two rear speakers, resulting in a four-speaker system.)

Dolby Surround got an upgrade in 1987, when the Dolby Pro Logic system was introduced. Dolby Pro Logic was a four-channel version of the early three-channel Dolby Surround technology, with a separate center channel, primarily used for dialogue. To this day, Dolby Pro Logic is the surround sound standard for prerecorded VHS videotapes.

Home surround sound went digital in 1995, with the introduction of the first Dolby Digital laserdisc. Like the cinema version, home Dolby Digital is a 5.1-channel system, with left front, center front, right front, left surround, and right surround, along with a separate LFE or subwoofer channel. The first Dolby Digital DVD was released in 1997; Dolby Digital is also the sound format of record for high-definition television broadcasts. The competing DTS 5.1-channel system was released in a home version in 1996.

The new millennium brought significant enhancements to all the major home surround sound technologies. Dolby Laboratories introduced the concept of rear channels to the mix, placed behind the listener, to supplement the normal side-firing surround speakers. The Dolby Digital EX system soon had a competitor with DTS ES, which added a single discrete rear channel. Going forward, Dolby and DTS have developed new surround sound formats for future high-definition DVDs. Dolby TrueHD and DTS HD both deliver lossless audio, for the best sound quality possible over an unlimited number of discrete channels.

ON THIS DAY: CONDENSED MILK PATENTED (1856)

On August 19, 1856, Gail Borden (name sound familiar?) received a patent for his process for condensed milk. Borden's slogan was "The milk from contented cows," and his mascot was Elsie the cow. The next time you're enjoy delicious holiday treats, remember Gail Borden helped make it possible.

BLOG OF THE WEEK: WILWHEATON.NET

Yeah, it's the same Wil Wheaton who starred on *Star Trek: The Next Generation* and in the movie *Stand By Me*. Turns out Wil is a bit of a tech geek, and a darned good writer. His blog is always an interesting read; find out for yourself at www.wilwheaton.net.

August 21, 2006

MONDAY

THIS WEEK'S FOCUS
TROUBLESHOOTING

TROUBLESHOOTING TIPS

When it comes to troubleshooting computer problems, my first piece of advice, stated simply, is—**DON'T PANIC!** In the grand cosmic scheme of things, a malcontent microcomputer just doesn't register on the significance scale. Besides, if anything major really went wrong, there's not much you could do about it *after* the fact, anyway. The thing to do is keep your cool and try to minimize the damage—and get things back to normal as soon as possible. (In fact, you might just want to walk away from your computer for a while, so you can approach your problem a little later with a clear head.)

After you've not panicked, you should try to reproduce the problem. If you're lucky, the problem won't occur again. And, after all, if it only happens once, it's not really that bad of a problem, is it? If you can't reproduce the problem, it means that the problem magically fixed itself (unlikely) or that your problem was caused by a one-time user error (quite common).

If you *can* reproduce the problem (too bad!), the next thing to do is to check your system's hookup and configuration. You need to examine every cable going into and out of your system unit, the installation of any add-in cards, the setup of specific software programs (including Windows), and anything else you can think of that might cause the problem.

If you still have problems, you should methodically and logically trace the steps that led to your problem. Did you do anything wrong—or different from normal? Have you recently added anything new to your system—or changed anything old? If you can isolate the cause of the problem, it's easy enough to remove the new peripheral or program and get your system back to pre-problem operating status.

Finally, if you can't fix the problem—and not all problems are easily fixed, unfortunately—use Windows' System Restore to restore you system to a point before the problem occurred. It's kind of the coward's way out (you're not really fixing what was wrong), but it gets you back up and running relatively quickly.

ON THIS DAY: ADDING MACHINE PATENTED (1888)

On this date in 1888, William Seward Burroughs received a patent on his adding machine. Sales were originally only a few thousand units a year, but by 1926 Burroughs had sold his one millionth machine.

FACT OF THE WEEK

Corzen (www.corzen.com) reports that the top U.S. job site, in terms of number of listings, is Careerbuilder (www.careerbuilder.com), followed by Monster (www.monster.com) and Yahoo! HotJobs (hotjobs.yahoo.com). Combined, these three sites saw a 61% increase in listings from 2003 to 2004.

August 22, 2006

TUESDAY

THIS WEEK'S FOCUS
TROUBLESHOOTING

DEALING WITH A DEAD COMPUTER

If your computer doesn't start when you flip on the power, check first for some obvious causes. First, be sure your computer's power cord is plugged in—and securely connected to the computer. Then make certain that the power outlet has power and that the wall switch is turned on. If you use a surge suppressor, check that it, too, is turned on.

If none of these recommendations work, then you really *do* have a problem—one that could have one of many potential causes. Just look at what can cause a dead computer:

- The cord from your computer to your power outlet might be bad. Try replacing your power cord with a cord from another computer.
- The power supply transformer in your computer might be faulty. Call a technician and get your bad power supply replaced (if you don't feel like doing it yourself).
- If your computer makes noise but nothing appears on your screen, something might be wrong with your monitor. Try adjusting your monitor's brightness and contrast settings or replacing your monitor with a monitor from another computer system.
- You might a major problem with your hard disk. It might have a damaged boot sector, it might not be connected properly, it might be missing the key system files, or it might even be completely dead. Try starting your system from the Windows installation CD or system restore CD, if you have them. (This problem might require calling a computer technician for more help.)

If none of this works, your computer is really and truly dead. It's time to call in a professional repairperson and let them tackle the problem!

ON THIS DAY: FIRST COMPUTER USER GROUP FOUNDED (1955)

On August 22, 1955, users of IBM's Model 704 computer founded SHARE, the first computer user group. SHARE enabled the growing community of IBM computer users (mainly aerospace companies on the west coast) to exchange information and programs.

GADGET OF THE WEEK: VENEXX PERFUME WATCH

I can't help it; the Venexx Perfume Watch reminds me of something the Joker would wear, although he'd probably fill it with laughing gas or acid or some other concoction that the Batman would find similarly sinister. For you and me, we get to fill this watch with perfume—yes, perfume. Buy it for $89.95 at www.venexxusa.com.

August 23, 2006

WEDNESDAY

THIS WEEK'S FOCUS
TROUBLESHOOTING

TROUBLESHOOTING IN SAFE MODE

Many Windows-related problems can be better troubleshooted (troubleshot?) by rebooting Windows into Safe mode. Safe mode is a special Windows operating mode that uses a very simple configuration which should run on all computer systems—even those that are experiencing operating problems. Safe mode is *not* your standard operating mode; it's a mode you use only when Windows won't work otherwise.

You enter Safe mode by rebooting your computer, and then carefully watching the onscreen messages as your computer goes through its startup procedure. When you see the line "Please select the operating system to start," press the F8 key. When you do this, you'll be presented with a Windows Startup menu. Select Safe Mode from the menu and Windows will start with just a bare minimum of drivers and utilities loaded. (You'll also get to experience Windows in glorious 640 × 480 VGA resolution—thrilling!)

When you're in Safe mode, you can examine all of Windows' configuration settings to find out what could be causing your problem. In particular, check your system's display and device settings, using the Device Manager tab in the System Properties dialog box. Correct any misconfigured settings and reboot your computer; if all is well, Windows will start in its normal mode. In a worst-case situation, you might need to uninstall a particular device and then reinstall it to reset the configuration.

While in Safe mode, you can also use the System Configuration Utility to selectively disable various startup programs that might be causing your problem. You open this utility by selecting Start, Run and entering `msconfig`. Go to the General tab and select Selective Startup, then go to the Startup tab to see all the stuff that gets loaded when you start Windows. You can troubleshoot startup problems by disabling items from this list.

ON THIS DAY: GALILEO DEMONSTRATES TELESCOPE (1609)

On this day in 1609, our pal Galileo first demonstrated his telescope. He was in Venice when he first heard of an invention that allowed distant objects to be seen as if they were nearby. Inspired, Galileo worked out the principle on his own, creating an eight-power telescope. He used this device to observe the heavens.

HARDWARE OF THE WEEK: MARATHON COMPUTER REPORTER

The RePorter is essentially a big ol' extension cable that plugs into the rear of your computer and replicates those ports on the opposite end. You get two USB ports, FireWire 400 and 800 ports, and audio in and out connections—all on a five-foot cable. You can buy it for $59 from www.marathoncomputer.com.

August 24, 2006

THURSDAY

THIS WEEK'S FOCUS
TROUBLESHOOTING

PAGE FAULT ERRORS

One of the most frustrating errors on a Windows system comes when you get the dreaded "blue screen of death," accompanied by a message about an invalid page fault—followed by an involuntary reboot. What the heck causes this problem—and what's a page fault, anyway?

In Windows, a page fault is an interrupt that occurs when a program requests data that is not currently in real memory. The interrupt triggers Windows to fetch the data from virtual memory on your hard disk, and then load it into RAM.

A page fault error occurs when Windows can't find the data in virtual memory. This usually happens when the virtual memory area on your hard disk becomes corrupt. The resulting error generates the "blue screen of death" and crashes your system.

If you run into this problem with some regularity, you need to flush out your system's virtual memory. In Windows XP, you do this by opening the Start menu, right-clicking My Computer, and selecting Properties. When the System Properties dialog box appears, select the Advanced tab, and then click the Settings button in the Performance section. When the Performance Options dialog box appears, select the Advanced tab and click the Change button in the Virtual Memory section. Check the No Paging File option, click Set, then click OK. Reboot your system, and then repeat the procedure. When you get to the Performance Options dialog box, check System Managed Size, click Set, click OK, and then reboot again. This should sweep out whatever problems were causing the page fault error.

If the problem persists, it's possible that you have some sort of hard disk problem. It might pay to run Windows Checkdisk or some similar disk repair utility to see what might be wrong. It's also possible that some software program (or combination of programs) is improperly accessing your system's virtual memory. Try to track down the offending program(s) by noting just what you're doing when these errors occur.

ON THIS DAY: WINDOWS 95 SHIPS (1995)

At midnight on August 24, 1995, millions of geeks around the country lined up to buy Microsoft's latest and greatest operating system, Windows 95. Ten years later, it's hard to believe that a new software release could be such a major event, but it was. Microsoft even paid a hefty sum to license The Rolling Stones' "Start Me Up" for their television commercials promoting the new OS. How things have changed!

SOFTWARE OF THE WEEK: PC CERTIFY LITE

When you're troubleshooting PC problems, it helps to have a good diagnostic program that can test, find, and sometimes fix what's wrong. PC Certify Lite does a good job of this, offering configuration reports, system diagnostic tests, and more. Buy it for $129 at www.protechdiagnostics.com/pccertifylite.htm.

August 25, 2006

FRIDAY

THIS WEEK'S FOCUS
TROUBLESHOOTING

CREATING AN EMERGENCY STARTUP DISK

If your PC came with the original Windows installation CD (or a similar "recovery" disk provided by your hardware manufacturer), you can use this CD to start your computer in case of a hard disk error or failure. Just insert the installation/recovery CD, reboot your computer, and the startup process will default to the CD before it accesses the hard drive. This lets you access an otherwise-inaccessible computer to initiate the troubleshooting process.

Unfortunately, most new PCs today don't come with Windows installation CDs. Computer manufacturers save money by not including the CDs, even though this puts you in a bit of a lurch if you have startup problems with your PC.

Fortunately, if you don't have an installation/recovery CD, you can create an emergency startup disk on a floppy disk, and then reboot with this disk inserted into your PC's 3.5-inch disk drive. You'll have to deal with the old-school MS-DOS operating environment, but it's better than being locked out completely.

To create an emergency startup disk, start by inserting a blank floppy disk into your computer's floppy drive. Open My Computer, and then select the floppy disk drive. Pull down the File menu, point to the name of the floppy drive, then click Format. When prompted, check the Create an MS-DOS Startup Disk option, and then click Start. The blank disk will now be formatted for MS-DOS bootup.

Troubleshooting from an MS-DOS disk can be a little tricky, especially if you've never used a command-line operating system. (Shame on you!) Unless you're an old fart like me who was raised on MS-DOS, you need some reference to the available commands—which you can find at www.computerhope.com/msdos.htm. Use these commands wisely, my son; they are very powerful tools.

ON THIS DAY: PARIS LIBERATED (1944)

On August 25, 1944, American troops accompanied Free French and French Resistance forces to liberate Paris from the German occupation. Okay, it's not really tech-related, but a little history is good for you!

WEBSITE OF THE WEEK: MICROSOFT KNOWLEDGE BASE

The advice might be a little obtuse at times, but Microsoft's Knowledge Base is the single best resource for tracking down Windows-related problems. Search for yourself at support.microsoft.com.

August 26/27, 2006

SATURDAY/SUNDAY

THIS WEEK'S FOCUS
TROUBLESHOOTING

PC SURVIVAL CHECKLIST

Want to keep your computer running in tip-top condition? Then make sure you take the following advice to heart:

- Keep your system in a well-ventilated area free of dust and smoke particles.
- Make sure that all your cables are firmly and properly connected.
- Plug your system into a surge suppressor—even better, into a universal power supply (backup battery).
- Keep all your original software and documentation—including your Windows installation CD, if you have one—in a safe yet accessible place.
- If you don't have a Windows installation CD, create an MS-DOS emergency startup disk. (See yesterday's article for instructions.)
- Buy an external hard disk and schedule daily full-disk backups.
- Institute a regular program of system maintenance, including defragmenting your hard disk, deleting unused files, and checking your disk for errors.
- Buy a third-party disk utility program—and keep it nearby for emergencies.
- Install and run anti-virus and anti-spyware utilities.
- Activate the Windows Firewall (or a similar third-party firewall utility) to protect your system from Internet-based intrusions.

ON THIS DAY: COMPAQ INTRODUCES PRESARIO (1993)

On August 27, 1993, Compaq announced its Presario family of personal computers. Presario computers were made for the mass market, intended to be low-priced and user-friendly. The basic Presario model sold for $1,399 (very cheap, back then) and included a monitor and modem.

BLOG OF THE WEEK: RESCUECOMP BLOG

An interesting blog with lots of helpful computer first aid, maintenance, and troubleshooting tips. Check it out at www.rescuecomp.net/blogs/.

August 28, 2006

MONDAY

THIS WEEK'S FOCUS
THE HISTORY OF VIDEO GAMES

FIRST GENERATION (1972-1977)

Like many technological advances, the idea for video games originated with the government. The U.S. military wanted some sort of device that would develop the reflexes of military personnel, and in 1966 Ralph Baer, an employee of defense contractor Sanders Associates, first came up with the concept of a "television gaming apparatus." This device included both a chase game and a video tennis game, and could be attached to a normal television set.

It took several years and numerous false starts, but in 1970 Baer showed the game to Magnavox, who signed a licensing agreement the following year. Then, on January 27, 1972, Magnavox launched Baer's "brown box" technology as the Odyssey video-game console—the world's first home video game system. Priced at $100, the Odyssey utilized simple black-and-white graphics, enhanced by plastic overlays for the television screen.

Also in 1972, inspired by an early peek of Baer's original video tennis game, Nolan Bushnell and his Atari company released an electronic arcade game called PONG, which became a huge success. In 1975, Atari partnered with Sears to release a home version of PONG under the Sears Tele-Games label. The $100 game system was Sears' best-selling item during the 1975 Christmas season, with sales of more than $40 million.

Following Atari's smashing success, several companies released PONG "clones" using General Instrument's "PONG-on-a-chip" integrated circuit. Chief among these competitors were Coleco's Telstar and Magnavox's Odyssey 100. None achieved the sales levels of the Sears and Atari-branded games, but all managed to sell a fair number of units during the 1976 Christmas season.

Then, in August 1976, Fairchild Camera and Instrument leveraged its position as the creator of the microchip to release the first programmable home video game system. Based on Fairchild's own 8-bit F8 microprocessor and displaying 16-color graphics, the Channel F Video Entertainment System sold for $169 and was capable of playing a variety of games as programmed by removable ROM cartridges. Although not a huge success, the Channel F was a harbinger of things to come.

ON THIS DAY: WORCESTER SAUCE INTRODUCED (1837)

You'll recognize most of these names. On this date in 1837, pharmacists John Lea and William Perrins of Worcester, England, began the manufacture of a flavoring sauce that they imaginatively called Worcester sauce.

FACT OF THE WEEK

The Entertainment Software Association (www.theesa.com) reports that the average game player is 30 years old and has been playing games for 9.5 years. In contrast, the average game *buyer* is 37 years old. (Parents buy games for their kids.) In 2005, 95% of computer game buyers and 84% of console game buyers were over the age of 18.

August 29, 2006

TUESDAY

THIS WEEK'S FOCUS
THE HISTORY OF VIDEO GAMES

SECOND GENERATION (1977–1982)

Fairchild had initiated the age of the programmable video game with its Channel F system, but its graphics weren't far removed from those of the original PONG game. Customers clamored for higher-resolution graphics, better game play, and more games, which Atari would soon deliver.

The so-called "golden age" of home video games was launched in October 1977, when Atari released its own programmable videogame system. Priced at $199, Atari's Video Computer System (VCS), later known as the Atari 2600, was based on an 8-bit Motorola 6507 microprocessor, with 256 bytes of RAM. On the market through 1990, the Atari VCS went on to sell more than 25 million units over its product life. Over the course of its production run, 40 different manufacturers created more than 200 different games for the system, selling more than 120 million cartridges of popular games such as *Space Invaders*, *Asteroids*, and *Pac-Man*.

One of Atari's first competitors in the programmable video game market was Bally, which launched the Bally Professional Arcade in 1977. Even though the Bally unit had better graphics than the Atari VCS, it sold at a much higher price ($350) and failed to catch on beyond a hard-core cult following who appreciated what the system had to offer. In 1981 Bally sold the rights to the Professional Arcade to Astrovision, who in 1982 marketed the then-dying system as the Astrocade.

Magnavox jumped into the programmable video game market with the Odyssey². Launched in 1978, the Odyssey² was more popular in Europe than in the United States, where parent company Philips Electronics marketed it as the Videopac. In the United States, it sold only about a million units—well below the market-leading Atari 2600.

Atari received a more serious competitor in 1980, when Mattel launched its Intellivision video game system. Intellivision featured better graphics than the VCS and was the first video game system to utilize a 16-bit microprocessor—the General Instruments 1600. Intellivision became known for its proprietary sports titles, such as *Major League Baseball*, *NFL Football*, *NHL Hockey*, and *NBA Basketball*, even though the lack of third-party games contributed to its second-place showing against the Atari 2600.

ON THIS DAY: FIRST RUSSIAN ATOMIC BOMB (1949)

The Russkies went nuclear on this date in 1949. This was when the U.S.S.R tested their first atomic device, code-named "First Lightning." (Or, as the Americans called it, "Joe Number One," for Joseph Stalin.)

GADGET OF THE WEEK: JAKKS TV GAMES

Jakks TV Games let you experience old-school gaming on any TV—no PC or game console required. These are portable, self-contained game systems; just connect the joystick to your TV and start playing. Buy them all for $20 apiece at www.jakkstvgames.com.

August 30, 2006

WEDNESDAY

THIS WEEK'S FOCUS
THE HISTORY OF VIDEO GAMES

THIRD GENERATION (1982-1984)

The third generation of video games became known as the "dark ages," due not to any new (or evil) technology, but rather to the precipitous drop in sales that started in 1982. The crash was caused by too many derivative or poor-quality game cartridges from too many manufacturers. Many third-party game developers went out of business during this period, and even established companies lost money on unsold inventory.

At the peak of the previous generation, the video game industry was grossing upwards of $3 billion a year in America alone; in 1985, at the end of the third generation, video game sales would only reach $100 million worldwide. The situation was so dire that Atari's stock dropped 32% in a single day (December 7, 1982), after it announced that VCS holiday sales would not meet company expectations.

The most prominent third-generation game system was Coleco's Colecovision. Launched in 1982 at a price of $199, Colecovision featured high-quality graphics and utilized an 8-bit Z-80A microprocessor with 8K RAM. Colecovision's main claim to fame is that it offered high-quality versions of arcade favorites *Donkey Kong*, *Defender*, *Frogger*, *Joust*, *Spy Hunter*, and *Zaxxon*.

Atari responded to Coleco by releasing the $299 Atari 5200 SuperSystem, which was based on the graphics and audio chips found in the Atari 400 personal computer. Games for the 5200 were essentially improved releases of older 2600 (VCS) games; this lack of new games failed to excite consumers, and the 5200 was lost amidst the overall market crash of 1982.

The crash of 1982-1984 didn't totally destroy the video game industry, however. The industry was soon to experience a rebirth, thanks to the introduction of a new generation of game units driven by two technological innovations—lower-cost memory chips and higher-power 8-bit microprocessors. These developments enabled game designers to produce home video game consoles that could successfully compete at a quality level equal to that of arcade machines. I'll talk about this fourth generation of video gaming tomorrow.

ON THIS DAY: FIRST AFRICAN-AMERICAN ASTRONAUT (1983)

In 1983, Guion S. Bluford, Jr., became the first black American astronaut to travel in space, flying aboard the shuttle Challenger 3 in the eighth Space Shuttle mission. In another first, Bluford and four colleagues blasted off from Cape Canaveral at night.

HARDWARE OF THE WEEK: CLASSICADE UPRIGHT GAME SYSTEM

The Classicade Upright is an honest-to-goodness replica of a classic arcade game, complete with wooden cabinet, 21-inch television display, and multi-function controllers. Each unit is a two-player machine and comes with 21 games pre-installed on a built-in PC. Buy one for a modest $4,095 at www.gamecabinetsinc.com.

August 31, 2006

THURSDAY

THIS WEEK'S FOCUS
THE HISTORY OF VIDEO GAMES

FOURTH GENERATION (1985-1989)

In 1983, Nintendo had released the Famicon ("family computer") video game system to the Japanese market. The console was a hit, selling 2.5 million units in its first year, and Nintendo began negotiations with Atari to distribute the system in the United States. Those talks fell through, however, and Nintendo decided to distribute the system itself in the United States, under the name Nintendo Entertainment System (NES).

The $199 NES was based on an 8-bit Motorola 6502 microprocessor and shipped with a version of the hit arcade game *Super Mario Bros*. Quantities of the NES were shipped into the New York market in time for Christmas 1985, and national distribution followed early in 1986. Nintendo sold more than 3 million NES units in its first two years of release; it is estimated that, over its entire product life, more than 65 million NES consoles were sold worldwide, along with 500 million cartridges.

Atari attempted to reverse its sliding fortunes by releasing the long-awaited Atari 7800 ProSystem in 1986. Unfortunately, the 7800 featured outdated technology (the unit was originally set for release in 1984 but was shelved when Warner Communications sold the company to Commodore founder Jack Tramiel) and did not compete effectively against newer fourth-generation game systems.

In 1989, Sega released its first game system in the U.S., the Sega Master System (SMS). The SMS had two cartridge ports; one in a standard cartridge configuration, and a second port that accepted small credit card-shaped cartridges. The system was capable of utilizing both ports at any given time, and Sega used this feature to produce plug-in 3-D glasses for use with certain games.

Also released in 1989 was the first programmable handheld game system, Nintendo's GameBoy. Priced at $100, the GameBoy featured a black-and-white LCD screen and came prepackaged with a *Tetris* cartridge. With more than 100 million units shipped in various configurations, the GameBoy holds the honor of being the world's all-time best-selling video game system.

ON THIS DAY: FIRST CAR ON THE MOON (1971)

On the last day of August, 1971, astronaut David Scott drove NASA's Lunar Roving Vehicle (AKA "moon buggy") during the Apollo 15 mission. For better traction across the lunar surface, each wheel had its own 0.1 hp electric motor.

SOFTWARE OF THE WEEK: ATARI 80 CLASSIC GAMES

If you're an old-school kind of guy, check out the Atari: 80 Classic Games in One CD-ROM. This collection lets you play classic Atari 2600 games on your home (or work) PC. Games include all your favorites: Asteroids, PONG, Missile Command, Breakout, Warlords, and the rest. Buy it for $19.95 at www.atari.com.

September 2006

September 2006

SUNDAY	MONDAY	TUESDAY	WEDNESDAY	THURSDAY	FRIDAY	SATURDAY
					1 — 1985 Titanic wreck located in the Atlantic Ocean, west of Newfoundland	**2** — 1752 Gregorian calendar adopted by England and its colonies and remains the system we follow today
3 — 1976 Viking II lands on Mars	**4** Labor Day — 1886 Geronimo, the last American Indian warrior, surrenders to U.S. military forces	**5** — 1885 First U.S. gasoline pump is sold in Fort Wayne, IN	**6** — 1892 First gasoline-powered tractor is sold to a farmer in Langford, SD	**7** — 1888 First baby incubator is used on Ward's Island, NY	**8** — 1930 Scotch Tape is invented	**9** — 1995 Orville Redenbacher dies
10 Grandparents Day — 1913 First coast-to-coast paved highway (the Lincoln Highway) is opened	**11** Patriot Day — 5-year anniversary of World Trade Center and Pentagon attacks	**12** — 1956 Construction completed on first U.S. coal pipeline	**13** — 1899 First American automobile fatality	**14** — 1886 Typewriter ribbon is patented in Memphis, TN	**15** — 1928 Sir Alexander Fleming discovers Penicillin	**16** — 1620 The Mayflower departs from England
17 Citizenship Day — 1976 NASA unveils its first space shuttle, the Enterprise	**18** — 1947 U.S. Air Force is founded	**19** — 1991 The Iceman, the oldest human body ever found, discovered in the Alps on the Italian-Austrian border	**20** — 1952 Scientists find heriditary data in DNA	**21** — 1937 J.R.R. Tolkien's *The Hobbit* published	**22** — 1893 Charles and Frank Duryea test drive America's first automobile in Springfield, OH	**23** First day of Autumn — 1846 Neptune discovered
24 — 1936 Jim Henson, creator of "The Muppets," is born	**25** — 1974 Scientists announce discovery of hole in ozone layer	**26** — 1996 U.S. and other countries sign nuclear weapons testing ban	**27** — 1825 First passenger transport train begins service in England	**28** — 1925 Seymour Cray, founder of the Cray Computing Company, born	**29** — 1914 Phonograph records patented	**30** — 1982 Cyanide-laced Tylenol kills six Chicago-area people

September 1, 2006

FRIDAY

THIS WEEK'S FOCUS
THE HISTORY OF VIDEO GAMES

FIFTH GENERATION (1989–1995)

The fifth generation of home video-game systems featured 16-bit processors, more detailed graphics, and more imaginative games. This generation was dominated by Nintendo and Sega.

Video gaming's fifth generation was ushered in by the 1989 American release of NEC's TurboGrafx-16. (The system was launched in Japan in 1988 as the PC Engine.) Although the TurboGrafx-16 was advertised as a 16-bit system, it actually used an 8-bit microprocessor, assisted by a 16-bit graphics chip and 64K RAM; it was notable as the first game console to have a CD player attachment. Unfortunately, a lack of games doomed the TurboGrafx-16 in the marketplace, and it was discontinued within four years of its initial release.

More formidable was the Sega Genesis game system (sold as the Mega Drive in Japan). Released to the U.S. market in 1989, Genesis was the first true 16-bit game system, using a Motorola 68000 microprocessor. Genesis was priced at $199 and ran excellent translations of Sega arcade hits; sales received a significant boost with the 1991 release of the *Sonic the Hedgehog* game.

To compete with the Sega Genesis, Nintendo launched its own 16-bit system in 1991. The Super NES (known as the Super Famicon in Japan) sold for $199 and included the *Super Mario World* cartridge. The initial U.S. production run of 300,000 units sold out overnight; over the course of its product life, more than 46 million Super NES units were sold worldwide.

In December 1993, a full six years since the release of its last game console, Atari attempted to reenter the fray by releasing a 64-bit video-game system. The CD-ROM–based Atari Jaguar promised to be a revolutionary machine but was hampered by a lack of game cartridges and practically non-existent marketing. In 1996, Atari officially killed the Jaguar—and dropped out of the video game market altogether—when it merged with JTS, a manufacturer of computer hard drives.

ON THIS DAY: VIRTUAL LIBRARY PROJECT STARTS (1994)

September 1, 1994, marked the launch of the United States Library of Congress "Virtual Library" project. The intent was to convert all of the LOC's materials into digital form over the Internet, where they would be accessible to students and researchers around the world.

WEBSITE OF THE WEEK: VIDEO GAME MUSEUM

Revisit all those classic game consoles and console games at the Internet's Video Game Museum. Browse by game system—they're all here, from Atari 2600 to Xbox. Check it out at www.vgmuseum.com.

September 2/3, 2006

SATURDAY/SUNDAY

THIS WEEK'S FOCUS
THE HISTORY OF VIDEO GAMES

SIXTH GENERATION (1995-1998) TO TODAY

The sixth generation of home video games featured high-powered microprocessors and dedicated graphics processors that enabled extremely realistic graphics and game play. These game consoles outperformed the much higher-priced personal computer systems of the day.

The Sega Saturn, released in May 1995, achieved its high graphics quality by using twin 32-bit microprocessors and CD-ROM-based games. Unfortunately, the Saturn's high $399 price and lack of third-party games led to its being overshadowed by Sony's upcoming game console.

In September 1995, Sony released its first video game system, the Playstation, to the U.S. market. The Playstation was priced at $299, $100 less than the competing Sega Saturn, and incorporated a 32-bit microprocessor designed to produce polygon graphics. Backed with a massive advertising campaign, the Playstation unseated both Nintendo and Sega to become the leading home video game system; to date, it has sold more than 50 million units worldwide.

In 1996, five years after the release of the Super NES, Nintendo released its own sixth-generation game system, the Nintendo 64. The Nintendo 64 was the first home system to utilize a 64-bit microprocessor (hence the name) and was priced at just $150, significantly lower than its competition. The launch was hugely successful, with 1.7 million units sold in the first three months of release.

The current generation of video games continues to push the envelope in terms of graphics, performance, and game play. The major players continue to be Sony (with the PlayStation 2) and Nintendo (with the GameCube), with one significant newcomer—Microsoft (Xbox). And the games continue!

ON THIS DAY: FERDINAND PORSCHE BORN

On September 3, 1875, Austrian automotive engineer Ferdinand Porsche was born. Yes, this is the same Porsche who designed the Volkswagen and whose name still graces today's finest high-speed cars. Ferdinand passed away on January 30, 1951.

BLOG OF THE WEEK: VIDEO GAME BLOG

Of all the video game blogs on the Web, I find this one the most interesting. Tom Zjaba is a classic video game buff (he writes the *Retrogramming Times* newsletter), and this blog covers his musings on video games from all eras. Give it a read at www.tomheroes.com/video_game_blog.htm.

September 4, 2006

MONDAY

THIS WEEK'S FOCUS
NOTEBOOK COMPUTERS

DIFFERENT TYPES OF NOTEBOOKS

Buying a notebook PC isn't the easiest thing in the world. That's because there are so many different types on the market today. At the very least, you'll have to choose from these types of models:

- **Ultracompact**—If you primarily use your notebook for email and running the occasional PowerPoint presentation, and not for heavy-duty typing or number-crunching, this might be the type of machine for you. An ultracompact has a smaller screen (10"–12") and cozier keyboard than larger laptops, and consequently is smaller and weighs less.

- **Business laptop**—For most non-executive road warriors, you actually need to do some real work when you travel—which means that a bigger screen and full-size keyboard are important. What we're talking about here is the traditional business laptop, with a 14"–15" screen, full-sized keyboard, combo CD/DVD drive, and enough computing horsepower for typical office applications.

- **Desktop replacement**—If power matters more to you than portability, you want a laptop that can function as a full-featured desktop PC. These so-called desktop replacement models use standard Pentium 4 processors (instead of the portable-oriented Pentium M) and provide extra-large 15" or larger screens (typically widescreen) and CD/DVD burning capabilities.

- **Tablet PC**—This is a subspecies of the notebook PC, with a touch screen you can write on—with a stylus, that is, not a real pen. Tablets are great for specific tasks, such as meter reading or checking inventory in a warehouse.

So, if you plan on using your notebook primarily as a desktop, spring for a desktop replacement model. If you use your notebook primarily to check email when you're on the road, go with an ultracompact. If you do a lot of heavy work when you're on the road, you need a business laptop. And if you're a UPS delivery person, you'll probably end up using a tablet PC. Got it?

ON THIS DAY: GERONIMO SURRENDERS (1886)

More history for you. On this day in 1886, Apache leader Geronimo surrendered to General Nelson A. Miles. Geronimo had led raids on white settlers for 10 years after the U.S. government attempted to move the Apache to a reservation.

FACT OF THE WEEK

Notebook PCs now account for nearly 27% of all PCs in use in the United States, reports the Computer Industry Almanac (www.c-i-a.com). Experts say that notebooks are becoming increasingly popular among home office users.

September 5, 2006

TUESDAY

THIS WEEK'S FOCUS
NOTEBOOK COMPUTERS

SHOPPING FOR A NEW NOTEBOOK

Whichever type of notebook computer you decide on, you'll want to make sure it meets some minimal performance specs. Assuming you don't go the Apple route (which I prefer, but you probably don't), you should look for a laptop that uses a Pentium M processor (runs cooler and uses less power than standard desktop Pentium 4 processors), has at least 512MB memory, offers built-in 802.11b/g WiFi wireless connectivity, and has a combo CD/DVD drive (typically with a CD burner).

By the way, machines billed as using Intel Centrino technology use Intel's Pentium M and WiFi chips, which is a sure-fire way to go—but not the only way. Some manufacturers use non-Intel WiFi chips, and even though they use the Pentium M processor, they can't use the official Centrino logo. No big deal, as long as you get some sort of Pentium M/WiFi combination.

Looking forward, we're just about due for notebooks that use Intel's next-generation Centrino technology, dubbed Sonoma. These notebooks will have improved performance thanks to a 533Mhz bus, DDR2 RAM, PCI Express, and improved audio and video. Sonoma will also support something called stack execution disabling, which will improve security on next-generation notebooks. Look for Sonoma-based notebooks to hit the market in the next year or so.

Back to notebook specs, you also need to consider how big a screen you want and how much battery life you need. The screen is kind of self-explanatory, other than noting that bigger screens take more power to run, which decreases battery life. That's not the only factor for battery life, however; the type of microprocessor used, the type and capacity of the battery itself, and other PC features all contribute to longer or shorter battery times. How you use the PC also matters; a lot of writing to hard disk and running a CD/DVD drive drain the battery faster than it might otherwise. Of course, if you're only using the notebook at home or in the office as a desktop replacement, that might not matter; if you use it on the road or away from home a lot, it's a different story.

ON THIS DAY: ON *THE ROAD* PUBLISHED (1957)

On this day in 1957, Jack Kerouac's *On the Road* was first published. The book, based on Kerouac's friendship with Neal Cassidy, became one of the best-known works of the Beat Generation.

GADGET OF THE WEEK: TARGUS DEFCON MDP MOTION SENSOR

The DEFCON MDP is a PC card you insert into your notebook computer to deter theft. The card contains a tiny tilt motion sensor which senses when a thief is making off with your PC. It then sounds a 110dB alarm and prohibits unapproved access. Buy it for $99.99 at www.targus.com.

September 6, 2006

WEDNESDAY

THIS WEEK'S FOCUS
NOTEBOOK COMPUTERS

NOTEBOOKS FOR COLLEGE

When your little one isn't so little anymore and is ready to move into a college dorm, it's time to think about what type of computer he or she will need at the university. This isn't always a simple decision.

First off, you have to decide between a desktop and a notebook PC. While a good desktop might cost you a little less and offer a little more in terms of multimedia features, I recommend going with a notebook. That's because a notebook is more flexible in the college environment; you're not limited to computing in the dorm room because you can take it anywhere. That means your little Joe or Jane College can take notes (and send instant messages) while in class, or do their research while sipping a cup of coffee at the student union or local coffeehouse. At least it gets them out of their rooms.

That said, you probably want to hold off on buying anything until your kid actually arrives on campus—or receives the official welcome notice from the college. That's because many colleges have specific requirements for their students' PCs. Some colleges include a computer as part of the tuition. (Of course, the cost of the PC is factored into the tuition—you're still paying for it, one way or another.) Other colleges specify the type of PC to buy, in terms of platform (Windows or Mac), wireless capabilities, and so on.

Even better, many colleges offer student discounts when you purchase a computer from their recommended vendor(s). This is often quite a good deal, as you'll pay substantially less than you would if you headed down to your local computer store.

Wherever you buy it, make sure you're looking at a notebook with built-in WiFi capability, so your kid can connect anywhere on campus. You should also consider a unit with a large screen and built-in DVD player and TV tuner, so the young 'uns can use a single device for both education and entertainment. That means you don't have to buy a separate TV and stereo system—which saves you money and saves your kid space in the dorm.

ON THIS DAY: MARGARET SANGER DIES (1966)

Margaret Sanger was a vocal champion of birth control and founded the first U.S. birth control clinic (in Brooklyn, in 1916). She was born on September 14, 1879, and died on September 6, 1966.

HARDWARE OF THE WEEK: LAPCOOL 2 NOTEBOOK COOLER

The LapCool 2 notebook cooler helps cool down hot-running notebooks, thanks to two ultra-quiet fans in the base. The LapCool has a flat 11.8" × 10.3" surface, so just about any notebook fits on top and stays cool while doing so. Buy it for $39.99 from www.vantecusa.com.

September 7, 2006

THURSDAY

THIS WEEK'S FOCUS
NOTEBOOK COMPUTERS

UPGRADING NOTEBOOK MEMORY

Upgrading memory on a portable PC is a little different than with a desktop system. On one hand, it's easier; memory is typically added through an easily accessible compartment on the bottom of the unit. On the other hand, it's more complicated; every manufacturer (and seemingly every model) uses different non-standard memory types and form factors.

When shopping for laptop memory, you do it by manufacturer and model number. Some portable RAM comes in modules, some on units that look like little credit cards. You have to get the exact type of memory used by your particular PC, whatever that might happen to be.

Installing the memory, however, should be a snap—literally. On most models you use a screwdriver to open a small compartment on the bottom of the unit, and then snap the new memory into place. All you have to do is power off your PC, open the memory compartment, insert the new memory module or card, refasten the cover, and then restart your computer. Your notebook should automatically recognize the new memory on startup.

Know, however, that you don't have an unlimited number of memory slots to work with. This means that you might need to remove an old lower-capacity memory card to add a new higher-capacity one.

How much memory is enough? While some low-priced notebooks still come with just 256MB, I recommend 512MB as the bare minimum. If you plan on doing video editing, photo editing, playing graphics-intensive PC games, or working with really big Excel spreadsheets or PowerPoint presentations, consider upgrading to 1GB of RAM; you'll notice the difference.

ON THIS DAY: FIRST BABY INCUBATOR (1888)

Way back in 1888, September 8 to be exact, little Edith Eleanor McLean became the first baby to be placed in an incubator. She was born at State Emigrant Hospital on Ward's Island, New York, and weighed just 2 pounds, 7 ounces. And here's something interesting: Originally, the incubator was called a "hatching cradle."

SOFTWARE OF THE WEEK: RMCLOCK

RMClock is a small utility that lets you monitor and reconfigure your notebook's CPU frequency, throttling, and load level, all on the fly. Download it for free at cpu.rightmark.org.

September 8, 2006

FRIDAY

THIS WEEK'S FOCUS
NOTEBOOK COMPUTERS

PRINTING ON THE ROAD

What do you do when you're on the road with your notebook and you need to print out an important document? Well, if you're lucky your hotel has a business center where you can borrow a printer for a few minutes—usually for a fee, of course. If you're not lucky and your hotel or motel doesn't have a printer for rent, you have to look around for other options.

One such option is to use the printers located at your local FedEx Kinkos store. Kinkos offers a print utility, called File, Print FedEx Kinkos (really, that's the name) that serves as a virtual printer on your notebook. When you select Print in your application, you select the File, Print FedEx Kinkos application from the list of available printers. Enter the appropriate information (such as your credit card number) and, as long as you're connected to the Internet, your file is whisked to the nearest Kinkos for printing. Drop by at your convenience to pick up your documents.

Kinkos offers other printing options, too. First, there's a web browser-based printing solution, located at print2.kinkos.com, which requires you to upload your document to the Kinkos site for printing. And, of course, you can always just drop by the Kinkos store and do the manual printing thing; printing is as simple as connecting your notebook to their printer.

Depending on where you are, you might also want to check out EFI's PrintMe service. This is a web browser-based service that lets you print documents at more than 1,800 locations across the U.S.—including printers located in various public locations, such as shopping malls and hotels. Check the www.printme.com site for more information on locations and pricing.

Another option, of course, is to carry a printer with you. So-called portable printers, such as the Pentax PocketJet, are small enough to fit in a briefcase and let you print from anywhere with no hassles at all. However, that's one more thing to carry around with you, plus you'll need to carry some blank paper, as well, so it's not a choice for everyone.

ON THIS DAY: FORD PARDONS NIXON (1974)

Tricky Dick got off the hook on this date on 1974, when his designated replacement, President Gerald Ford, pardoned him for any "crimes he committed or may have committed." The issue was a big factor in Ford losing the 1976 election to Jimmy Carter.

WEBSITE OF THE WEEK: NOTEBOOKREVIEW.COM

Find out everything you need to know about the latest notebook PCs at the NotebookReview.com site, located at (you guessed it) www.notebookreview.com. Tons of notebook reviews, as you might expect, as well as user comments and discussions.

September 9/10, 2006

SATURDAY/SUNDAY

THIS WEEK'S FOCUS
NOTEBOOK COMPUTERS

STRETCHING NOTEBOOK BATTERY LIFE

I hate it when I'm on the road, in the middle of writing something important, and get that little beep that signals the impending death of my notebook PC battery. There are, fortunately, several things you can do to extend your notebook's battery life.

First, turn down the brightness of your computer screen. Going from full brightness to a mid-range setting gets me an extra half-hour or so per charge, believe it or not. Keeping the screen bright takes a lot of juice.

Next, minimize your drive use. You might need to configure your notebook's power management settings to turn off your hard disk and CD/DVD drives, but if you're not using them, it's worth it. If you use your PC as a portable music player (thus accessing the hard disk or spinning the CD drive somewhat constantly) you'll drain your battery lickety-split. The less work your notebook has to do, the longer the battery will last.

Along the same lines, turn off your notebook's WiFi function if you're not using it. Even if you're not connected, your notebook will periodically scan for available wireless networks—and all that scanning takes a bit of juice to do.

Another option is to use two batteries in your notebook. If your PC has a removable media drive, chances are you can purchase a second battery to fit in the media drive bay. Of course, you won't be able to use your CD/DVD drive when you're using the second battery, but two batteries mean twice the life before your PC runs down.

Finally, if your notebook is getting on in years, consider buying a new battery. The available charge on a battery decreases with use; that new battery that lasted three hours when new might run out in two hours a year later. If you want maximum battery life, go with a newer battery.

ON THIS DAY: LINCOLN HIGHWAY OPENS (1913)

On September 10, 1913, America's first paved coast-to-coast highway opened for traffic. Dubbed the Lincoln Highway, it took a northern route from New Jersey to California, roughly following (in order) what became U.S. Highways 1, 30, 530, 50, and 40.

BLOG OF THE WEEK: LAPTOP REVIEW

Find out about all the latest notebook PC models and upcoming technology at the Laptop Review blog. It's the single best source for all things portable PC-related; check it out at www.breakingnewsblog.com/laptop/.

September 11, 2006

MONDAY

THIS WEEK'S FOCUS
COMPUTERS FOR KIDS

ENCOURAGE SAFE COMPUTING

Although there are technical means to monitor your kids' online activities, the best thing you can do is promote safe-surfing habits among your younger family members. Here are some good guidelines:

- Make sure your children know never to give out any identifying information (home address, school name, telephone number, and so on) or to send their photos to other users online.
- Provide each of your children with an online pseudonym so they don't have to use their real names online.
- Don't let your children arrange face-to-face meetings with other computer users without parental permission and supervision. If a meeting is arranged, make the first one in a public place and be sure to accompany your child.
- Teach your kids that people online might not always be who they seem; just because someone says that she's a 10-year-old girl doesn't necessarily mean that she really is 10 years old, or a girl.
- Consider making Internet surfing an activity you do together with your younger children—or turn it into a family activity by putting your kids' PC in a public room (such as a living room or den) rather than in a private bedroom.
- Set reasonable rules and guidelines for your kids' computer use. Consider limiting the number of minutes/hours they can spend online each day. And if they misbehave, cut off the Internet!
- Monitor your children's Internet activities. Ask them to keep a log of all websites they visit; oversee any chat sessions they participate in; check out any files they download; even consider sharing an email account (especially with younger children) so that you can oversee their messages.
- Install content-filtering software on your PC, and set up one of the kid-safe search sites as your browser's start page.

ON THIS DAY: FIRST REMOTE COMPUTATION (1940)

On this date in 1940, the first remote computation took place. Computer pioneer George Stibitz set up a terminal at Dartmouth College that allowed users to perform remote computation via telegraph wire with the Complex Computer located in New York City.

FACT OF THE WEEK

Hitwise (www.hitwise.com) reports that for the first quarter of 2005, the top five most-visited reference sites on the Web were (1) Dictionary.com (www.dictionary.com), (2) Wikipedia (www.wikipedia.com), (3) About.com (www.about.com), (4) Answers.com (www.answers.com), and (5) MSN Encarta (encarta.msn.com).

September 12, 2006

TUESDAY

THIS WEEK'S FOCUS
COMPUTERS FOR KIDS

SAFE SEARCHING FOR KIDS

As should be blatantly obvious, not every site on the Web is especially kid-friendly—especially traditional search engines, which are more likely than not to turn up adult-oriented sites in their search results. Fortunately, several sites on the Web offer kid-safe searching, including some of the major search sites (via content filtering options).

Here are some of the best kid-safe search sites on the Web:

- **AltaVista—AV Family Filter** (www.altavista.com; go to the Advanced Search Settings page and click the Family Filter link)
- **Ask Jeeves for Kids** (www.ajkids.com)
- **Awesome Library** (www.awesomelibrary.org)
- **Fact Monster** (www.factmonster.com)
- **Family Source** (www.family-source.com)
- **Google SafeSearch** (www.google.com; go to the Preferences page then choose a SafeSearch Filtering option)
- **KidsClick!** (www.kidsclick.org)
- **MSN Kids Search** (search.msn.com/kidz/)
- **Net Nanny Kids Directory** (search.netnanny.com/?pi=nnh1&ch=kids)
- **SearchEdu.com** (www.searchedu.com)
- **SuperKids** (www.super-kids.com)
- **Yahooligans!** (www.yahooligans.com)

And here's something else to know. These kids-safe search sites are good to use as the start page for your children's browser because they can serve as launching pads to guaranteed safe content. You can't control everything they find online, but at least you can start them off clean!

ON THIS DAY: TEST OF FIRST INTEGRATED CIRCUIT (1958)

On September 12, 1958, Jack Kilby successfully tested the first integrated circuit in his lab at Texas Instruments. His test proved that resisters and capacitors could co-exist on the same piece of semiconductor material.

GADGET OF THE WEEK: HASBRO VIDEONOW COLOR

Hasbro's VideoNow Color is a portable video player for kids. Movies come on mini 4.25" DVD discs that hold up to 25 minutes of programming and are displayed on the unit's color LCD screen. Buy it for $59.99 from www.hasbro.com/videonow/.

September 13, 2006

WEDNESDAY

THIS WEEK'S FOCUS
COMPUTERS FOR KIDS

ONLINE LIBRARIES

When I was a kid, I used to love going to the library. Today, the traditional library is moving online, as witnessed by these terrific (and extremely kid-friendly) library sites:

- **Internet Public Library** (www.ipl.org). "The first public library of the Internet," with a huge assortment of online reference works and collections.
- **Library of Congress** (www.loc.gov). America's national library online, complete with online collections, texts of many of the library's publications, and access to bibliographic catalogs and legislative information (THOMAS).
- **LibDex** (www.libdex.com). This site lets you browse the catalogs of close to 18,000 libraries.
- **Literature.org** (www.literature.org). A mini-library that offers the full and unabridged texts of classic works of literature, from Rene Descartes to Charles Dickens to Mark Twain.
- **New York Public Library Digital Library Collections** (digital.nypl.org). Numerous online collections and exhibitions from the NYPL.
- **Questia** (www.questia.com). A large online collection of books and articles from various traditional and academic publishers.
- **refdesk.com** (www.refdesk.com). A family-oriented collection of online reference materials.
- **Smithsonian Institution Libraries** (www.sil.si.edu). Digital collections, electronic journals, and other online services from the Smithsonian Institution.

And don't forget your neighborhood bricks-and-mortar public library. Many public libraries have their own websites, where you can check on availability, find out about upcoming events, and so on. Go to Publiclibraries.com (www.publiclibraries.com) for links to public libraries across the United States!

ON THIS DAY: FIRST AUTO ACCIDENT DEATH (1899)

Cars don't kill people—well, actually, cars *do* kill people. The first vehicle-related fatality occurred in 1899, when H.H. Bliss was run over by a car as he alighted from a streetcar in New York City. The driver of the car, one Arthur Smith, was arrested and held on $1,000 bail.

HARDWARE OF THE WEEK: LOGITECH FOOTBALL MOUSE

Okay, it's cool enough that this is a mouse that looks like a football. But the really neat thing is that this is a mouse designed for little kids—it's just 2/3 the size of a standard mouse, perfect for small hands. Buy it for $14.95 at www.logitech.com.

September 14, 2006

THURSDAY

THIS WEEK'S FOCUS
COMPUTERS FOR KIDS

HELP WITH HOMEWORK

One of primary reasons for K-12 students to go online is to do research for school reports and get help with their homework. Here's a good starter list of sites your kids can use when it's homework time:

- **AOL@SCHOOL** (www.aolatschool.com). Resources for primary, elementary, middle school, and high school students and educators.
- **B.J. Pinchbeck's Homework Helper** (school.discovery.com/homeworkhelp/bjpinchbeck/). A terrific list of homework-related links compiled by real 16-year-old B.J. Pinchbeck.
- **Homework Center** (www.factmonster.com/homework/). Part of the Fact Monster site, specially designed for K-12 homework problems.
- **Jishka** (www.jiskha.com). Homework help from experts, as well as forums where you can ask questions of other users.
- **Kid Info** (www.kidinfo.com). Homework help and links to curriculum guides, lesson plans, reference sites, and teaching aids.
- **MadSci Network** (www.madsci.org). Focuses on K-12 science education; scientists answer questions submitted by users, and all the answers are archived at this site.
- **Open Door** (www.knockonthedoor.com). Reference material for students aged 9-17, with a focus on biology, chemistry, physics, and history.
- **Webmath** (www.webmath.com). Hone your math skills with thousands of practice problems.
- **Word Central** (www.wordcentral.com). Access Merriam Webster's student dictionary, a kid's Word of the Day, and various word games.

Great sites, all—and a lot more help than was available when I was still doing homework!

ON THIS DAY: FIRST LOBOTOMY (1936)

The first prefrontal lobotomy surgery was performed on this date in 1936. Surgeons J.W. Watts and Walter Freeman performed the procedure on a 63-year-old woman at George Washington University Hospital. For most of us, lobotomies make us think of Jack Nicholson and Nurse Ratchet in *One Flew Over the Cuckoo's Nest*, a chilling portrayal of abusive and downright brutal treatment of mental patients in the first half of the twentieth century.

SOFTWARE OF THE WEEK: MICROSOFT STUDENT 2006

Microsoft Student is productivity software for middle- and high-school students. The Learning Essentials component adds useful tutorials, toolbars, and templates to Microsoft Word, Excel, and PowerPoint; it also includes tons of homework help for a range of subjects. Buy it for $99.99 at www.microsoft.com/student/.

September 15, 2006

FRIDAY

THIS WEEK'S FOCUS
COMPUTERS FOR KIDS

KIDS' SPORTS

Kids love their sports, and there are a lot of websites designed to support kids' sports. Here's a short list:

- **Amateur-Sports.com** (www.amateur-sports.com). A site serving young athletes, sports teams, coaches, and parents.
- **InfoSports** (www.infosports.net). A virtual community for players, coaches, and parents associated with youth sports.
- **Kids Sports Network** (www.ksnusa.org). Promoting non-school sports and fitness for children 3–19.
- **Kids' Sports Bulletin Board** (www.coachjerry.com). A place for kids and coaches in youth sports to interact.
- **MomsTeam** (www.momsteam.com). For sports moms (and dads), promoting a safer, saner, less-stressful youth sports experience.
- **Play Football** (www.playfootball.com). The official NFL site for kids.
- **Sports Illustrated for Kids** (www.sikids.com). An online magazine for young athletes.
- **Youth Sports Instruction** (www.youth-sports.com). Focusing on fundamentals and sportsmanship in a variety of youth sports, including t-ball, soccer, football, and more.

In addition, don't forget the websites for your local professional sports teams. All MLB, NBA, NFL, and NHL teams have their own websites, most of which are very kid-friendly. Go team!

ON THIS DAY: ACM FOUNDED (1947)

The Association for Computing Machinery (ACM) is the world's oldest computing society, and it was founded on this date in 1947. Today, ACM has more than 80,000 members and organizes a variety of conferences and educational workshops.

WEBSITE OF THE WEEK: KIDS' SPACE

Kids' Space is a website created for kids, by kids. It's an award-winning site that encourages and facilitates creative activities, complete with galleries of artwork and stories. Direct your kids to www.kids-space.org.

September 16/17, 2006

SATURDAY/SUNDAY

THIS WEEK'S FOCUS
COMPUTERS FOR KIDS

COMPUTERS MAKE KIDS DUMBER?

I always thought the opposite, but that might not be the case. According to a study of 100,000 pupils in 31 countries, kids that use computers do worse in school than those that don't. Or so found two researchers from Munich University, working for the U.K.'s Royal Economic Society.

The researchers used PISA tests to measure the skills of 15-year-olds. They found that students perform significantly worse if they have computers at home. In contrast, children with access to 500 or more books at home performed significantly better on the tests. So, reading books is better than fiddling with computers. Not really a surprise, I suppose.

The implication is that kids with computers tend to neglect their homework more than non-techie kids. I can understand that; it's really easy to waste time at the computer keyboard, especially if you have the Internet nearby for some needless browsing.

Another upshot of this research is that kids with computers aren't nearly as creative as kids used to be. The computer enforces a literalness on activities that detracts from the users' ability to analyze and imagine. In other words, computers don't stimulate your imagination. It's better to develop imagination and critical thinking than it is to hone information access skills. In other words, facts are no substitute for knowledge.

So if you want to make your kids smarter, make sure you supplement their computer activities with other activities that better stimulate their imagination and their analytical skills. Appearances to the contrary, using a computer too often is a spectator activity rather than a participatory one.

ON THIS DAY: RCA EXITS COMPUTER MARKET (1971)

RCA used to be a big player in the mainframe computer market. Used to be is the operative phrase, as the computer business became a big money-loser for RCA in the late 1960s, causing the company to officially shut down its computer business on September 17, 1971.

BLOG OF THE WEEK: "HEY THAT SMELLS GREAT"

The "Hey That Smells Great" blog, subtitled Cooking for My Kids, is chock-full of great-tasting, healthy recipes for kids of all ages. Just reading this blog will make you hungry! Check it out at www.heythatsmellsgreat.com.

September 18, 2006

MONDAY

THIS WEEK'S FOCUS
HOME THEATER

AUDIO/VIDEO RECEIVERS

Home theater sound all starts with the audio/video receiver, to which you connect all your audio and video components and then switch between inputs with a single remote control. The A/V receiver also serves as the main processor/amplifier for your system's audio. Surround sound sources (either broadcast or DVD) are fed into the receiver, which decodes and amplifies the surround sound signal using the appropriate technology.

A/V receivers are available at a variety of price points, and if you can't tell the difference between a $200 and a $6,000 receiver, you need an ear exam. It's all about the sound—and, of course, the control. While all A/V receivers perform similar functions, the higher-priced models simply provide better-quality sound and more flexibility in terms of control and component switching. Consider the following variables:

- **Power**—In general, you're better off getting as much power as you can afford, within limits. But don't sweat 10–20 watt per channel differences between models; the difference won't likely be noticeable.

- **Number of channels**—At the most basic, you want a receiver capable of amplifying five channels (for a 5.1-channel surround sound system). Many higher-end receivers have seven channels of amplification (for 7.1-channel systems), and some even over nine or more channels so you can power speakers in a separate room.

- **Inputs and outputs**—Make sure there are enough—and the right kinds of—input and output jacks to connect all your different components. Video I/O varies significantly between lower-priced and higher-priced models; the very best receivers include component video, DVI, and HDMI inputs and outputs.

- **Control**—You operate an A/V receiver with its remote control unit—and there are big differences in remote controls. Look for a universal or learning remote that can be programmed to control all the components in your system.

ON THIS DAY: JIMI HENDRIX DIES (1970)

Rock guitar god Jimi Hendrix died of drug-related causes on this day in 1970. He was just 27. In the 36 years since his death, Hendrix's work has influenced nearly every musician to pick up a guitar. Hendrix is best known for tracks such as "Purple Haze" and "Are You Experienced." He was also known for playing the guitar with his teeth, his on-stage otherworldly gyrations, and for igniting his guitars with lighter fluid and matches.

FACT OF THE WEEK

Jupiter Research (www.jupiterrearch.com) reports that 55% of all households currently own a DVD player, 22% own a surround-sound home theater system, 8% own an HDTV television or set-top box, 6% own a DVD recorder, and just 4% own a digital video recorder (DVR).

September 19, 2006

TUESDAY

THIS WEEK'S FOCUS
HOME THEATER

DIGITAL VIDEO RECORDERS

A hard disk recorder—sometimes called a digital video recorder (DVR) or personal video recorder (PVR)—is simply a little computer with its own hard disk. The video signal comes into the DVR and is recorded, digitally, onto the hard disk. When you play back the recording, you're reading the stored file off the hard disk. It's actually nothing too fancy, if you're used to computers; in the world of consumer electronics, however, this is really gee-whiz stuff.

What makes DVRs so appealing is the accompanying electronic program guide (EPG). After all, a DVR doesn't do anything that a VCR didn't do (except with much better picture quality, of course), so why is everyone all of a sudden raving about being able to record their favorite television programs? Trust me on this one—it's all about the EPG, which makes it *way* easier to schedule a recording than it was in the VCR era.

The most notable EPG is TiVo, which costs you $12.95 a month to use, and is only available with specific TiVo-compatible units. Other DVRs offer other EPGs, such as the one offered by TV Guide, most of which are free. I still prefer the versatility and functionality of TiVo, but I also understand that zero dollars a month is a lot more attractive to most folks than $12.95 per month—especially when the basic hard disk recording functions are the same, regardless of which EPG is used.

When you're shopping for a DVR, make sure you like the EPG and that you have a big enough hard disk for all the programs you want to record. Don't settle for the basic 40GB models, which will only hold about 12 hours of programming in high-resolution mode (or about 40 hours in lower-resolution mode); I recommend at least an 80GB model, larger if you're a packrat or record a lot of HDTV programming.

Even more fun are those combination DVR/DVD recorders that let you record either to hard disk or to blank DVDs. Learn more about these puppies tomorrow.

ON THIS DAY: FIRST UNDERGROUND NUCLEAR TEST (1957)

On this date in 1957, the United States conducted its first underground nuclear test. Named the Rainier Event, the blast took place at Area 12 of the Nevada Test Site. It was detonated in a horizontal tunnel about 1,600 feet into and 900 feet beneath the top of the mesa.

GADGET OF THE WEEK: LOGITECH HARMONY 880

The Harmony 880 is a home theater remote that you can program by filling in a series of forms on Harmony's website. It's also activity-based (as opposed to equipment-based); press the Watch DVD button and the remote turns on your TV, switches to the proper video input, turns on your DVD player, and starts playback. Buy it for $249.99 from www.logitech.com.

September 20, 2006

WEDNESDAY

THIS WEEK'S FOCUS
HOME THEATER

SHOPPING FOR A DVD PLAYER

The DVD player is an essential component of any home theater system. Whether you choose a basic sub-$100 player or an uber-expensive $3,500 one (yes, they exist—and one is my Leo's Pick in this category!), you want a progressive scan player that can handle both DVD movies and CD audio discs, as well as all the important subcategories, such as DVD-R and CD-R discs.

The more money you have to spend, the better the performance and the more features you get. For example, that $3,500 player is as solid as the rock of Gibraltar, and also plays DVD-Audio and SACD discs. You can also splurge and go for a DVD megachanger, so you can store your entire movie collection in a single machine, no disc-swapping necessary.

Also popular these days are DVD recorders. A DVD recorder lets you record your favorite television programming direct to DVD +/ R/RW discs, so you have a permanent copy of the program. It's like using a VCR, except with DVDs instead of tapes.

Or you can go the next step and get a combo DVD/DVR machine. This is the ultimate way to go, for all types of recording and playback. Record your television programming onto the hard disk (using the electronic program guide for assistance), and then when you find a program you really want to keep, burn it from the hard disk to DVD. This type of unit definitely offers the most flexibility; use the hard disk for short-term storage and burned DVDs for your permanent library.

The best DVD/DVRs let you do some degree of editing of your content before you burn the DVD discs. At the very least, it's nice to edit out commercials and edit in some chapter stops when you burn a disc. However, some manufacturers are in the pockets of advertisers, and don't let you edit broadcast programming you've recorded. I try to avoid DVD recorders that put these sort of restrictions on my fair use—if I want to skip the commercials, let me skip the commercials!

ON THIS DAY: FIRST FORTRAN PROGRAM (1954)

On September 20, 1954, the first successful FORTRAN program was run. The FORTRAN programming language (which stood for FORmula TRANslator) was devised at IBM by a group led by John W. Backus.

HARDWARE OF THE WEEK: SONOS DIGITAL MUSIC SYSTEM

The Sonos Digital Music system consists of one main ZonePlayer unit (which connects to your main stereo speakers) and as many auxiliary ZonePlayers as you want in additional rooms. Each ZonePlayer streams music from your PC and is controlled by an iPod-like remote controller. Systems start at around $1,000 from www.sonos.com.

September 21, 2006

THURSDAY

THIS WEEK'S FOCUS
HOME THEATER

CHOOSING A SPEAKER SYSTEM

Choosing the right speakers is essential to creating the best possible home theater experience; if you have any spare money in your home theater budget, there's no better place to spend the bucks than in upgrading your system's speakers! There are three primary types of home theater speakers:

- **Floor-standing speakers**—Floor-standing speakers typically are larger than other types of speakers, reproduce a wider range of frequencies (including deep bass), and are quite efficient, producing more volume per watt. They're the best type of speakers for pure music reproduction.

- **Bookshelf speakers**—With bookshelf speakers, you get smaller speakers that take up less space (and can be mounted on stands or on shelves), good performance, and a smaller price tag. Some bookshelf speakers don't have a lot of oomph on the low end, and benefit from being paired with a powered subwoofer.

- **Satellite speakers**—Think of these as mini-bookshelf speakers, if you like; they're small enough to be mounted or placed just about anywhere in the room. Despite the small size, some of these satellite speakers deliver surprisingly good performance, which can be enhanced by a separate subwoofer.

(There are also in-wall and in-ceiling speakers, but these function more or less like bookshelf speakers. The same advice applies.)

If you're using bookshelf or satellite speakers (or even some floor-standing speakers), you'll want to include a separate subwoofer in your system. The subwoofer is a powered speaker (it contains its own power amplifier) that reproduces the very lowest bass frequencies. In a Dolby Digital or DTS soundtrack, the subwoofer is the ".1" of the 5.1-channel system and is fed a separate low frequency effects (LFE) audio channel.

ON THIS DAY: GALILEO ENDS MISSION (2003)

On September 21, 2003, NASA's Galileo space probe ended its eight-year mission to Jupiter. Its work done, the Jet Propulsion Laboratory in Pasadena, California, steered the craft into Jupiter's atmosphere, where it heated up and vaporized.

SOFTWARE OF THE WEEK: MEEDIO ESSENTIALS

Meedio Essentials is kind of like Windows Media Center Edition, except that it runs on any type of computer. You get the ability to store, manage, and play back digital audio, video, and photographs, with an interface that reads well in the living room environment. Buy it for $49.99 from www.meedio.com.

September 22, 2006

FRIDAY

THIS WEEK'S FOCUS
HOME THEATER

SETTING UP YOUR SYSTEM

As you plan your home theater layout, you'll need to position your video display, speakers, and furniture carefully to create the best possible viewing and listening experience. In general, you want to center your seating area between the side walls on which you mount your surround speakers. The distance from your viewing screen to the center of your seating area should be roughly two times the diagonal measure of a widescreen TV. So if you have a 60-inch diagonal TV, set your couch or chair about 10 feet (120 inches) from the screen.

Probably the most challenging part of the setup process is connecting all the speakers. The front speakers are easy enough, but running wires to surround speakers is never fun or easy. You might need to run speaker wire under the carpet or feed it up through a basement or crawl space or down from the attic. When you get into feeding the wire through the wall and out again, you're into a lot of work—which is probably necessary, but a chore nonetheless.

Position the center speaker either directly above or directly below the television screen. The left and right front speakers should be positioned immediately to the left and right of the television screen. The left and right surround speakers should be positioned to the left and right sides of the main listening area—not to the rear. And, because the subwoofer is nondirectional, where you put it is unimportant—although placing it in a corner or under a stairway can add a little extra oomph to your system's low end.

Finally, if all this positioning and cabling makes you nervous, bite the bullet and call a professional installer. They do this sort of thing for a living, have all the proper tools, and know all sorts of shortcuts and workarounds that you'll never stumble across. In addition, a professional installer can help you create a great-looking system, especially if you like the look of built-in components and custom furniture. It'll cost, but the results are worth it!

ON THIS DAY: PEACE CORPS AUTHORIZED (1961)

On this date in 1961, the U.S. Congress formally authorized the creation of the Peace Corps, which was created in March of that year by an executive order by President Kennedy.

WEBSITE OF THE WEEK: AUDIOREVIEW.COM

AudioREVIEW.com compiles links to all the reviews of audio equipment that appear on the Web. It's a great resource when you're shopping for new equipment; check it out at www.audioreview.com.

September 23/24, 2006

SATURDAY/SUNDAY

THIS WEEK'S FOCUS
HOME THEATER

UNIVERSAL REMOTE CONTROLS

With all the various components in a typical home theater system, you're bound to end up with a coffee table full of remotes. There's the one for the TV, one for the A/V receiver, one for the DVD player, one for the digital video recorder, one for the Media Center PC, one for the cable box or satellite dish, and on and on and on. How do you deal with all those remotes?

Well, the easiest way to deal with remote control clutter is to do a little consolidation. The key is to combine all your operating functions into a single universal remote control unit. Most universal remotes have codes for the most popular audio/video components preprogrammed; other codes can be "learned" from the old remote. Once you have it programmed, the new remote can control four or more components, just by pressing the right buttons.

The best universal remotes feature some sort of LCD touchscreen display. Typically, this display varies depending on which component you're trying to operate. Press the button for TV, and the touchscreen changes to display the television controls. Press the button for DVD, and the screen displays the DVD's controls. And so on.

Even better are those remotes that let you program their functionality via your PC. It's really quite easy (and very cool) to design your own custom remote control layout on your PC, using the remote's supplied software, and then download that layout to your remote via a USB or serial connection. Some remotes, like Philips' Pronto line, even let you add your own custom graphics; go online to find all sorts of custom screens and logos to use.

The best remotes include separate on/off codes for your components, instead of the normal toggle codes. For example, the single "power" button on a remote typically toggles a component on and off. But even if the remote doesn't have separate "on" and "off" buttons, your component probably has separate "on" and "off" states. If your universal remote has individual codes for both these states, it's easier to program macros that turn components on or off. Without separate on and off commands, you run the risk of turning a component off if it's already on when you send an on/off toggle command in a macro. Better to send a separate "on" command that won't turn anything off by mistake!

ON THIS DAY: DIRIGIBLE DEMONSTRATED (1852)

On September 24, 1852, Frenchman Henri Gifard demonstrated his new invention, a semi-rigid airship he called a dirigible. He installed a small steam engine in the gondola of a 147-foot long balloon filled with coal gas; the engine turned an 11-foot propeller that propelled the craft at 5 miles per hour over a 17-mile distance.

BLOG OF THE WEEK: EHOMEUPGRADE

This is a terrific blog about all manner of digital lifestyle topics, focusing on home theater and the connected home. It's on my daily must-read list; find out why at www.ehomeupgrade.com.

September 25, 2006

MONDAY

THIS WEEK'S FOCUS
SHOWCASING YOUR HOME THEATER

HDTV PROGRAMMING

This week I'm going to discuss some of the best television programs, DVDs, and CDs you can use to show off your new home theater system. Today, the focus is on HDTV programming that will put your high-definition television through its paces.

For the best HDTV programming, turn no further than the DiscoveryHD, HDNet, and INHD networks, available on many cable and satellite systems. All the programming on these channels is designed to satisfy the needs of viewers hungry for high-definition entertainment. Especially impressive are the various nature shows and bikini contests—which, come to think of it, are kind of nature shows in their own way.

On network television, you have a lot of HDTV programming to choose from, although not all of it looks that impressive. The best of the lot, IMHO, is CBS's *CSI: Crime Scene Investigation*. All the shows in the franchise look good, but I'm partial to the original Las Vegas-based program. The coolly lit CSI labs look really hip in high definition, and the gory computer-generated graphics take full advantage of the increased resolution of the HDTV format. (Some viewers disagree and find *CSI: Miami* more visually appealing; it's all that Florida sun, I guess.)

Also on network TV, I recommend catching *Late Night with Conan O'Brian*, which is now broadcast in HDTV. *Late Night* does a very good job with its HDTV telecasts, with extra-sharp camera work and excellent lighting. (In fact, the picture's so sharp you can see which of Conan's guests need to change makeup artists....) This can't be said by Conan's lead-in, *The Tonight Show with Jay Leno*. Even though *The Tonight Show* is in high-definition, the picture doesn't look nearly as good as the later show; the lighting is barely adequate and the sharpness leaves much to be desired. Conan's a much better demo, even if you don't like his sophomoric brand of humor.

Finally, I'll admit to one high-definition guilty pleasure. The *Low Carb Cookworx* show, shown on many PBS stations, looks simply glorious on my HD television. The sets are sharp, the food is colorful, and the whole thing is extremely attractive.

ON THIS DAY: IBM ANNOUNCES MCA

On September 25, 1989, IBM announced a Micro Channel Architecture for its PS/2 line of personal computers. MCA could transfer data about eight times faster than the standard PC architecture, and IBM hoped it would become the industry standard. (It didn't.)

FACT OF THE WEEK

The Digital Entertainment Group (www.dvdinformation.com) reports that 2004 was a record year for DVD sales with 1.5 billion discs shipped. The DEG also handed out a bunch of awards for excellence, including Theatrical DVD Title of the Year (*Lord of the Rings: The Return of the King, Extended Edition*), Catalog DVD Title of the Year (*Gone with the Wind, Collector's Edition*), and TV on DVD Title of the Year (*Seinfeld: Seasons 1–2*).

September 26, 2006

TUESDAY

THIS WEEK'S FOCUS
SHOWCASING YOUR HOME THEATER

SHOWCASE DVDS (PART I)

Today we'll look at the best DVDs to show off your television's picture. For this you want a movie that's sharp and colorful, with a wide range of brightness and contrast. And, while eye-popping visuals are good, there's something to be said for subtle shadings (in both light and dark scenes), as well. Here are the discs I recommend:

- *Finding Nemo*, which is—to my eyes, anyway—the best-looking DVD, period. This film is a glorious celebration of color that looks its best on a high-quality, well-calibrated widescreen display. For a good demo, try the "first day of school" scene, as Marlin and Nemo sweep through an amazingly colorful coral reef.
- *Alien, Collector's Edition*, one of the best DVDs for testing your TV's capability to render detail in dark scenes.
- *Amélie*, a French film that's a candy-colored treat for the eyes. The film's color palette is wonderfully vibrant, ranging from richly monochromatic to vividly rainbow-like.
- *Blue Crush*, a kind of guilty pleasure, but it's great eye-candy and the ocean never looked this good on disc before. (It's also a great disc for surround sound; the underwater scenes are especially realistic.)
- *Crouching Tiger, Hidden Dragon*, a DVD of extraordinary beauty; the picture is very sharp, with an outstanding sense of dimension. Colors are exceptionally rich, and the night scenes will test your system's capability to display shadow detail.
- *The Hulk*, not necessarily the best comic-book film ever made, but the transfer to DVD presents a reference-quality picture. It may, in fact, be the best-looking live action movie on disc today.
- *Seabiscuit*, a horseracing flick with plenty of opportunities to view fine details. Pay particular attention to the scene where Seabisuit races War Admiral; the detail in the close-ups is stunning.
- *Citizen Kane*, because black and white can be just as impressive as color—especially on this disc, which sports a reference-quality transfer with vivid contrasts between dark and light and incredibly sharp focus throughout.

ON THIS DAY: JEAN HOERNI BORN (1924)

Jean Hoerni invented the planar processor, which led to the creation of the modern integrated circuit. The planar process enabled the placement of complex electronic circuits on a single chip. Hoerni was born on this date in 1924.

GADGET OF THE WEEK: LASERPOD LIGHT

Throw out your lava lamp, kids, something cooler has finally arrived. The Laserpod uses three lasers and three blue and purple LEDs, projected through a crystal diffuser, to create one heck of a light show in your dorm room. Buy it for $100 from www.1aserpod.com.

September 27, 2006

WEDNESDAY

THIS WEEK'S FOCUS
SHOWCASING YOUR HOME THEATER

SHOWCASE DVDS (PART II)

Now let's look—or rather, listen—to some DVDs that will show off the audio component of your home theater system:

- *Saving Private Ryan.* This may be the most impressive-sounding movie on DVD, especially when you listen to the DTS version. What makes this soundtrack so good is its accurate sound steering around the channels; you'll hear bullets zing from one channel to another, with each and every effect perfectly placed in the soundscape.
- *Master and Commander.* Pay particular attention in the battle and storm sequences; you'll hear cannonballs and shrapnel move from channel to channel, crashing waves that wash right over you, and thundering explosions will give your subwoofer a real workout.
- *Das Boot.* Listen to this film on a good surround sound system and you'll feel as if you're actually inside the submarine, with creaks and groans and drips coming from all around you; the effect is actually somewhat claustrophobic.
- *Apollo 13.* If you want the single best showcase for your system's subwoofer, you have to get the IMAX edition of this disc; go directly to the launch scene for the deepest bass you'll find on DVD.
- *Days of Heaven.* This is the perfect disc for a more understated demonstration, a gentle film with an equally gentle soundtrack. There are no big explosions here; it's all subtle, realistic effects, like crickets in the surround channels and gentle breezes blowing back and forth.

Of course, another good way to show off your system's sound is with music. Here we have a number of fine concert DVDs to choose from, all of which reproduce not only high-fidelity music but also the full live concert experience, using all the surround channels. To my ears, the best-sounding concert DVDs include *James Taylor: Live at the Beacon Theater*, *Peter Gabriel: Growing Up Live*, *The Eagles: Hell Freezes Over*, *Concert for George*, and *Alison Krauss and Union Station: Live*. It goes without saying, you should listen to the DTS soundtrack on these discs, when available; DTS has a much wider dynamic range than does Dolby Digital.

ON THIS DAY: KEVIN MITNICK INDICTED (1996)

On this day in 1996, criminal hacker Kevin Mitnick was indicted on charges resulting from a 2 1/2-year hacking spree. He was accused of stealing software worth millions of dollars from major computer corporations.

HARDWARE OF THE WEEK: LOGITECH Z-5500 SPEAKERS

The Z-5500 is one fine-sounding PC speaker system, perfect for both music and games. What you get here is a THX-certified 5.1 system with 505 watts of total amplification, controlled via the Digital SurroundTouch outboard controller. Buy it for $399.95 from www.logitech.com.

September 28, 2006

THURSDAY

THIS WEEK'S FOCUS
SHOWCASING YOUR HOME THEATER

SHOWCASE DVDS (PART III)

Now let's look at those DVDs that deliver the whole ball of wax—terrific picture *and* sound, in a totally immersive experience. Here are the best of the best:

- *Kill Bill: Vol. 1*. This film is a visually and aurally impressive experience from start to finish. My favorite bit runs from when Uma Thurman purchases her ticket to Japan (accompanied by the theme from the old *Green Hornet* TV series) through the Crazy BBs fight, through the final duel with Lucy Liu in the Japanese garden.
- *The Lord of the Rings: The Return of the King*. This is another one of those discs where the colors are lush, the picture is sharp enough to note fine costume details, and the sound simply surrounds you. This film's big battles are simply amazing on a good widescreen TV, accompanied by room-filling sound.
- *The Matrix*. For many viewers, this is the ultimate demo disc. The best version of this film is in *The Ultimate Matrix Collection*; my standard demo involves the final shootout between Neo and Agent Smith, with bullets whizzing around the room—in slow motion, of course.
- *Sky Captain and the World of Tomorrow*. A visually stunning film, all computer generated (except for the actors, of course), with enough room-shaking action to impress even the most jaded viewers. My favorite demo scene is at the beginning, when the giant robots are rampaging through Manhattan; Jude Law's fighter plane veers from front to rear to front again, giving your surround speakers a real test.
- *Spider-Man 2*. Check out the Superbit version of this DVD and you'll find an exceptional transfer of a visually and aurally dynamic film; the film's bright comic book colors virtually pop off the screen.
- *Star Wars Episode II: The Attack of the Clones*. This movie was created entirely in the digital domain, which makes for a superb transfer to DVD. Colors are superbly rendered, and the detail in some of the very crowded scenes makes you want to pause your DVD player and just gawk.

ON THIS DAY: SEYMOUR CRAY BORN (1925)

Seymour Cray was the engineering whiz who pioneered the concept of the supercomputer. For more than three decades he built the fastest computers in the world, first with Control Data Corporation and then for his own companies. Cray was born on this date in 1925.

SOFTWARE OF THE WEEK: READERWARE

ReaderWare is a set of programs designed to catalog your music, movie, and book libraries. Just enter or scan in the item's bar code and the program builds a custom database for you. Buy the three-media bundle for $75 at www.readerware.com.

September 29, 2006

FRIDAY

THIS WEEK'S FOCUS
SHOWCASING YOUR HOME THEATER

SHOWCASE CDS (PART I)

Enough with the video, it's time to showcase the audio aspect of your home theater system. So let's take a peek at some of the best-sounding CDs available—the ones that will tell you whether you have a true high-fidelity system or not. Today we'll focus on some of the mellower CDs, those that show off the subtle qualities of your systems.

Critics seldom agree on anything, but most of them concur that possibly the best-sounding CD of recent years is Norah Jones' Grammy-winning *Come Away With Me*. You have to hear this disc through a really high-end system to appreciate its true quality. It's a small-group affair, with all the instruments closely-mic'd but with lots of presence. Jones' voice sounds as if she's in the room with you—it's like she's right beside you, singing softly into your ear. This CD has tremendous "air" and clarity; there's a reason it garnered so much acclaim. (For a single demo track, listen to "Don't Know Why"; it's the best thing on the album.)

Other top choices for audio subtlety include Bob Dylan's *Blood on the Tracks* (especially the SACD version), Janis Ian's *Breaking Silence*, and Lucy Kaplansky's *Every Single Day*. And, for jazz fans, you should check out Diane Reeves' *In the Moment* (recorded live in the studio), Shirley Horne's *You Won't Forget Me*, *Night and the City* by Charlie Haden and Kenny Barron, and *Meant to Be* by Ramsey Lewis and Nancy Wilson.

The one thing that all these CDs have in common is that they perfectly reproduce the softest sounds of acoustic instruments. Soft is more difficult to capture than loud, which is why these recordings will show off any defects in your audio system. All of these discs require a quiet listening room and a top-notch speaker system to hear all the fine details.

ON THIS DAY: ENRICO FERMI BORN (1901)

Enrico Fermi was the Italian-born American physicist who was one of the chief architects of the nuclear age. Born on this date in 1901, Fermi developed the mathematical statistics required to clarify a large class of subatomic phenomena, discovered neutron-induced radioactivity, and directed the first controlled chain reaction involving nuclear fission.

WEBSITE OF THE WEEK: METACRITIC

Now here's an interesting site. Metacritic lists critic's reviews of music, films, books, and games; it might be the best source on the Web for CD reviews. Check it out for yourself at www.metacritic.com.

September 30, 2006

SATURDAY

THIS WEEK'S FOCUS
SHOWCASING YOUR HOME THEATER

SHOWCASE CDS (PART II)

Now let's look at some showcase CDs that rock a little harder than the ones we discussed yesterday—and that show off the full dynamic range of your home theater system. Here are some of the most effective demo discs you can play:

- *10,000 Hz. Legend* (Air). Air is a unique electro-pop duo and this is kind of sort of a concept album. But what's important is the sound—which, while highly synthesized, will pull every last drop of low bass and high treble out of your speakers.
- *Brothers in Arms* (Dire Straits). This was one of the first recordings issued in the then-new CD format, and Mark Knopfler took particular care in the album's production. Highlights are the finger-picked guitar on "Why Worry," and the way the sound grows and envelopes the listener on the title track.
- *Casino Royale* (Soundtrack). During the vinyl era, this soundtrack album was a sought-after treasure, gracing audiophile top-10 lists for decades. The standout track is "The Look of Love"; Dusty Springfield's breathy vocal is so intimate, you can hear the slightest catch in her voice.
- *Strange Little Girls* (Tori Amos). A terrifically recorded CD, ranging from tender moments to full-on rockers. I particularly like Tori's version of Tom Waits's *Time*, with close-mic'd vocals and one of the best-sounding grand pianos on disc.
- *The Gershwin Connection* (Dave Grusin). This was one of the first digital recordings made (back in 1991), and Grusin and friends paid particular attention to the recording quality. My favorite track is "Fascinating Rhythm"; everything sounds extra crisp and clear, from the punchy drums to the lush piano chords.

Make sure you have plenty of amplifier power when listening to these discs; the sound is so good you'll want to turn up your system extra-loud to hear all the details!

ON THIS DAY: FIRST NUCLEAR SUB COMMISSIONED (1954)

On September 30, 1954, the USS Nautilus was commissioned at Groton, Connecticut. The Nautilus was the world's first nuclear submarine; a nuclear reactor eliminated the diesel engines that had previously limited a sub's range and speed. In 1958, the Nautilus became the first vessel to travel under the Arctic ice and cross the North Pole.

BLOG OF THE WEEK: JAZZ & BLUES MUSIC REVIEWS

This is a great blog for jazz lovers. It focuses primarily (but not exclusively) on reviews of jazz artists and CDs, although there's a little blues and alt rock thrown in for good measure. It's all good stuff; check it out at www.jazzandblues.blogspot.com.

October 2006

October 2006

SUNDAY	MONDAY	TUESDAY	WEDNESDAY	THURSDAY	FRIDAY	SATURDAY
1 1982 First compact disc player unveiled in Japan	**2** 1866 First tin can with a key opener patented	**3** 1941 First aerosol can used in a commercial application patented	**4** 1968 Film rating system adopted by the Motion Picture Association	**5** 1582 Gregorian calendar introduced in Italy	**6** 1942 Xerography (the process for duplicating documents) is patented	**7** 1959 The dark far side of the moon first photographed
8 1906 First hair permanent is demonstrated	**9** Discovery Day (Columbus Day) 1946 First electric blanket patented	**10** 1933 Plasticized PVC, better known as vinyl, patented	**11** 1983 Last hand-cranked telephones in the U.S. taken out of service in Bryant Pond, Maine	**12** 1928 The artificial respirator is first demonstrated	**13** 1860 First successful aerial photograph of the U.S. taken over Boston	**14** 1947 Chuck Yeager travels faster than the speed of sound
15 1990 Killer bees reach the U.S. by way of Texas	**16** 1985 Intel introduces the first 32-bit processor, the 386	**17** 1989 Earthquake rocks San Francisco, killing 63	**18** 1931 Thomas Edison dies	**19** 1992 Intel introduces the Pentium processor	**20** 1906 First three-element electrical vacuum tube announced	**21** 1879 Thomas Edison invents incandescent electric light
22 1797 First parachute jump is made over Paris	**23** 1991 *Star Trek* creator Gene Roddenberry dies	**24** 1836 Fire matches, then called Lucifers, patented	**25** 1881 The air brush is patented	**26** 1948 A killing industrial smog blanketed the small town of Donora, PA	**27** 1904 New York City subway begins operating	**28** 1955 Bill Gates born
29 1958 First coronary angiogram is performed	**30** 1938 Orson Wells' fictional War of the Worlds radio broadcast causes panic	**31** Halloween 1992 Vatican admits erring for over 359 years in formally condemning Galileo				

October 1, 2006

SUNDAY

THIS WEEK'S FOCUS
RIPPING AND BURNING CDS

CD RIPPING AND BURNING SOFTWARE

If you want to store your CDs on your computer's hard disk, for playback on your computer or transfer to a portable media player, you need a CD ripping program. Alternatively, if you want to take those digital audio files stored on your hard disk and copy them to a CD you can play in other CD players, you need a CD burning program. For what it's worth, many programs do both ripping and burning, so you can do the whole two birds with one stone thing.

The most popular CD ripping/burning programs include

- **#1 CD Ripper**, a CD to MP3 ripper with "anti-cackle" feature. Buy it for $25 at www.maskbit.com.
- **Accoustica MP3 CD Burner**, a combination CD ripper/burner program. Buy it for $24.95 from www.accoustica.com.
- **AltoMP3 CD Ripper**, which rips CDs to MP3 files. Buy it for $19.95 from www.yuansoft.com.
- **Audiograbber**, a CD ripper that can also capture audio from turntables, cassette decks, and other devices. Download it for free from www.audiograbber.com-us.net.
- **Blaze Media Pro**, a combination audio and video CD/DVD editor/converter/burner program. Buy it for $50 from www.blazemp.com.
- **CD to MP3 Gecko**, another CD to MP3 ripper. Buy it for $29.95 from www.cdtomp3gecko.com.
- **Cute CD DVD Burner**, a CD/DVD burner that also lets you do CD-to-CD copying (cloning). Buy it for $29.95 from www.cute-cd-dvd-burner.com.
- **Easy Music CD Burner**, a CD burner with drag-and-drop interface. Buy it for $19.95 from www.easydvdcdburner.com/music/features.htm.
- **Media Jukebox**, an all-in-one ripper/burner/audio player program. Buy it for $24.98 from www.mediajukebox.com.
- **Winamp Pro**, a classic digital audio player that also features CD ripping and burning. Buy it for $14.95 from www.winamp.com.

ON THIS DAY: CONCORDE BREAKS THE SOUND BARRIER (1969)

On the first day of October 1969, the prototype of the French-built Concorde hit Mach 1 for the first time. The plane's inaugural flight had taken place on March 2nd, but things really got up to speed on this flight seven months later. The whole point of the Concorde, of course, was to fly at supersonic speeds—thus shortening long flights, such as those between Europe and the United States.

BLOG OF THE WEEK: ONLINE MUSIC BLOG

As the name implies, this is a blog all about music—new CD releases, digital music news, and the like. It has a bit of an indie/alternative music slant and is a great place to learn about new and upcoming artists. Check it out at www.onlinemusicblog.com.

October 2, 2006

MONDAY

THIS WEEK'S FOCUS
RIPPING AND BURNING CDS

CONFIGURING WINDOWS MEDIA PLAYER FOR RIPPING

If you have a decent compact disc collection and a CD-ROM drive in your computer system, you can make your own digital music files from the songs on your CDs. You can then listen to these files on your computer, transfer the files to a portable music player for listening on the go, share them with other users via the Internet, or use these files to burn your own custom mix CDs.

The process of copying files from a CD to your hard disk is called *ripping*. You can use Windows Media Player to quickly and easily rip all your CDs to hard disk, but first you need to tell WMP what file format you want to use for your ripped music files. In addition, you can choose a quality level for recording, by selecting a specific *bit rate*. (The higher the bit rate, the better the sound quality—and the larger the file stored on your hard disk.)

To set the format and sound quality from within WMP 10, start by pulling down the Tools menu and selecting Options. When the Options dialog box appears, select the Rip Music tab. Select the file format you want from the Format list; your choices include MP3, WMA, WMA Lossless, and WMA Variable Bit Rate. Then you can adjust the Audio Quality slider to select your desired bit rate. Move the slider to the left to create smaller files at lower quality; move the slider to the right to create larger files at better quality.

What's a good quality level? Well, if you want to preserve the full original fidelity of the CD, use WMA Lossless format and the quality level choice is moot. (In fact, it's blanked out.) If you don't mind sacrificing a little sound quality for smaller files, choose the WMA format and a quality level of 128Kbps or 160Kbps.

ON THIS DAY: ENIAC RETIRED (1955)

After 11 years of hard work, the ENIAC computer was retired on October 2, 1955. The machine was capable of running 5,000 operations per second using a system of pluggable boards, switches, and punch cards.

FACT OF THE WEEK

The Recording Industry Association of America (www.riaa.com) reports that unit sales of compact discs rose 2.8% in 2004, to 766.9 million units. (That translates to $11.5 billion in revenue, if you're counting—which the record labels certainly are.) For what it's worth, 1.3 million vinyl LPs/EPs were sold in 2004; they're still hanging on! And who says the MP3 revolution is hurting music sales?

October 3, 2006

TUESDAY

THIS WEEK'S FOCUS
RIPPING AND BURNING CDS

RIPPING CDS WITH WMP

After you've chosen the desired file format and sound quality (see yesterday's article), it's really easy to start ripping. Just insert the CD you want to rip into your PC's CD-ROM drive, and then follow these easy instructions.

In Windows Media Player, select the Rip button to show the contents of the CD. Put a check mark by the tracks you want to copy, and then click the Rip Music button.

I told you it was easy!

WMP now begins to copy the tracks you selected, in the format you selected and at the quality level you selected. Unless you specify otherwise in the Options dialog box, the tracks are recorded into your My Music folder, into a subfolder for the artist, and within that in another subfolder for this particular CD.

By the way, you should make sure your computer is connected to the Internet before you start ripping. That's because WMP will download album and track details from an online database. If you don't connect, you won't be able to encode track names or CD cover art.

As for the legality of burning CDs, while some record labels might disagree, we believe it's perfectly okay to burn your legally purchased CDs. The trouble comes in when you share those music files with others who haven't purchased the same music, or participate in peer-peer file swamping, such as is possible on file trading services, such as Kazaa and Blubster. The Recording Industry Association of America (www.riaa.com) certainly frowns on such activities. Rip and share at your own risk. Of course, legal music files can be purchased (generally for $1 each) from a variety of online retailers, such as iTunes, Musicmatch, and Napster.

ON THIS DAY: TRANSISTOR PATENTED (1950)

On this date in 1950, the United States Patent Office issued a patent to John Bardeen, Walter Brattain, and William Shockley for the transistor. The three researches worked at AT&T Bell Laboratories.

GADGET OF THE WEEK: DARTH TATER

I love it. Darth Tater is a set of accessories that turns Mr. Potato Head into the Dark Lord of the Sith. You even get a spud-sized light saber! Buy it for $7.99 at www.hasbro.com/playskool/.

October 4, 2006

WEDNESDAY

THIS WEEK'S FOCUS
RIPPING AND BURNING CDS

BURNING YOUR OWN MUSIC CDS

Unlike ripping songs from a CD, burning digital music to CD doesn't require you to set a lot of format options. That's because whatever format the original file is in, when it gets copied to CD it gets encoded into the CD Audio (CDA) format. All music CDs use the CDA format, so whether you're burning an MP3 or WMA file, your CD burner software translates it to CDA before the copy is made.

There are no quality levels to set, either. All CDA-format files are encoded at the same bit rate. So you really don't have any configuration to do, other than deciding which songs you want to copy.

The easiest way to burn a CD full of songs is to assemble a playlist beforehand (in your music player or CD burner program), and then copy that entire playlist. You can record up to 74 minutes (650MB) worth of music on a standard CD-R disc, or 80 minutes (700MB) on an enhanced disc.

After you've decided which songs to copy, load a blank CD-R disc into your computer's CD-R/RW drive, launch your CD burner software, and then follow the program's instructions to start translating and copying the song files. After the ripping begins, the digital music files on your hard drive are converted and copied onto a blank CD-R in standard CD Audio format.

And about those disc formats—which should you use? Well, if you want to play your newly ripped CD in a regular (non-PC) CD player, record in the CD-R format and use a blank CD-R disc specifically labeled for audio use. That's because CD-RW discs will not play in most home CD players.

ON THIS DAY: JOHN ATANASOFF BORN (1903)

October 4, 1903, was the day that computer pioneer John Vincent Atanasoff was born. While teaching at Iowa State University in the 1930s, Atanasoff developed the idea of a calculating machine that would use binary arithmetic and electronic calculating switches. Together with doctoral student Clifford Berry, he build the Atanasoff-Berry computer, which preceded the construction of the better-known ENIAC computer.

HARDWARE OF THE WEEK: POCKET CDRW

Add CD burning to any computer, even notebooks, with Addonics' Pocket CDRW. It weighs around 14 ounces and connects (and is powered) by USB. It's also small enough to fit in your pocket, almost, at 5.6" × 5.3" × 0.9". Buy it for $179 at www.addonics.com.

October 5, 2006

THURSDAY

THIS WEEK'S FOCUS
RIPPING AND BURNING CDS

BURNING CDS WITH WINDOWS MEDIA PLAYER

Just as you used Windows Media Player to rip CDs to hard disk, you can also use WMP to burn CDs from songs stored on your hard disk. Here's what you need to do.

Start by inserting a blank CD-R disc into your computer's CD-ROM drive. Then, from within WMP, select the Burn tab. Pull down the Items to Burn list (on the left) and select All Music, and then check those songs you want to burn to CD.

If you don't mind doing a little planning beforehand, a better method is to first create a playlist of the songs you want to burn. Once you assemble the playlist, all you have to do is select that single playlist on the Burn tab. It's a lot easier than assembling songs to burn on the fly.

However you select your tracks, you now pull down the Device list (on the right), select your CD-R/RW drive, and then select Audio CD. Click the Start Burn button and WMP inspects the files you want to copy, converts them to CDA format, and copies them to your CD. When the entire process is done, WMP displays a Closing Disk message for the final track on your playlist. The burning is not complete until this message is displayed.

By the way, if you're trying to copy too much music to the CD, WMP will display a May Not Fit message in the Status column of the final tracks. WMP will try to fit these tracks on the CD, but there might not be room for them. Better to time your tracks before you start burning!

ON THIS DAY: FIRST NON-STOP PACIFIC FLIGHT (1931)

On this date in 1931, Clyde Pangborn and Hugh Herndon completed the first nonstop flight across the Pacific Ocean. They arrived in Washington state about 41 hours after taking off from Japan.

SOFTWARE OF THE WEEK: EASY MEDIA CREATOR

I've talked it about several times already, but if you want a good all-around CD/DVD burner program, go with Roxio's Easy Media Creator. It burns music CDs, movie CDs, and even DVDs, all with a very user-friendly interface. Buy it for $99.95 at www.roxio.com.

October 6, 2006

FRIDAY

THIS WEEK'S FOCUS
RIPPING AND BURNING CDS

COPYING ENTIRE CDS

Making a copy of a CD is pretty much a combination of ripping and burning. That is, you rip the files from your original CD to your hard disk, and then burn those files to a new CD. Just make sure you rip the files to WAV or WMA Lossless format, to retain the original audio fidelity. (If you rip to regular WMA or MP3 format, you'll suffer a degradation in audio quality, which will make your resulting CD sound compressed.)

An even easier solution is to use a program specifically designed for CD copying, such as Roxio's Easy Media Creator. This program makes copying an audio CD pretty much a one-button operation; just click the Disc Copier button and follow the onscreen instructions.

Then there's the issue of CD copy protection. Fortunately, very few CDs have any sort of copy protection, so the rip-to-burn lets you copy your CDs without any problems. However, the big music labels are constantly trying out new copy-protection schemes, with the expressed purpose of reducing bootleg CDs—and in the process, of course, stopping you from ripping or copying those CDs that you legally purchase. If these guys ever find something that works (and isn't easily hackable), expect an armed uprising from legitimate CD owners everywhere.

The latest copy protection technology takes advantage in the slight differences in specifications for RedBook audio CDs and YellowBook data CDs to make the CDs unplayable on computer CD-ROM drives. Various schemes have consisted of adding additional information to the table of contents area on the CD, or adding errors to the track that an audio CD player skips over but a CD-ROM reader has trouble with. Unfortunately, this approach causes problems with some high-end audio equipment that use CD-ROM read mechanisms and with Media Center PCs used for legitimate CD playback—which is why it hasn't seen widespread adoption. If you get a copy-protected CD that won't play on your equipment, ask for your money back!

ON THIS DAY: THOR HEYERDAHL BORN (1914)

Thor Heyerdahl was the Norweigan ethnologist and adventurer who led the famous Kon-Tiki and Ra expeditions. Both voyages were intended to prove the possibility of ancient transoceanic contacts between distant civilizations. Heyerdahl was born on October 6, 1914, and died on April 18, 2002.

WEBSITE OF THE WEEK: FREEDB.ORG

This website is a free database of albums, artists, and songs. The database contains track listings for more than 1.7 million CDs; you can search the database online or download the whole thing for your own personal use. Go searching at www.freedb.org.

October 7/8, 2006

SATURDAY/SUNDAY

THIS WEEK'S FOCUS
RIPPING AND BURNING CDS

WRITING TO A DATA CD

Creating a data CD is a bit different than creating a music CD. If you're using Windows XP, the process is similar to copying a file from one disk to another.

Start by opening My Documents or My Computer and navigating to the files you want to copy. Right-click the files, and then right-click Send To > *drive*, where *drive* is the drive letter for your CD-ROM drive.

Contrary to what you might think happens next, the file doesn't get copied to your CD-ROM drive. Instead, Windows moves it into a temporary storage area on your hard drive, from which it can then be written to CD. Look for a "you have files waiting to be written to the CD" message in the system tray (click it), or just open My Computer and double-click the icon for your CD-ROM drive. The files you have waiting are now displayed. Insert a blank CD-R or CD-RW disc into your CD-ROM drive, select the files in My Computer, and select Write These Files to CD from the CD Writing Tasks pane. Burning then takes place.

If you have a pre-XP computer, the process is a bit more difficult because older versions of Windows didn't support direct-to-CD copying. The best solution is to invest in a CD copying program such as Roxio's Easy Media Creator, and use that program to burn files to CD.

If you only want to read the data on the PC on which it was originally stored, or if you want to rewrite additional data to the same disc, you can use CD-RW (rewritable) discs. On the other hand, if you want to share this data with other computer users, you need to use CD-R (recordable) discs. While CD-R discs are almost universally compatible with other CD drives and players, they are write-once discs—which means you can burn to each disc only one time, and then it's set. CD-RW discs, in contrast, let you rewrite each disc an unlimited number of times—but chances are, other PCs won't be able to read the discs you burn. Choose the right type of disc for your particular use.

ON THIS DAY: CARBON PAPER PATENTED (1806)

On October 7, 1806, Englishman Ralph Wedgwood secured a patent for an "apparatus for producing duplicates of writings." This apparatus, which we know as carbon paper, was a byproduct of another invention, the Stylographic Writer—a machine designed to help the blind write through the use of a metal stylus.

BLOG OF THE WEEK: POSTSECRET

PostSecret is a blog that lets you post your deepest, darkest secrets for everyone on the Internet to read. Users mail in their secrets, anonymously, on one side of a postcard; PostSecrets scans in the postcards and posts the secrets. (One of my favorites: "I don't take my kids to the playground because I don't like talking to the other moms.") Check it out at postsecret.blogspot.com.

October 9, 2006

MONDAY

THIS WEEK'S FOCUS
MOBILE PHONES

CELLULAR NETWORK TECHNOLOGIES

A cell phone's a cell phone's a cell phone, right? Well, not quite. As you've no doubt learned if you've ever switched cell phone providers, there are a number of different cellular network technologies in use today, and none of them seem to be compatible with the others. Which technology you use depends on what's offered by your cellular provider. The most common technologies include

- **FDMA** (frequency division multiple access)—FDMA is an older analog technology that assigns each call on the network its own frequency; essentially, each phone call functions as its own radio station. It's not very efficient and isn't used in today's all-digital networks.
- **CDMA** (code division multiple access)—This is a digital technology that assigns a unique code to each call. A single call can be broken up and spread over all available frequencies, and multiple calls can be overlaid on a single frequency. CDMA networks operate in either the 800MHz or 1900MHz frequency band. In the United States, Sprint and Verizon networks use CDMA technology.
- **TDMA** (time division multiple access)—Instead of a single call spread over multiple frequencies, TDMA splits a single narrow frequency into three separate and complete calls. Each call is assigned a specific portion of time on the designated frequency; it's essentially a form of time sharing. TDMA networks operate in either the 800MHz or 1900MHz frequency bands.
- **GSM** (global system for mobile communications)—GSM is a form of TDMA and is the international standard in Europe, Australia, and much of Asia and Africa. In Europe and Asia, GSM operates in the 900MHz and 1800MHz frequency bands; in the United States, GSM is assigned the 1900MHz band. Cingular and T-Mobile networks use GSM technology.

ON THIS DAY: CALLIOPE PATENTED (1855)

On this date in 1855, Joshua C. Stoddard of Worcester, Massachusetts, received a patent for his calliope. This first instrument consisted of 15 whistles of graduated sizes, attached in a row to the top of a small steam boiler.

FACT OF THE WEEK

The Computer Industry Almanac (www.c-i-a.com) reports that sales of smartphones are eating into PDA sales—a lot. In Western Europe, smartphone sales surpassed PDA sales for the first time in 2004. In the United States, which lags in mobile phone technology, smartphone sales are only about a quarter of PDA sales.

October 10, 2006

TUESDAY

THIS WEEK'S FOCUS
MOBILE PHONES

DUAL BAND AND DUAL MODE PHONES

If you find all the cellular technologies and frequency bands confusing, you're not alone. This is why many manufacturers offer cell phones that utilize more than one technology and more than one frequency band.

Here are the most common options available:

- **Dual-band CDMA**—These phones can switch between 800MHz and 1900MHz CDMA frequencies used by different networks.
- **Dual-band TDMA/GSM**—These phones can switch between the 800MHz and 1900MHz TDMA/GSM frequencies used by different networks (in the United States only).
- **Tri-band GSM**—These so-called "international" phones can switch between the 800MHz and 1900MHz GSM networks used in the United States, as well as the 1800MHz networks used in Europe and elsewhere around the globe.
- **Dual-mode**—These phones can switch between analog FDMA and digital CDMA or TDMA/GSM networks.
- **Dual-band/dual-mode**—These phones can not only switch between analog and digital networks, but also between digital networks operating at both 800MHz and 1900MHz frequencies.

If you have a dual- or tri-band phone, the frequency switching is done automatically. The phone first tries to connect to the default frequency used by your cellular service provider, but then switches bands if it can't access the default network. These phones are popular among business users who travel in areas of the country or the world that use different networks.

ON THIS DAY: BILLIARD BALL PATENTED (1865)

It wasn't just any billiard ball but rather the first billiard ball made from a composite material resembling ivory that was patented by John Wesley Hyatt on this date in 1865. Hyatt also invented celluloid and opened the way for the creation of the modern plastics industry. Billiard balls were originally made of wood. Later, they were made from ivory, which comes from elephant tusks, meaning thousands were needlessly slaughtered before the advent of Hyatt's composite billiard balls.

GADGET OF THE WEEK: FREEPLAY FREECHARGE

When your cell phone battery runs out, what do you do? Well, with the FreePlay FreeCharge, you crank it up—that is, you turn the crank to recharge the battery. The FreeCharge provides emergency power to your phone; you get 2–3 minutes of talk time per 45-second crank. Buy one for $65 from www.freeplayenergy.com.

October 11, 2006

WEDNESDAY

THIS WEEK'S FOCUS
MOBILE PHONES

CHOOSING A MOBILE PHONE

Choosing a cell phone is a major lifestyle decision. Do you prefer a flip phone or a candy bar phone? Do you use it strictly for conversations, or do you do text messaging? Do you want to play games on your phone or watch videos? How about taking pictures? And do you prefer a silver case or a blue one?

More important—well, equally important—is the cell phone provider, and the network technology used. As you learned a few days ago, different providers use different cell phone technologies for their networks. This means that a phone that works with one provider won't work with another provider; it also means you can't switch providers and expect to keep using your old phone.

Which providers use which technologies? Sprint and Verizon use 800MHz or 1900MHz CDMA technology; Cingular and T-Mobile use 1900MHz TDMA/GSM technology. And that's just within the United States. Overseas (over any sea, actually), GSM technology is standard. However, because European and Asian providers use the 900MHz and 1800MHz bands, these systems are incompatible with the United States providers' 1900MHz GSM systems.

Confused yet?

It gets better, especially when you discover that different cell phone manufacturers produce different models for different systems. You might see a particular cell phone you like, but then find out it isn't available for your particular service provider. So you have to pick a model that works with your particular provider, and vice versa. Darned confusing, if you ask me.

So the best approach is probably to pick your cellular service provider first, and then find out which phones that provider offers. If a provider doesn't offer the particular phone you want, tough luck!

ON THIS DAY: H.J. HEINZ BORN (1844)

On October 11, 1844, Henry John Heinz was born. The distant relative of Teresa Heinz Kerry's first husband, Henry founded H.J. Heinz and Co., came up with the "57 varieties" slogan, and started building the family fortune.

HARDWARE OF THE WEEK: VERBATIM STORE 'N' GO PRO

Verbatim's Store 'n' Go Pro is one of those little USB flash memory drives that everybody is using these days. It's faster than most and ultra slim, which makes it easier to plug into stacked USB ports on your PC. Capacities range from 256MB to 2GB; prices range from $50 to $170.

October 12, 2006

THURSDAY

THIS WEEK'S FOCUS
MOBILE PHONES

SMARTPHONES

A smartphone is a combination PDA and cell phone; that is, it's a device that makes cellular calls and stores contact and scheduling information. Most smartphones offer a larger display than a normal cell phone (more like a PDA display), as well as a mini-QWERTY keyboard (for data entry).

Why would you want a smartphone? Well, if you carry both a PDA and a mobile phone, a smartphone lets you cut your number of portable gadgets in half—that's if you don't mind the compromises inherent in such a combo device, of course. You see, the typical smartphone is somewhat larger than a typical cell phone, shaped and sized more like a PDA. This makes for a somewhat awkward phone, but if you think of it as a PDA plus, then you're okay.

My favorite PDA/phone is the Treo 650, which is a bit of a brick to haul around, but it's extremely functional, with its built-in QWERTY keyboard. (It uses the Palm OS operating system for all its PDA functions.) And if you like the Treo, you should also check out the 7100, which has a similar form factor and features. Either will do the job for serious mobile professionals.

Other smartphones look and feel like traditional cell phones but add some degree of PDA functionality. Most of these type of smart phones use Microsoft's Windows Mobile operating system, which you'll either love or hate—or love to hate, as the case may be. These smartphones are easier to carry around than the larger PDA/phones but more difficult to use, PDA-wise. You typically have a smaller screen and no keyboard, which makes entering data challenging, at best.

In any case, before you purchase any smartphone, you should give it a full try-out. Be sure it does everything you need it to do, in a way that's intuitive and comfortable to you. And definitely be sure you like the size and heft; whichever model you choose, you'll be using it a lot!

ON THIS DAY: NEXT COMPUTER INTRODUCED (1988)

On this date in 1988, Steve Jobs introduced the NeXT Computer, his first project since leaving Apple Computer. Although the NeXT ultimately was a market failure, it introduced several features new to personal computers, including optical digital storage and an object-oriented programming language. Of course, Jobs later returned to Apple where he's still exceptionally successful.

SOFTWARE OF THE WEEK: HITCHHIKER'S GUIDE TO THE GALAXY GAME

Hitchhiker's Guide to the Galaxy: Vogon Planet Destructor is a Java-based game that brings the Hitchhiker's world to your Java-equipped mobile phone. It's a cool way to waste time while waiting in line or on the train; learn more at www.atatio.com.

October 13, 2006

FRIDAY

THIS WEEK'S FOCUS
MOBILE PHONES

CHOOSING A HEADSET

If you use your cell phone for extended periods at a time, you know how uncomfortable it can get. It's also not terribly safe (and, in some states, illegal) to use a handheld phone while driving. And then there's the whole issue of cell phone radiation frying your brain; if you're like me, your brain doesn't need to get any more fried than it already is.

All of these are good reasons to use some sort of headset with your cell phone. That way you don't have to hold the handset to the side of your head all the time. They're very popular among people who work all day on the phone, such as call center professionals. They're also great for using a cell phone in the car, which you really shouldn't be doing anyway, although I know you do.

Until recently, all headsets attached to the phone via a long cord—easy to connect, if somewhat inconvenient. Today, many new headsets attach cordlessly, thanks to Bluetooth wireless technology. If your phone is already Bluetooth-enabled (and more and more are), just synch a Bluetooth headset with your phone and you're ready to go. If you don't have a Bluetooth phone, you'll have to attach a Bluetooth adapter to it to use a wireless headset.

When you're shopping for a headset, whether wired or wireless, the main thing to look for is comfort. Do you like the way it hangs on your ear? You should also check the performance; those mini-mics don't always work that well, especially if you're a quiet speaker. You might have to evaluate several models to find one you really like. And then there's the style issue; some of these puppies are ultra-stylish, others look like giant plastic bugs growing out of your ear canal. Style is in the eye of the beholder (and the ear of the beholden), so choose accordingly.

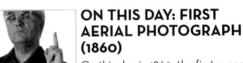

ON THIS DAY: FIRST AERIAL PHOTOGRAPH (1860)

On this day in 1860, the first successful aerial photograph in the United States was taken over Boston. The photographer was James Wallace Black, and he was riding in a hot air balloon tethered 1,200 feet above the city.

WEBSITE OF THE WEEK: THE MOBILE PHONE DIRECTORY

The Mobile Phone Directory is the online resource for everything cell phone related—including detailed specs on hundreds of mobile phones. Check it out at www.mobile-phone-directory.org.

October 14/15, 2006

SATURDAY/SUNDAY

THIS WEEK'S FOCUS
MOBILE PHONES

FAVORITE CELL PHONES

With all this talk of cell phones (or "mobiles," as our European friends call them), I thought I'd share my thoughts on some of the coolest phones on the market today.

With the current trend in cell phones being bigger and bulkier with more and more useless features, it's nice to find a stylish phone in an ultra-small form factor—which precisely describes this season's hottest model, Motorola's Razr V3 (www.motorola.com). At just a half-inch thick, it's small enough to fit in any pocket. And the anodized aluminum case and cool-blue touch keypad are the height of 21st-century high-tech. Yeah, you'll pay around $600 for it, but can you really put a price tag on coolness?

Of course, not all phones have to flip open; some swivel. My favorite swivel phone is the Sony Ericsson S710z (www.sonyericsson.com), which hides its keypad until you swivel it open. You also get a built-in digital camera, unique five-way directional button to control the phone's operations, slot for Memory Stick DUO flash memory, and Bluetooth wireless connectivity.

Nokia's N90 (www.nokia.com) is another swivel phone, but this one swivels in a much different fashion. The top portion of the clamshell twists and turns to turn the phone into a digital camera or camcorder-like device.

The N90 features a 2 megapixel digital camera with Carl Zeiss lens and integrated flash, 352 × 288 pixel video camera, Bluetooth wireless, and (surprise!) a USB connection.

Finally, we have the Siemens S66 (www.siemens-mobile.com), a good-looking candy bar phone with strong multimedia features. You get a 1.3 megapixel digital camera, integrated video recorder, video player, and digital music player—all in a compact package. Quite a performer—and quite stylish, as well.

ON THIS DAY: EDISON ELECTRIC LIGHT COMPANY FOUNDED (1878)

On October 15, 1878, Thomas Edison founded the Edison Electric Light Company in New York City, backed by a syndicate of leading financiers, including J.P. Morgan and the Vanderbilts. Edison's company became a part of General Electric 14 years later.

BLOG OF THE WEEK: MOBILE BURN

The Mobile Burn blog focuses on mobile phone news and reviews, compiled by industry veterans. Check it out at www.mobileburn.com.

October 16, 2006

MONDAY

THIS WEEK'S FOCUS
COMPUTER VIRUSES

ARE YOU INFECTED?

A computer virus is a malicious software program designed to do damage to your computer system by deleting files or even taking over your PC to launch attacks on other systems. A virus attacks your computer when you launch an infected software program, launching a "payload" that often is catastrophic.

Whenever you share data with another computer or computer user, you risk exposing your computer to potential viruses. There are many ways you can share data and many ways a virus can be transmitted, including opening an infected file attached to an email message, launching an infected program file downloaded from the Internet, sharing a floppy disk that contains an infected file, or sharing a computer file over a network that contains an infected file.

How do you know whether your computer system has been infected with a virus? In general, whenever your computer starts acting different from normal, it's possible that you have a virus. You might see strange messages or graphics displayed on your computer screen or find that normally well-behaved programs are acting erratically. You might discover that certain files have gone missing from your hard disk or that your system is acting sluggish—or failing to start at all. You might even find that your friends are receiving emails from you (that you never sent) that have suspicious files attached.

If your computer exhibits one or more of these symptoms—especially if you've just downloaded a file from the Internet or received a suspicious email message—the prognosis is not good. Your computer is probably infected. (But don't panic—you can get rid of the infection by using a good antivirus program!)

ON THIS DAY: FIRST U.S. BIRTH CONTROL CLINIC (1916)

On this date in 1916, Margaret Sanger and her sister Ethel Byrne opened the first birth control clinic in the United States The clinic was located at 46 Amboy Street in Brooklyn, New York, but was soon closed by the police because disseminating birth control was against the law back then.

FACT OF THE WEEK

Sophos (www.sophos.com) reports that the Zafi-D virus accounted for 46.6% of all virus reports in April 2005. Netsky-P was the number-two virus, with 20.6% of all reports.

October 17, 2006

TUESDAY

THIS WEEK'S FOCUS
COMPUTER VIRUSES

PRACTICING SAFE COMPUTING

There's no way to be 100% safe from the threat of computer viruses, save completely isolating your computer from any contact with other computers and computer users. (And what's the fun in that?) There are, however, some steps you can take to reduce your risk of infection:

- Don't open email attachments from people you don't know—or even from people you *do* know, if you aren't expecting them. (That's because some viruses can hijack the address book on an infected PC, thus sending out infected email that the owner isn't even aware of.)
- Don't click links sent to you in spam email messages.
- Download files only from reliable websites, such as Tucows or Download.com.
- Don't execute programs you find in Usenet newsgroups.
- Don't click links or download files sent to you from strangers via instant messaging or in a chat room.
- Share disks and files only with users you know and trust.
- Use antivirus software.

These precautions—especially the first one about not opening email attachments—should provide good insurance against the threat of computer viruses. Remember, whenever you open a file attached to an email message, you stand a good chance of infecting your computer system with a virus—even if the file was sent by someone you know and trust. The bottom line is that no email attachment is safe unless you were expressly expecting it.

ON THIS DAY: MAE JEMISON BORN (1956)

Mae C. Jemison was born on this date in 1956. Jemison is a physician who became the first African-American woman in space when she lifted off aboard the space shuttle Endeavor on September 12, 1992. She has since left NASA and is now the director of The Jemison Institute for Advanced Technology in Developing Countries, which researches, designs, and implements cutting-edge technology for developing nations.

GADGET OF THE WEEK: SCANGAUGE AUTOMOTIVE COMPUTER

The ScanGauge is a multi-function diagnostic computer that works with any automobile. It can check for and interpret diagnostic trouble codes, turn off the "check engine" light, display a variety of digital gauges, and function as a trip computer for up to three trips at a time, tracking average speed, fuel economy, maximum RPM, and the like. Buy it for $129.95 from www.scangauge.com.

October 18, 2006

WEDNESDAY

THIS WEEK'S FOCUS
COMPUTER VIRUSES

ANTIVIRUS UTILITIES

Antivirus software programs are capable of detecting known viruses and protecting your system against new, unknown viruses. These programs check your system for viruses each time your system is booted and can be configured to check any programs you download from the Internet. They're also used to disinfect your system if it becomes infected with a virus. Bottom line: You need to install an antivirus utility on every computer you own.

The most popular antivirus utilities include

- AVG Anti-Virus (www.grisoft.com)
- Kaspersky Anti-Virus Personal (www.kaspersky.com)
- McAfee VirusScan (www.mcafee.com)
- NOD32 (www.nod32.com)
- Norton AntiVirus (www.symantec.com)
- Panda TruPrevent Personal (www.pandasoftware.com)
- PC-cillin (www.trendmicro.com)

Whichever antivirus program you choose, you'll need to go online periodically to update the virus definition database the program uses to look for known virus files. As new viruses are created every week, this file of known viruses must be updated accordingly. Your antivirus software is next to useless if you don't update it at least weekly. An outdated antivirus program won't be capable of recognizing—and protecting against—the very latest computer viruses.

LEARN MORE ABOUT PROTECTING YOUR COMPUTER

To learn more about protecting your computer from viruses, spam, spyware, and other Internet nasties, we recommend picking up a copy of Que's *Absolute Beginner's Guide to Security, Spam, Spyware, and Viruses*, by Andy Walker. (By the way, Andy is my co-host on *Call for Help*, which airs on TechTV Canada, so you know he knows what he's talking about.)

ON THIS DAY: BABBAGE, BINET, AND EDISON DIE

Not all in the same year, of course. Computer pioneer/mathematician Charles Babbage died on October 18, 1871; psychologist/testing pioneer Alfred Binet died on this date in 1911; and noted inventor Thomas Alva Edison died on this date in 1931.

HARDWARE OF THE WEEK: WD PASSPORT POCKET HARD DRIVE

USB flash drives are nice, but sometimes you just need more storage than you get with those little gizmos. The answer is to carry around a portable hard drive, like the 80GB WD Passport, an 80GB model that's small enough to fit in your pocket. Buy it for $199.99 from www.wdc.com.

October 19, 2006

THURSDAY

THIS WEEK'S FOCUS
COMPUTER VIRUSES

REPAIRING AN INFECTED COMPUTER

If your computer has been infected with a computer virus, all hope is not lost. Assuming that your PC is still up and running—slowly, perhaps, or exhibiting some unusual behavior—you're in good shape; all you have to do is scan your system to see if it really is infected, and then (if the news is bad) remove the infection. It's as simple as that.

Start by using your antivirus software to run a manual scan of your system. If infected files are found, make note of the type of infection, and then try to clean or disinfect those files. If an infected file can't be cleaned, delete the file. Then reboot your system.

Next, you want to go online to your antivirus software's website and search for information about the type of virus identified during the scan. Follow any additional instructions given on the website for completing the removal of that specific virus. (For example, you might be instructed to delete or edit certain entries in the Windows Registry.)

In some instances, there might be a specific "fix file" available from your antivirus software's website. These programs are specifically designed to remove a particular virus from your system, and supplement the normal virus-removal operation of your antivirus software.

Finally, if you were forced to delete any document files that were infected, you should now restore those files from a backup copy. (Which is yet another good reason to make regular backups of all your important data.)

If you take all these steps and are still infected, the only recourse might be to completely reformat your computer's hard disk. If you're forced to take this extreme step, you'll need to reinstall all your programs from their original CDs and restore your document files from backup copies. (You did make backup copies, didn't you?)

ON THIS DAY: SIR THOMAS BROWNE BORN—AND DIED

Sir Thomas Browne was a noted physician and writer, best known for his book of reflections, *Religio Medici*. His forte was natural history, and he kept a notable collection of bird eggs, maps, and medals. He was born on October 19, 1605, and died on the same date in 1682.

SOFTWARE OF THE WEEK: NOD32 ANTIVIRUS

My personal favorite antivirus utility is NOD32 from Eset. NOD32 runs unobtrusively without interfering with other programs (sometimes a problem with this kind of software), and automatically updates itself with new definitions. Buy it for $39 from www.nod32.com.

October 20, 2006

FRIDAY

THIS WEEK'S FOCUS
COMPUTER VIRUSES

FINDING NEW VIRUSES

With new viruses being discovered every day, how do the antivirus companies keep up? Well, these companies have several ways they can search for new viruses in the wild, and then add the newly discovered virus signatures to their definition databases. Among the most popular methods are the following:

- **User reporting**—If you stumble across a previously unknown virus on your computer, you're encouraged to submit a sample of that virus to your antivirus company. Company researchers will analyze the virus to determine just what it is that you found. If it's really a new virus, they'll decode its signature and add it to their virus definition database—so that future attacks can be prevented.

- **Research analysis**—The researchers at the antivirus companies are constantly analyzing new virus samples, looking not only for new viruses but also for new infection techniques. Fortunately, this research is used for good, not evil, as what they discover in the lab can be added to their product's virus definition database.

- **Web searching**—Some antivirus companies take the proactive approach and go actively looking for new viruses. In most cases this search takes the form of a web crawl, with "spider" software sent across the Web to look for specific signatures or behaviors. For example, Symantec's Bloodhound system is essentially a Java-based web crawler that looks for virus-like behavior on the Internet. When it finds something suspicious, it sends it back to the Symantec AntiVirus Research Center (SARC) to be analyzed—and possibly added to the company's virus definition database.

It goes without saying that all this effort the companies devote to finding new viruses is totally wasted if you don't keep your antivirus program updated with the latest virus definitions. Configure your program to download new definitions at least once a week, so your computer will be protected from the newest nasties that the antivirus companies discover.

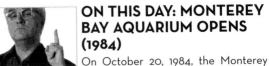

ON THIS DAY: MONTEREY BAY AQUARIUM OPENS (1984)

On October 20, 1984, the Monterey Bay Aquarium opened. It is the largest artificial environment for marine life, built on Cannery Row on the site of the old Hovden sardine cannery. It was financed with a $40 million grant from David Packard (of Hewlett-Packard fame) and houses 6,500 marine animals from more than 525 different species.

WEBSITE OF THE WEEK: VIRUS BULLETIN

Virus Bulletin offers up a compelling mix of news, articles, and other information about the latest computer viruses and virus trends. The site includes a massive malware directory, free for your searching. Check it out at www.virusbtn.com.

October 21/22, 2006

SATURDAY/SUNDAY

THIS WEEK'S FOCUS
COMPUTER VIRUSES

SIGNATURE SCANNING

The primary method of detection used by virtually all of today's antivirus programs is called signature scanning or pattern matching. This method compares a scanned file with the characteristic signature (individual bytes of program code) of a known virus stored in the antivirus utility's virus definition database. If the scanned file matches a known signature (contains the same code pattern), the program marks the file as infected and takes appropriate action.

Of course, signature scanning is only as good as the data in the database. What happens when there's a new virus in the wild, with a signature that doesn't yet appear in the virus definition database?

Fortunately, as soon as an antivirus company finds out about a new virus, it immediately updates its virus definition database. But it might take days or weeks to realize that a new virus is making the rounds—and even longer before all users download the updated virus definitions.

During that period between the creation of the virus and the updating of your antivirus program, your computer is at risk of catching the new virus—simply because your copy of the antivirus software doesn't yet know about the new virus. If your antivirus company identified the virus quickly, and you update your virus definitions frequently, you won't be at risk for long. However, if your antivirus company is slow on the uptake, or if you *don't* update your virus definitions on a regular basis, you're at major risk of being hit by the new infection.

The big problem is that, by default, most users only update their virus definitions once a week. In fact, many users don't update near that frequently, and a large number of users don't update their programs *at all*. This is why you need to configure your antivirus utility to update your definitions at least once a week, if not more frequently. The longer you go without updating, the more at risk you'll be!

ON THIS DAY: FIRST TRIMLINE TELEPHONE (1963)

On October 21, 1963, the first trimline telephone was placed into service by the Michigan Bell Telephone Company. The first phone to integrate the dial and hang-up button into the handpiece, it was available to all customers for an additional monthly charge of $1.

BLOG OF THE WEEK: WORM BLOG

No, this blog isn't about creepy crawly earthworms, but rather about the types of worms distributed by computer viruses. This site features all the latest virus news; check it out at www.wormblog.com.

October 23, 2006

MONDAY

THIS WEEK'S FOCUS
FUN WITH PHOTOSHOP

FIXING ONLY PART OF THE PICTURE

It's possible to take a picture that's only partially perfect. That is, part of the picture looks great, but another part—maybe it's a particular object, maybe it's a larger area—needs a little work. When you need to edit only a particular part of a picture, you first need to select that area; any edits you make are then restricted to the area you've selected.

If you're really, really fortunate, the area you want to edit is rectangular in shape, such as a sign or the side of a building. In this situation, you can easily select a rectangular area, using Photoshop's Rectangular Marquee tool.

When you need to select an irregularly shaped area, the first tool to try is the Magic Wand tool. This is an interesting little tool, in that it selects an area based on its color, in contrast to surrounding areas. So, if you need to select a red object against a green background, or a person with a white shirt framed against a dark wall, or even a gray automobile against a deep blue sky, the Magic Wand will do a pretty good job.

If the area you need to select blends in with its surroundings, the Magic Wand isn't so magic. In these instances, you have to manually draw around the area you want selected. The easiest way to do this is to use Photoshop's Magnetic Lasso tool. This tool lets you trace roughly over the area you want selected, and then it snaps its border more precisely to the selection, based on how the edge of the area contrasts with the surrounding area. Although the Magnetic Lasso requires some skill to use, I've found it a useful tool for selecting difficult-to-draw areas.

By the way, if you're drawing a border and the Magnetic Lasso doesn't recognize a point on the edge of the selection area, you can click the mouse to manually add a fastening point there. If you need to undo any fastening points while you're selecting, press the Backspace key (Mac: Delete) to delete points in reverse order.

ON THIS DAY: MR. TORNADO BORN (1920)

Tetsuya Theodore Fujita was born on this date in 1920. Fujita was a Japanese-American meteorologist who specialized in severe storms. His nickname was "Mr. Tornado," and he devised the Fujita scale for measuring the intensity of tornados based on the amount of their damage.

FACT OF THE WEEK

Adobe (www.adobe.com) reports that Photoshop has an installed base of more than four million users. Not surprisingly, more than 90% of creative professionals have Photoshop on their desktops; it's that ubiquitous.

October 24, 2006

TUESDAY

THIS WEEK'S FOCUS
FUN WITH PHOTOSHOP

SOFTEN YOUR EDGES

When you're selecting an item in Photoshop, either to edit or to copy and paste, the result sometimes looks a little too hard-edged and artificial. For a more natural effect, it's often better to soften the border, so the edited area blends in better with its surroundings.

All of Photoshop's marquee and lasso tools (but not the Magic Wand tool) let you soften the edges of your selection via *feathering*. To soften the edge of your selection, all you have to do is enter a value (in pixels) into the Feather box on the Options bar.

When you feather the edge of a selection, you blur the transition from one area to the next. The amount of blurring is specified in pixels; the more pixels you feather, the softer the edge becomes.

If you made a selection with the Magic Wand tool, you can feather after the fact by selecting Select, Feather. This opens the Feather Selection dialog box; enter a pixel value, and then click OK. (You can also use this approach to soften the edges of any selected area, including those selected by the marquee and lasso tools.)

LEARN MORE ABOUT FIXING BAD DIGITAL PICTURES

Interested in turning your average digital photographs into masterpieces? Want to rescue a good picture gone bad? Look no further than Que's *Bad Pics Fixed Quick*, written by Michael Miller, the co-author of this book. Here, you'll find a bevy of practical, expert advice for rescuing digital pictures. The best part: You don't have to be a Photoshop expert or a professional photographer to create beautiful digital pictures.

ON THIS DAY: FIRST NYLON STOCKINGS (1939)

On October 24, 1939, nylon stockings went on sale for the first time to employees of DuPont's Wilmington, Delaware, nylon factory. Previous to this, all women's stocking were made of silk.

GADGET OF THE WEEK: BOSE NOISE CANCELING HEADPHONES

The Bose QuietComfort 2 is, without a doubt, the quietest headphone on the market today—which is why you see so many of them in use on long business flights. Bose's Acoustic Noise Canceling technology electronically identifies and then reduces unwanted noise. Buy a pair for $299 from www.bose.com.

October 25, 2006

WEDNESDAY

THIS WEEK'S FOCUS
FUN WITH PHOTOSHOP

REMOVING RED EYE

Probably the most common problem for digital photographers is the red eye effect. You know what I'm talking about: You shoot with a flash, and your subjects end up with glowing red eyes. Unless your friends or family members actually have glowing red eyes (in which case you have bigger problems than just a bad picture), you need to eliminate the effect to achieve a realistic photo.

Because red eye is such a common problem, Photoshop makes it easy to fix. Photoshop CS2 includes a new tool just for fixing red eye, called the Red Eye Removal tool. It's a fix so easy that just about anyone can do it—and a lot more effective than using similar tools in previous versions of the program.

Before you fix the red eye, it helps to zoom in on the area you want to edit. Use the Zoom tool to get the eye big enough to work with comfortably.

Now you can select the Red Eye Removal tool. Position the cursor over the red area of the pupil, and then click once. That should remove the red; repeat this step for each eye you need to fix.

And if the one-click approach doesn't remove all the red eye (which it sometimes doesn't), there's another technique you can use. With the Red Eye Removal tool selected, click and drag the cursor to draw a rectangle over the entire red area of the eye. When you release the mouse button, all the red should be gone.

By the way, Photoshop CS2's Red Eye Removal Tool is such a big improvement over the similar-but-less-effective tool in CS1 that it almost makes the upgrade worthwhile for this tool alone. If you've tried fixing red eye in previous versions of Photoshop, you know what I'm talking about; believe me when I tell you that this new tool really works!

ON THIS DAY: FIRST MICROWAVE OVEN SOLD (1955)

On this date in 1955, Tappan sold the first domestic microwave oven. The Radarange was first demonstrated by Raytheon in 1947; Tappan's consumer model sold for a whopping $1,300—and that's in 1955 dollars!

HARDWARE OF THE WEEK: GRAPHIRE BLUETOOTH TABLET

The Graphire Bluetooth is a drawing tablet without the wires—it operates via Bluetooth technology. It's nice to sit on your couch with the tablet in your lap, while you draw and paint and whatever at your convenience. Buy it for $249.95 from www.wacom.com.

October 26, 2006

THURSDAY

THIS WEEK'S FOCUS
FUN WITH PHOTOSHOP

REMOVING WRINKLES

Ever wonder how those 40- and 50-something Hollywood stars keep looking so young? Well, part of it is makeup and part of it is plastic surgery, but a big factor in their eternal youthfulness is photo retouching. In other words, they really don't look that young in person; their youthfulness only comes through in their digitally retouched photos.

The easiest way to remove wrinkles is to use Photoshop's Healing Brush tool. This tool works by "cloning" an unwrinkled area of the face over the wrinkled area, and then blends the cloned area into the surrounding area. The result is surprisingly natural.

Start by zooming into the area you want to fix. Select the Healing Brush tool, and then select a relatively small brush. You should also make sure that the Normal mode and Sampled options are selected.

Now you need to select the area of the face you want to clone. Find an unwrinkled area of the face that is near the wrinkle, hold down the Alt key (Mac: Option) and click once in this area.

After you've selected the area to clone, you can start "painting" the cloned area over the wrinkled area. To do this, click and drag the cursor over the wrinkles. As you drag over the area, the area you previously selected is "cloned" under the cursor, effectively painting smooth skin over the wrinkled skin. Note, however, that the full effect isn't visible until you release the mouse button; this is when Photoshop completes the healing process by blending the cloned area into the surrounding area.

ON THIS DAY: C.W. POST BORN (1854)

Charles William Post was born on this date in 1854. He created Grape-Nuts cereal (based on a similar concoction he discovered while a patient at Dr. John Harvey Kellogg's sanitarium) and founded the Post Cereal Company. Post Cereal later became General Foods Corporation.

SOFTWARE OF THE WEEK: ADOBE PHOTOSHOP ELEMENTS

If you don't want to spring for the full, extremely expensive, and difficult-to-use Photoshop CS package, consider going with Photoshop Elements instead. Elements is essentially a subset of Photoshop CS, containing the most-used tools for hobbyists and casual users, in an easier-to-use interface. It costs a lot less, too; buy it for just $99.99 from www.adobe.com.

October 27, 2006

FRIDAY

THIS WEEK'S FOCUS
FUN WITH PHOTOSHOP

REMOVING SKIN BLEMISHES

Some people are particularly sensitive about skin blemishes, from freckles and acne to moles and scars. Your subjects will be impressed when you use Photoshop to remove these types of blemishes—and make them look even more gorgeous than they are in person.

The easiest way to remove small moles and blemishes is with Photoshop's Spot Healing Brush tool. Using the Spot Healing Brush is a one-click operation that replaces the blemish with a new patch of skin, automatically calculated from the surrounding skin area.

Start by zooming in on the area with the blemish, and then select the Spot Healing Brush tool. Pull down the Brushes menu and select a soft-edged brush; the brush size should be slightly larger than the blemish. You should also make sure that the Type is set to Proximity Match.

Fixing the blemish is as easy as positioning the blemish and then clicking once. Remember, you only click once—you don't try to paint over the blemish!

This technique does a really good job 90% of the time—and a horrible job 10% of the time. Sometimes Photoshop will "heal" the area with a much darker or different-textured spot; if this happens to you, press Ctrl+Z to undo the operation and try another technique, instead.

Note that despite their similar names, the Spot Healing Brush tool is a totally different tool from the Healing Brush tool we discussed yesterday. The Spot Healing Brush is a one-click tool best used to fix small areas (like moles and freckles), while the regular Healing Brush uses a "painting" process to fix larger areas (like wrinkles).

ON THIS DAY: ARPANET CRASHES (1980)

October 27, 1980, saw the first large-scale crash of a computer network. The ARPANET, predecessor of today's Internet, was brought down by a redundant single-error detecting code; the network was down for a total of four hours.

WEBSITE OF THE WEEK: PLANET PHOTOSHOP

Learn how to do lots of cool stuff with Photoshop at Planet Photoshop. This site is chock-full of information and tutorials that help you get the most out of the Photoshop program. It's located at www.planetphotoshop.com.

October 28/29, 2006

SATURDAY/SUNDAY

THIS WEEK'S FOCUS
FUN WITH PHOTOSHOP

CREATE A GLAMOUR GLOW

Here's one of my favorite Photoshop tricks, ideal for creating glamorous close-ups with a kind of soft-focus glow—without losing the overall picture sharpness. The technique involves using multiple layers, so make sure you're familiar with Photoshop's layer feature.

You start by making a copy of the current layer by selecting Layer, Duplicate Layer. You'll want to blur this new layer, so select it in the Palette Bin and then select Filter, Blur, Gaussian Blur. When the Gaussian Blur dialog box appears, adjust the Radius control until the picture is sufficiently blurry (probably somewhere between 3 and 6 pixels). Click OK when you're done.

You now need to blend this blurry layer with the still-sharp original layer, so go back to the Layers palette and adjust the Opacity control to 50. Your picture should now look only slightly blurred.

Next, you need to sharpen the non-skin areas of the picture, which you do by erasing some of the duplicate layer. So select the Eraser tool (with a soft-edged brush), and click and drag the mouse over those areas of the face you want to sharpen—eyebrows, eyes, teeth, lips, hair, and so on. What you're doing is erasing parts of the blurred layer so the sharp original layer shows through. Avoid dragging the Eraser tool over open areas of skin.

(If you want the background of your photo to be really sharp, you'll also want to "erase" the blur from the background area. On the other hand, many portraits feature equally blurry backgrounds, which argues for not erasing the blur there. It's your choice.)

The result is a very professional-looking portrait, with the kind of soft-focus effect you'd pay a professional glamour photographer big bucks for!

ON THIS DAY: STATUE OF LIBERTY UNVEILED (1886)

On October 28, 1886, the monument of Liberty Enlightening the World (also known as the Statue of Liberty) was unveiled in New York Harbor. It was a gift from the people of France to commemorate the 100th anniversary of American Independence. The statue was assembled on Liberty Island (formerly Bedloe Island) and rises 151 feet high on a 155-foot high granite pedestal.

BLOG OF THE WEEK: THE UNOFFICIAL PHOTOSHOP BLOG

Wow, this blog has tons of information! It's all about Photoshop—everything from the latest Adobe news to various tips and tricks on how to better use the program. Find it all at photoshop.weblogsinc.com.

October 30, 2006

MONDAY

THIS WEEK'S FOCUS
URBAN LEGENDS AND HOAXES

THE WAR OF THE WORLDS

Perhaps the biggest hoax of all time (assuming the moon landing was legitimate—or was it?) was Orson Welles's *The War of the Worlds* broadcast on October 30, 1938—the night before Halloween. Welles decided to adapt H.G. Welles's famous novel for his weekly Mercury Theatre radio play. However, Welles made one important change in adapting the book for radio. His troupe performed the play so that it sounded like a real news broadcast—a broadcast about an invasion from Mars.

The broadcast started with realistic sounding dance music, which was interrupted a number of times by fake news bulletins about a "huge flaming object" that had dropped on a farm near Grovers Mill, New Jersey—a real location. In spite of several announcements about the broadcast being a fictional radio play, tens of thousands of listeners thought that what they were listening to was real. People hid in their cellars, loaded guns, even packed the roads to escape the phony "invasion." Widespread panic ensued.

Despite the mass hysteria, it should have been clear that Welles was simply using his radio forum to pull a good old-fashioned Halloween prank. Here's how he ended the broadcast:

"This is Orson Welles, ladies and gentlemen, out of character to assure you that *The War of The Worlds* has no further significance than as the holiday offering it was intended to be: The Mercury Theatre's own radio version of dressing up in a sheet and jumping out of a bush and saying 'boo!'

"Starting now, we couldn't soap all your windows and steal all your garden gates by tomorrow night... so we did the best next thing. We annihilated the world before your very ears, and utterly destroyed the CBS. You will be relieved, I hope, to learn that we didn't mean it, and that both institutions are still open for business.

"So goodbye everybody, and remember please, for the next day or so, the terrible lesson you learned tonight. That grinning, glowing, globular invader of your living room is an inhabitant of the pumpkin patch, and if your doorbell rings and nobody's there, that was no Martian... it's Halloween."

ON THIS DAY: BALLPOINT PEN PATENTED (1888)

On this date in 1888, the first United States patent for a ballpoint pen was issued to John L. Loud of Weymouth, Massachusetts. The patent described a pen having a spheroidal marking point capable of revolving in all directions, held down by three smaller anti-friction balls.

FACT OF THE WEEK

In 1950, just 8.8% of all households owned a TV; today that number is somewhere north of 98%. In 1950, just 1% of those TV households owned multiple sets. Today, 76% of TV households have more than one set. How things change!

October 31, 2006

TUESDAY

THIS WEEK'S FOCUS
URBAN LEGENDS AND HOAXES

VIRUS HOAXES

What's almost as bad as having your PC infected with a computer virus? How about someone telling you your computer is infected, when it's really not—and causing you to run around like a chicken with your head cut off trying to fix something that's not broken.

Virus hoaxes have been around pretty much since the first computer virus itself. They're perpetuated by pranksters with nothing better to do than annoy other users—or, sometimes, by virus creators looking for yet another way to spread their damaging creators.

The most benign type of virus hoax is an email that tells you there's a new and particular dangerous virus making the rounds. The email tells you to be afraid, very afraid, and to forward this message to all your friends and colleagues. If you do so, you perpetuate the hoax. These bogus warnings are sent purely to frighten and mislead you and other users; there's typically no such virus, or if there is the antivirus companies have it well under control.

A more dangerous type of virus hoax is one that not only tells you about a bogus virus, but purports to offer a software patch to protect you against it. As you might suspect, the attached file doesn't include the promised fix; instead, it contains a virus-infected file, that if you click will infect your computer. As before, the best recourse when you receive this type of message is to delete it—and definitely *don't* open the attached file!

ON THIS DAY: VATICAN ADMITS GALILEO WAS RIGHT (1992)

This might count as too little, too late, but on October 31, 1992, the Vatican admitted that it had erred (for almost 360 years!) in formally condemning Galileo for saying the Earth revolves around the sun. Galileo was placed under house arrest for the last eight years of his life. Turns out Galileo was right (duh!), and Pope John Paul II himself (after 13 years of inquiry on the matter) met with the Pontifical Academy of Sciences to set the record straight.

GADGET OF THE WEEK: SUUNTO SMART WATCH

Suunto's N3i looks like a normal wristwatch with a big LCD display. The display is for all the information you'll receive from Microsoft's MSN Direct service, which delivers news, weather, sports, and stock information, as well as instant messages from your online buddies. You'll pay $9.99 to subscribe to MSN Direct; the watch itself costs $299.99 and is available from www.suuntowatches.com.

November 2006

November 2006

SUNDAY	MONDAY	TUESDAY	WEDNESDAY	THURSDAY	FRIDAY	SATURDAY
			1 — 1952 U.S. explodes the first hydrogen bomb	**2** — 1988 First computer "worm" unleashed by a Cornell University graduate student	**3** — 1983 First computer virus created by Len Adleman	**4** — 1922 The entrance to King Tutankhamen is discovered in Thebes, Egypt
5 — 2001 Nintendo releases the GameCube	**6** — 1923 Patent issued to Col. Jacob Schick for the first electric shaver	**7** — 1908 Ernest Rutherford isolated a single atom of matter	**8** — 2001 Microsoft releases the Xbox	**9** — 1934 Carl Edward Sagan born	**10** — 1989 The Berlin Wall comes down, unifying East and West Germany	**11** — Veteran's Day — 1851 Telescope patented in Cambridge, MA
12 — 1926 First airplane bombing takes place in Williamson County, IL, during a feud between rival liquor factions	**13** — 1927 Holland tunnel opens, connecting Jersey City and New York City	**14** — 1665 First blood transfusion is made between two dogs in Oxford	**15** — 1492 Christopher Columbus notes the use of tobacco among American Indians	**16** — 1841 First patent life preserver issued	**17** — 1970 First patent issued for the computer mouse	**18** — 1978 Rev. Jim Jones and 916 of his cult followers found dead in Jonestown, Guyana
19 — 1954 First automated toll booth went into service on New Jersey's Garden State Parkway	**20** — 1985 Windows 1.0 released	**21** — 1620 Mayflower lands at the tip of Cape Cod	**22** — 1620 Mayflower lands at the tip of Cape Cod	**23** — Thanksgiving Day — 1897 Pencil sharpener patented	**24** — 1903 First U.S. patent for an automobile electric self-starter	**25** — 1867 Dynamite invented
26 — 1867 First U.S. patent for refrigerated railroad cars issued	**27** — 1963 First flight of a space vehicle powered by a liquid hydrogen and liquid oxygen fuel	**28** — U.S. Mariner 4 completes first mission to Mars	**29** — Leo Laporte's birthday	**30** — 1869 Mason jar patented		

November 1, 2006

WEDNESDAY

THIS WEEK'S FOCUS
URBAN LEGENDS AND HOAXES

HOW URBAN LEGENDS BECOME LEGENDARY

An urban legend is a tale—often cautionary, in some sense—passed along by those who believe the events befell someone they knew personally, or perhaps someone who is a friend of a friend or family member. By definition, all urban legends are false; whatever was supposed to happen never happened.

Urban legends take place in the recent past, and in places familiar to all of us. The typical urban legend involves "this guy my brother used to know," or someone similar, and takes place at the local shopping mall, college dormitory, or some such place. The point is that whatever was supposed to have happened to "that guy" in the urban legend could just as easily happen to you.

The thing is, none of these urban legends ever happened—at least, not precisely as told. In fact, some of the tales told in urban legends are actually reworkings of similar tales that date back a century or more—with the details updated to contemporary times, of course. The date that ends with the hook from the escaped mental patient embedded in the roof of the car is an urban legend, as is the story about gang members killing the drivers of cars who flash their headlights at them. Good stories, all, highly entertaining, but completely untrue.

The Internet has been a boon to the spread of these urban legends. Instead of a friend talking to a friend talking to a friend, urban legends are now spread to dozens—if not thousands—of people at once via email. It's easy for urban legends to spread over a wide geographic area, and for the legends to mutate and take on additional details with each retelling.

One last point. A small handful of urban legends are actually based on true incidents. However, the nugget of truth in the story tends to get obscured by the storyteller's need to personalize the story. So maybe one kid died from ingesting Dust Off compressed air (Kyle Williams was his name, and he died in 2005), but it didn't happen to your cousin's best friend's brother.

ON THIS DAY: FIRST H-BOMB (1952)

On the first day of November 1952, the United States tested the first thermonuclear device, a hydrogen bomb dubbed "Mike." The bomb was exploded at Eniwetok Atoll, 3,000 miles west of Hawaii. The blinding white fireball was more than three miles across; the blast was a thousand times more powerful than the atomic bomb that destroyed Hiroshima.

HARDWARE OF THE WEEK: LACIE BIGGER DISK EXTREME

For you storage hounds, LaCie offers the Bigger Disk Extreme, a one-box solution containing multiple hard disks in a RAID array. Capacity is truly humongous; the Big Disk Extreme comes in 1 terrabyte, 1.6 terrabyte, and 2 terrabyte versions, all with high-speed FireWire 400 and 800 connections. Prices range from $979 to a budget-busting $2,299.

November 2, 2006

THURSDAY

THIS WEEK'S FOCUS
URBAN LEGENDS AND HOAXES

FAVORITE URBAN LEGENDS

Today I'll relate some of my favorite urban legends. You've probably heard some of them yourself. Just remember: None of these stories are true, so don't pass them on!

- According to some official at the Environmental Protection Agency, when the temperature of the artificial sweetener Aspartame exceeds 86 degrees, it converts to formaldehyde and then formic acid, which causes people to behave as if they have multiple sclerosis. (Yuck!)
- A Florida policeman warns that someone has been attaching hypodermic needles to the underside of gas pump handles. These needles are covered with HIV positive blood, thus infecting unsuspecting customers who prick themselves when filling up their gas tanks.
- A friend warns that seven women have died after smelling a free perfume sample that was mailed to them. The product was, of course, poisonous. Because the government is afraid this might be a terrorist act, they're not announcing it to the news media. Don't want to cause panic!
- A health department official warns that three women in Chicago have died from some sort of toxic poisoning. It appears that they were all bitten by a South American Blush Spider, hiding under the toilet seat at a fast food restaurant.
- A friend relates that a friend of his was on vacation in Mexico and brought back a cactus plant as a souvenir. A few months later, the plant started shivering, and then exploded—spewing forth hundreds of baby tarantulas. It seems the cactus had been full of spider eggs, and eggs eventually hatch....

Any of these sound familiar? It doesn't matter—they're all bunk!

ON THIS DAY: GEORGE BOOLE BORN (1815)

Logician George Boole was born on this date in 1815. Boole wrote the book *Mathematical Analysis of Logic* and was responsible for the development of Boolean logic, which as you might know is what Web search engines, such as Google, use to join simple terms with AND, OR, and NOT. Using a Boolean search allows you to limit or qualify your search.

SOFTWARE OF THE WEEK: WALLPERIZER

Tired of the same old desktop background? Then let Wallperizer change your background for you—automatically, on a set schedule, using any graphics file you have stored on your PC. Download it for free at www.georgegabrielhara.org.

November 3, 2006

FRIDAY

THIS WEEK'S FOCUS
URBAN LEGENDS AND HOAXES

MORE FAVORITES

There are so many urban legends floating around the Internet, it's hard to narrow my favorites down to just a few. So here are some more of my favorite urban legends—just remember, none of them are true!

- A Department of Homeland Security official reports that hundreds of UPS uniforms have been purchased over the past month or so, by person or persons unknown. The official warns that subjects might try to gain access to sensitive areas or instigate terrorist attacks while dressed as UPS drivers.
- A friend of a friend says that groups of teenagers have been caught playing a new and dangerous game they call Spunkball. This game consists of a group of teens in a car pulling up to a stoplight, shouting "Spunkball," and then tossing into the window of a nearby car a gasoline soaked rag wrapped in aluminum foil and attached to a lit firecracker. The firecracker explodes, sending foil shrapnel through the air and catching the car on fire.
- A college researcher claims that KFC doesn't use real chickens, but rather creates genetically manipulated organisms without beaks, feathers, and feet to use in their tasty fried "chicken" meals. (Yummy!)
- The president of Old Navy (or the Gap, or Bed Bath and Beyond, or some similar retail chain) promises you a free gift card if you forward this email to 10 of your friends.
- Remember the Mikey kid from the old Life cereal commercials? A friend of a friend of a friend says that he died from ingesting six bags of Pop Rocks candy, followed by a six-pack of Pepsi. The result—a deadly fizz. (This is why, the email goes on to explain, Pop Rocks were taken off the market in the early 1980s.)

Are these stories fact or urban legend? The latter, of course!

ON THIS DAY: FIRST FROZEN FOOD (1952)
On this date in 1952, Clarence Birdseye marketed the first frozen peas. He learned the technique of flash freezing from Labrador Intuit while a field naturalist near the Arctic.

WEBSITE OF THE WEEK: SNOPES
Not sure whether someone is telling you the truth or repeating an urban legend? Then check out Snopes, a website that compiles virtually all known urban legends and hoaxes—and debunks them. Do your own debunking at www.snopes.com.

November 4/5, 2006

SATURDAY/SUNDAY

THIS WEEK'S FOCUS
URBAN LEGENDS AND HOAXES

ICE ON YOUR CELL PHONE

Here's an item that many think is a hoax, but really isn't. In fact, it's a pretty good idea!

Many users have received an email (or had a friend who supposedly received the email) from a paramedic, encouraging them to store emergency contact information in their cell phone's address book, under the name "ICE." In this instance, ICE stands for "in case of emergency," and is designed to help ambulance or hospital staff notify your next of kin in case of an emergency. Storing under the common entry ICE is advised, as otherwise personnel have no way of knowing just who it is in your address book should be notified.

As I said, this one isn't a hoax. It's actually a somewhat organized campaign launched by Bob Brotchie, who works as a paramedic for the East Anglian Ambulance NHS Trust in England. He came up with the idea after treating numerous accident victims and having the normal difficulty of finding next-of-kin to notify. He figured if he could turn on a victim's cell phone and dial the ICE entry, it would solve a lot of problems—and notify family members a lot faster that their loved one was hurt. While it's not a perfect solution to the identification problem (what if you lose your cell phone or it gets damaged in an accident?), it's still a good idea.

Accompanying this real message, however, are a number of hoaxes. One warns you that adding an ICE entry to your mobile phone will trigger premium charges or some sort of automated text message; it won't. Another hoax warns that the original ICE email is phase one of a phone-based virus; it isn't.

But here's the plain truth: There's no harm at all in adding an ICE entry to your phone's address book. In fact, it's a smart thing to do!

ON THIS DAY: ANCIENT BEER DISCOVERED (1992)

On November 5, 1992, archeologists discovered chemical evidence of beer inside a 5,000 year-old pottery vessel in the Zagros mountains of Iran. Beer was the preferred fermented beverage of the ancient Sumerians; the clay jar had apparently been used for beer fermentation or storage. Unfortunately for the Sumerians, pretzels weren't invented for a long, long time.

BLOG OF THE WEEK: UFO NEWS BLOG

Want to find out about the latest UFO sightings, alien abductions, and crop circles? Then check out the UFO News Blog, located at www.cosmicparadigm.com/ufonews/. The truth is out there!

November 6, 2006

MONDAY

THIS WEEK'S FOCUS
WEB PAGE TRICKS

CYCLE YOUR BACKGROUND COLORS

Here's a trick that's bound to amaze (and possibly annoy) visitors to your web page. Instead of using a static background color, you program the page to cycle through a series of different colors; the colors change automatically every few seconds.

For what it's worth, changing background colors is a relatively easy thing to do in JavaScript, thanks to the document.bgColor property. Throw in a few lines of script to randomize the background color, and you have everything you need to drive users insane. All you have to do is insert the following code after the <body> tag in your HTML document:

```
<script language="JavaScript">
<!--

function ChangeColor()
{color = "#";
for (i = 0; i < 6; i++)
{hex = Math.round(Math.random() * 15);
if (hex == 10) hex = "a";
if (hex == 11) hex = "b";
if (hex == 12) hex = "c";
if (hex == 13) hex = "d";
if (hex == 14) hex = "e";
if (hex == 15) hex = "f";
color += hex;}
document.bgColor = color;
setTimeout("ChangeColor();",2000);}

ChangeColor();

//-->
</script>
```

You can change the cycling time between colors by changing the setTimeout value. The default value (2000) is about two seconds.

Know, however, that while a background that constantly changes colors does an effective job of catching the user's attention, it's rotten for actually reading any text on the page. But it is attention-getting!

ON THIS DAY: MICROSOFT CONTRACTS WITH IBM (1980)

Microsoft entered the big time on this date in 1980, when it signed a contract with IBM to develop an operating system for its upcoming IBM Personal Computer. The operating system was known as the Microsoft Disk Operating System, or MS-DOS—perhaps you've heard of it?

FACT OF THE WEEK

Forty-two percent of all game players say they play games online for at least an hour each week, reports the Entertainment Software Association (www.theesa.com). In addition, 34% of heads of households play games on a wireless device, such as a cell phone or PDA, up from 20% who did so in 2002.

November 7, 2006

TUESDAY

THIS WEEK'S FOCUS
WEB PAGE TRICKS

ADD A WATERMARK TO YOUR WEB PAGE

Many web pages incorporate a watermark—a graphic background image that stays set in place, even when a visitor scrolls down the page. This sort of static background is a good way to impress an image on website visitors. For example, you might want to use a screened-back version of a company logo and have it stay static in the center of the browser window; this technique is how you do that. For that matter, any dominant image—a picture of a company's headquarters on a corporate page, a picture of a dog on a canine page, and so on—can be watermarked in this fashion.

It's relatively easy to define a background image that always stays centered in the browser window. All you have to do is create a style sheet that positions a fixed, untiled, nonscrolling image in the center of the window's background. Just add the following code into the head of your HTML document:

```
<style type="text/css">
<!--
body{background-image:url("filename");
background-attachment:fixed;
background-repeat:no-repeat;
background-position:center;}
-->
</style>
```

Of course, you'll need to replace *filename* with the complete URL and filename of the background image.

And if you're using FrontPage to design your website, the whole process is even easier. Just select Format, Background and select the Watermark option.

ON THIS DAY: TACOMA NARROWS BRIDGE COLLAPSES (1940)

Engineers don't always get it right. At approximately 11:00 a.m. on November 7, 1940, the first Tacoma Narrows suspension bridge collapsed due to wind-induced vibrations. Situated in Puget Sound, near the city of Tacoma, Washington, the bridge had been open for traffic only a few months.

GADGET OF THE WEEK: PALMONE LIFEDRIVE

This new Palm OS gizmo is more than a simple PDA. The LifeDrive contains a 4GB hard drive that lets it store and transport PC files and play digital audio and video files. It also has a big 320 × 480 pixel color display, built-in 802.11b WiFi and Bluetooth wireless, an SD/MMC expansion slot, and built-in voice recorder with microphone. Buy it for $499 from www.palmone.com.

November 8, 2006

WEDNESDAY

THIS WEEK'S FOCUS
WEB PAGE TRICKS

MAKE THE PAGE BACKGROUND SCROLL—AUTOMATICALLY

A nonscrolling watermark is cool enough, but it's even cooler to use a background image that scrolls slowly down the browser window, all by itself. This sort of scrolling effect is achieved by rapidly changing the position of the background graphic, incrementing the vertical position by 1 pixel, over and over.

Doing this is a two-step process. First, define the background image in the <body> tag of your HTML document, like this:

```
<body background="filename.jpg">
```

Then insert following code after the <body> tag:

```
<script language="JavaScript">
<!--

var GraphicPosition = 0;
var GraphicObject = eval('document.body');

function BackgroundScroll(maxSize)
{GraphicPosition = GraphicPosition + 1;
if (GraphicPosition > maxSize) GraphicPosition =
➥0;
GraphicObject.style.backgroundPosition = "0 " +
➥GraphicPosition;}

var ScrollRate =
➥window.setInterval("BackgroundScroll(307)",
➥10);
//-->
</script>
```

You can change the speed of the scrolling by setting a new value for the last variable in the window.setInterval command. The default is 10; set a larger number to scroll slower or a smaller number to scroll faster. Note, however, that if you set too slow a scroll rate, you won't actually see the image scroll; it will look more like an intermittent redraw of the background at a slightly lower position on the page.

ON THIS DAY: JACK KILBY BORN (1923)

Jack Kilby was an engineer at Texas Instruments when he came up with the idea of the integrated circuit. Kilby was born in Jefferson City, Missouri, on this date in 1923.

HARDWARE OF THE WEEK: DELL ULTRASHARP 24" LCD DISPLAY

This Dell UltraSharp LCD display is a 24" widescreen model that also functions as a remote hub with four USB ports and a 9-in-1 flash card reader. Maximum resolution is 1920 × 1200 (HDTV quality), and the picture is first-rate, with a 1000:1 contrast ratio. Buy it for $1,199 at www.dell.com.

November 9, 2006

THURSDAY

THIS WEEK'S FOCUS
WEB PAGE TRICKS

DISPLAY A MOUSEOVER ALERT

What do you do if you want to warn someone about something when they click a link on your web page? It's simple: You create a mouseover alert, a dialog box that appears when a visitor clicks a link. All you have to do is insert the following code into the head of your HTML document:

```
<script language="JavaScript">
<!--

function linkAlert(messageText)
{newMessage = messageText
alert (newMessage);}

//-->
</script>
```

Now use the following code within the body of your document to associate an alert box with a specific link; replace *message* with the desired text for your dialog box.

```
<a href="url" onMouseOver="linkAlert
➥('message')">linktext</a>
```

When a user hovers over the link, the dialog box will appear and stay visible until the user clicks OK. What's *really* annoying is that the user can't ever click the link—the dialog box keeps popping up and getting in the way!

If you've coded a link to a page that isn't active yet, you can alert users to this fact by displaying this type of mouseover alert box. The only problem with this trick is that it's somewhat disruptive—it prevents users from actually clicking on the selected link. If you actually have a link that you don't want users to click, think about removing it completely rather than going through this extremely frustrating procedure.

ON THIS DAY: CARL SAGAN BORN (1934)

Billions and billions of years ago... well, actually, it wasn't quite that long ago, in cosmic terms. In fact, it was November 9, 1934, when Mr. and Mrs. Sagan gave birth to little Carl, who would grow up to be a famed astronomer, writer of popular science books, and host of the television series *Cosmos*. He passed away on December 20, 1996.

SOFTWARE OF THE WEEK: JAVASCRIPT MENU MASTER

This is a nifty little shareware editing program that lets you create all manner of JavaScript menus for your web pages, no programming experience necessary. Try it for free or buy it $14.95 from www.harmonyhollow.net/jmm.shtml.

November 10, 2006

FRIDAY

THIS WEEK'S FOCUS
WEB PAGE TRICKS

OPEN LINK IN NEW WINDOW

With all the complicated JavaScript-based web tricks out there, it's nice to find a simple HTML-based trick that anyone can apply. It's doubly nice when that trick is a useful one, like this one is.

This trick forces a new window to open when someone clicks on a link. Normally, you click on a link and the browser diverts to the linked page. This is fine if you don't mind visitors leaving your site, and is less desirable if you want to hang on to your visitors. With this trick, you can have your cake and eat it too; the current page stays open (you keep the visitor on your site) while the linked-to page opens in a brand-new browser window.

The great thing about this trick is how easy it is to implement. All you have to do is include a new target attribute to the standard HTML link tag. The code looks like this:

```
<a href="url" target="_new">linktext</A>
```

That's it; the "_new" value for the target attribute is what does the trick. When the link is clicked, a new window will open, pointed to the designated URL.

An alternative approach to this effect—one that allows for more control over the newly opened window—is possible by creating pop-up windows in JavaScript. This approach requires you to insert the following code in the head of your HTML document:

```
<script language="JavaScript">
<!--
function genericPopup(popupAddress)
{new_window =
window.open(popupAddress,'windowName')}
// -->
</script>
```

You then call the code (open the window) from within the main body of your document. The easiest way to do this is by creating a link to pop-up the new window, like this:

```
<a href="javascript:genericPopup('popup-url');">linktext</a>
```

Obviously, you have to supply the URL for the pop-up window (*popup-url*), as well as assign a name to the windows (*windowName*). Past that, it's pretty straightforward.

ON THIS DAY: FIRST MOTORCYCLE (1885)

On this date in 1885, Gottlieb Daimler introduced the world's first motorcycle. The frame and wheels were made of wood, and a leather belt transferred power from the engine to large brass gears mounted to the rear wheel.

WEBSITE OF THE WEEK: JAVASCRIPT.COM

Want to find other JavaScript tricks to add to your web pages? Then check out JavaScript.com (www.javascript.com), a library filled with hundreds of JavaScript goodies.

November 11/12, 2006

SATURDAY/SUNDAY

THIS WEEK'S FOCUS
WEB PAGE TRICKS

ADD MUSIC TO YOUR WEB PAGE

Want to provide some soothing background music for your web page visitors? Or just subject them to a loud and annoying audio clip? Then here's a trick for you. Most web page background music is in the MIDI format. MIDI files will play through just about any modern web browser and can be set to play automatically when your web page loads or when a link is clicked.

To add background music to your web page, all you have to do is insert this code just after the <body> tag in your HTML document:

```
<embed src="filename" autostart=true loop=true
➥hidden=true>
<noembed>
<bgsound src="filename" loop=infinite>
</noembed>
```

Obviously, you replace *filename* (in both places) with the full URL and filename of the audio file you want to play. You can use this code to play MIDI, WAV, MP3, AIFF, AU, and other formats of audio files.

The code as-written plays the designated audio file in a continuous loop—which is great for background music, less desirable for other types of audio effects. To make the audio file play just once, change the loop=true variable (in the <embed> tag) to loop=false, and the loop=infinite variable (in the <bgsound> tag) to loop=1.

If you'd rather not force your visitors to listen to the audio, you can give them the option of listening when they click a text link. Just enter the following code at the appropriate place in your HTML document:

```
<a href="filename">Click here to play background
➥music</a>
```

ON THIS DAY: JACK RYAN BORN (1926)

No, not the hero of Tom Clancy's novels, but rather the inventor who created the Barbie doll, Hot Wheels, and other cool toys. Before joining Mattel, Ryan was an engineer for the Pentagon, where he designed Sparrow and Hawk missiles. Ryan was born on November 12, 1926, and died on August 13, 1991. For what it's worth, Harrison Ford, also known as Jack Ryan, was born July 13, 1942.

BLOG OF THE WEEK: THE JAVASCRIPT WEBLOG

Keep up on all the latest JavaScript-related news and developments at The JavaScript Weblog. Read it here: javascript.weblogsinc.com.

November 13, 2006

MONDAY

THIS WEEK'S FOCUS
POWERPOINT PRESENTATIONS

CREATE A SELF-RUNNING PRESENTATION

You know that PowerPoint is great for creating "live" presentations—but it's also great for creating self-running presentations. Self-running presentations are perfect for trade shows, displaying in public kiosks, or running on a computer in your company's lobby. You can program a self-running presentation to be entirely hands-off (it just repeats over and over, with no user interaction required) or to be interactive (requiring user input to advance to a particular slide).

To create a self-running show, select Slide Show, Set Up Show to display the Set Up Show dialog box. In the Show Type section of the dialog box, select the Browsed at a Kiosk (Full Screen) option. If you want the presentation to advance at a specified pace without user interaction, select the Using Timings, If Present option; if you want users to advance from slide to slide manually, select the Manually option. And if you want the presentation to include a prerecorded narration, make sure the Show Without Narration option is *not* selected.

When you run a kiosk show, your presentation will loop continuously, over and over, until the Esc key is pressed.

You also need to tell PowerPoint how long to hold on each slide before advancing. You do this through slide timings. The easiest way to deal with slide timings is to rehearse your presentation manually, and let PowerPoint record how long you stay on each slide. You can then edit the timing for any particular slide manually, if necessary.

To record slide timings, select Slide Show, Rehearse Timings. PowerPoint now switches to Slide Show mode, displaying a Rehearsal toolbar in the corner of the screen. Advance through your presentation as if you were giving it live; PowerPoint will record the timing as you progress. When you finish the presentation, PowerPoint will tell you the total time for the time show; if you're happy with this, tell PowerPoint to record the slide timings and use them when you view the slide show.

ON THIS DAY: ARTIFICIAL SNOW PRODUCED (1946)

On this day in 1946, artificial snow from a natural cloud was produced for the first time. The cloud was located over Mount Greylock, Massachusetts; an airplane spread small pellets of dry ice (frozen carbon dioxide) for three miles at a height of 14,000 feet. Unfortunately, the snow evaporated as it fell through dry air and never reached the ground.

FACT OF THE WEEK

Hall & Partners (www.hall-and-partners.com) reports that the most popular types of files stored on USB flash drives are music (62%), graphics (52%), and video (46%) files. The average USB drive user is male, in his mid-30s to mid-50s, and earns more than $50,000 per year.

November 14, 2006

TUESDAY

THIS WEEK'S FOCUS
POWERPOINT PRESENTATIONS

CREATE A BRANCHING PRESENTATION

It's nice when you can get double-duty out of a single presentation—that is, use the same PowerPoint slideshow for several different events, or for different options in a self-running kiosk. All you have to do is create a self-running presentation, add objects or action buttons at the right places, and then assign actions to those objects or buttons that link to another slide elsewhere in your presentation—or to another custom presentation within your overall presentation.

Let's work through an example where the opening slide directs users to different spots in the presentation, and assume you've already created the bulk of the presentation. You want to add a new first slide to the presentation, with pictures or text buttons that refer to the different areas of the presentation. (For a self-running kiosk, you probably want to insert the text "Click here for more information" under each button.)

One at a time, select each object or button, and then select Slide Show, Action Settings to display the Action Settings dialog box. Select the Mouse Click tab, and then check the Hyperlink To option, pull down the list, and select Custom Show; this displays the Link to Custom Show dialog box. For each object or button, select the corresponding custom show, and then click OK.

When you're finished programming the action items, select Slide Show, Set Up Show to display the Set Up Show dialog box. Select the following options: Browsed at a Kiosk (Full Screen); Loop Continuously Until 'Esc'; and Using Timings, If Present. Click OK, and your branching presentation is now ready to run.

When you start the slideshow, you or your kiosk viewers can click on the appropriate picture or button to go to the custom presentation they want.

ON THIS DAY: FIRST BBC BROADCAST (1922)

On this day in 1922, the BBC officially began its daily domestic radio service, broadcasting the 6:00 p.m. news read by Arthur Burrows. The London station, 2LO, broadcasted from a 100 watt transmitter on the top floor of Marconi House.

GADGET OF THE WEEK: DELL AXIM X50V

Dell's Axim X50v is one fine PocketPC PDA. You get a beautiful 3.7" color screen and built-in WiFi and Bluetooth wireless connectivity. Even better, this is simply one of the most stylish PDAs on the market today. See for yourself at www.dell.com; it'll set you back $424 if you want to buy one.

November 15, 2006

WEDNESDAY

THIS WEEK'S FOCUS
POWERPOINT PRESENTATIONS

ADDING MUSIC TO YOUR PRESENTATION

Adding a sound to accompany a special effect or an important point can really catch the attention of your audience. In addition, a little background music—soft and low to underscore your speech, or big and dramatic to emphasize an important point—can make your presentation stand out from everyone else's. (Just remember that overuse of audio in your presentation—or the overuse of *any* special effect—can draw attention away from your content.)

You insert an audio clip on a specific slide. You can elect to have the sound or music start automatically when you advance to the slide, or to have it start when you click an icon on your slide. You can also use PowerPoint's animation settings to have the sound play in sequence with other animations on a specific slide. When you're talking music, PowerPoint supports various audio file formats, including .WAV, .AU, and MIDI (.MID).

To insert an audio file, display the slide to which you want to add the sound, and then select Insert, Movies and Sounds, Sound From File to display the Insert Sound dialog box. Navigate to and select the audio file you want to insert, and then click OK. (Alternatively, you can insert a song directly from a CD by selecting Insert, Movies and Sounds, Play CD Audio Track.)

PowerPoint now displays the following message: How do you want your sound to start in the slide show? If you want the sound to play when you advance to the slide, click Automatically. If you want to start the sound manually via a mouse click, click When Clicked.

If you'd rather not see that annoying sound icon on your slide (especially if you program the sound to play automatically when you advance to the slide), right-click the sound icon and select Edit Sound Object. When the Sound Options dialog box appears, check the Hide Sound Icon During Slide Show option.

ON THIS DAY: TOBACCO DISCOVERED (1492)

As you might be able to tell from the date, it was Christopher Columbus who made the first written reference to tobacco. The notation in his journal was dated November 15, 1492.

HARDWARE OF THE WEEK: WACOM TOUCHSCREEN DISPLAY

Wacom's Cintiq 21UX combines a standard 21" LCD display with a pen tablet—which means you can "draw" directly onscreen with a pressure-sensitive pen. This is ideal for graphics artists, who can use the display as a drawing tablet for illustrations, CAD blueprints, and the like. Buy it for $2,999 from www.wacom.com.

November 16, 2006

THURSDAY

THIS WEEK'S FOCUS
POWERPOINT PRESENTATIONS

ADD DEPTH TO A 2-D CHART

Have you ever wanted some dimensionality in a chart—but without employing a complicated 3-D chart type? The easiest way to do this is to apply a drop shadow to your chart elements, which gives your chart an appearance of depth.

To add a shadow to bars or columns in a 2-D chart, start by double-clicking on any series of bars, columns, or slices. This displays the Format Data Series dialog box. From this dialog box, select the Patterns tab. In the Borders section, check the Shadow option. Then click OK to close the dialog box. Repeat this procedure for every series in your chart.

An even neater effect is to add a drop shadow to a borderless bar or column chart. You must select the Shadow option *before* you format the bars or columns to have no borders. (Which you do by opening the Format Data Series dialog box and selecting None in the Borders section.) If you select None (for borders) and then Shadow, selecting Shadow will toggle None to Automatic. You gotta go in order!

Of course, you can always display a two-dimensional chart in three dimensions, which gives each bar or column real depth. You might think that 3-D charts are only for displaying data that you need charted in three dimensions (using three axis), but the reality is that you can use any chart type to display any type of data. When you choose a 3-D chart, you display big thick bars and columns, even if they're all aligned in a single row. PowerPoint lets you adjust the depth of each chart element, so you can make your 3-D bars very thin or very thick, or somewhere in between—which is a lot more versatile than simple drop shadows.

ON THIS DAY: GENE AMDAHL BORN (1922)

On this date in 1922, Gene Amdahl was born in Flandreau, South Dakota. Amdahl helped design the IBM System/360 computer, which marked IBM's transition from discrete transistors to integrated circuits. Amdahl went on to found his own company, Amdahl Computer Corporation, which developed the first IBM-compatible mainframe computers.

SOFTWARE OF THE WEEK: CRYSTAL GRAPHICS POWERPLUGS

Crystal Graphics manufactures a wide variety of PowerPoint plug-ins, which it dubs "PowerPlugs." I'm talking about moving backgrounds, fancy transitions, professional-looking charts, stock photographs, and more. Check out the selection at www.crystalgraphics.com.

November 17, 2006

FRIDAY

THIS WEEK'S FOCUS
POWERPOINT PRESENTATIONS

PUBLISH YOUR PRESENTATION TO THE WEB

As an alternative to printing your presentation, you can create web pages from your slides and publish these pages either on the public Internet or on your company's private intranet. You can choose to publish your entire presentation, a custom show, selected slides, and even slides with notes.

To publish your presentation to the Web, select File, Save as Web Page; this displays a special version of the Save As dialog box. Enter a name for your presentation in the File Name box, make sure that Web Page is selected in the Save as Type list, and then select a location for the Web-based presentation. To publish to your company's intranet, pull down the Save In list, select Network Neighborhood, and navigate to the appropriate server and folder. To publish to a server on the Internet, pull down the Save In list and select either a web folder or an FTP location.

Next, click the Publish button to display the Publish as Web Page dialog box. To publish all the slides in your presentation, select the Complete Presentation option. To publish a range of slides, select the Slide Number option and select a starting and ending slide. To publish a custom show, select the Custom Show option, and then select a show from the accompanying list.

To set additional options, click the Web Options button to display the Web Options dialog box. From here, select the General tab to add navigation controls to your slides, show slide animation while browsing, and resize graphics to fit the browser window; select the Files tab to choose various file options; select the Pictures tab to set the target monitor size for your presentation; or select the Encoding tab to encode your pages for other languages. Click OK to return to the Publish as Web Page dialog box.

With all the options selected, you're ready to publish. Confirm the title and location of your presentation, and then click Publish to publish the presentation. Your presentation is now converted to HTML format and published as instructed. When people view your presentation in a web browser, they use the navigation buttons at the bottom of each page to change from slide to slide.

ON THIS DAY: MOUSE PATENTED (1970)

No, not the mammal, but the computer variety. On November 17, 1970, Doug Engelbart received a patent for an "X-Y Position Indicator for a Display System." Engelbert called it a mouse because of its tail-like cable. The first mouse was a hollowed-out wooden block with a single pushbutton on top.

WEBSITE OF THE WEEK: CLIPART CONNECTION

Spruce up your PowerPoint presentations—and your web pages—with clip art images from Clipart Connection. We're talking thousands and thousands of free pictures, all available at www.clipartconnection.com.

November 18/19, 2006

SATURDAY/SUNDAY

THIS WEEK'S FOCUS
POWERPOINT PRESENTATIONS

PRINTING PROFESSIONAL HANDOUTS

I've seen a lot of great presentations in my time, almost all of which were accompanied by some kind of hard-copy "takeaway." In almost all cases, the printed version of the presentation was inferior to the live presentation; in some cases, the printed piece smelled of afterthought.

Here's the thing. You need to give as much attention to your printed material as you do to your live presentation. Naturally, you won't be able to capture your stellar oratorical style on paper, but you can—and should—capture all the four-color glitz of your slides, and thus transmit the same look and feel on paper as you did live.

Most speakers, however, select File, Print and spit out either black-and-white screen prints (one per page, landscape orientation) or one of PowerPoint's unimaginative handouts or notes pages (also in black and white). No doubt you've received dozens of similar printouts; no doubt you were duly underwhelmed by what you received.

At the very least, your printouts should match your presentation in terms of color. If you're giving a color presentation (and you are, of course), then that presentation should be accompanied by a four-color "takeaway" piece. This piece can be printed on your home or office four-color printer—although this is a less-than-ideal option, both quality- and practicality-wise. A better option is to use a high-quality color LaserJet printer, like those found at FedEx Kinko's and other professional printers.

Finally, ask yourself if providing full-page landscape versions of your slides is really the best "takeaway" for your audience. They might be better served by handouts with 2, 3, 4, or 6 slides per page, or by notes pages that mix pictures of your slides with blank lines for note-taking. If you want to go this route, consider exporting your notes or handouts pages to Microsoft Word, where you can use all of Word's powerful formatting tools to create a really nice-looking printout.

ON THIS DAY: VITAMIN C WARDS OFF COLDS (1970)

On November 18, 1970, Nobel Prize winner Linus Pauling declared that large doses of vitamin C could ward off the common cold. He proposed that adults take somewhere between 2.3 and 10 grams of vitamin C daily, which is far higher than the official recommended daily allowance.

BLOG OF THE WEEK: BEYOND BULLETS

Yep, there's a blog devoted exclusively to PowerPoint presentations. It's called Beyond Bullets, and it's pretty interesting. Check it out at www.beyondbullets.com.

November 20, 2006

MONDAY

THIS WEEK'S FOCUS
SATELLITE RADIO

HOW SATELLITE RADIO WORKS

If you're on the road a lot, you know how frustrating it is to have radio stations fade out as you drive out of their broadcast areas. Satellite radio offers the much better alternative of fade-free nationwide coverage—you're never out of range. Plus, you get more than 100 channels of entertainment, with a good mix of music and talk, all with high-quality digital sound.

In the United States, there are two similar but competing satellite radio services: SIRIUS Satellite Radio and XM Radio, both of which work in a similar, deceptively simple, manner. In essence, they bounce digital signals off orbiting satellites; the signals are then received and decoded by compatible satellite receivers back on Earth. Both SIRIUS and XM signals are transmitted in the 2.3GHz S-band, reserved by the FCC in 1992 for digital audio radio service (DARS) transmissions.

XM uses two Boeing HS 702 satellites, dubbed "Rock" and "Roll." (Cute, eh?) The two satellites are in geostationary orbit, approximately 22,000 miles above the Earth. (XM has a third satellite already built and awaiting emergency launch, just in case something happens to the two orbiting satellites.)

Instead of geosynchronous satellites, SIRIUS uses three Space Systems/Loral satellites in an inclined elliptical constellation. This configuration puts at least one satellite over the United States at all times; each satellite spends about 16 hours a day over the continental United States. Naturally, the SIRIUS and XM systems are incompatible with each other.

Interestingly, both systems broadcast digital audio and a digital data stream for each channel. The data stream provides information about the music that's playing, including artist and song information. To display this information, most SIRIUS and XM radio receivers have big multi-line LCD displays.

ON THIS DAY: EDWIN HUBBLE BORN (1889)

Noted American astronomer Edwin Hubble was born on this date in 1889. He is considered the founder of extragalactic astronomy, and identified the first known objects outside the Milky Way galaxy. The Hubble Space Telescope was named in his honor.

FACT OF THE WEEK

As of mid-2005, XM claims 4.3 million subscribers and expects to close out the year with 5.5 million users; they're signing up close to 200,000 new subscribers a month. Competitor SIRIUS claimed 1.1 million subscribers at the beginning of the year, but hopes to close out the year with 2.5 million users. These services are getting popular!

November 21, 2006

TUESDAY

THIS WEEK'S FOCUS
SATELLITE RADIO

XM SATELLITE RADIO

XM Radio (www.xmradio.com) is the most popular satellite radio system in the United States. Basic programming runs $12.95 per month, which gives you more than 150 total channels, including close to 70 commercial-free music channels.

The programming divides up this way. There are 7 channels devoted to pop music by decade ('40s, '50s, '60s, '70s, '80s, and '90s); 7 country, folk, and bluegrass channels; 9 pop music channels (including adult contemporary and showtunes); and 3 Christian music channels. You also get 14 rock music channels (from classic to alternative to heavy metal to jam bands), 7 hip-hop and urban channels (including classic soul), 5 jazz and blues channels, 3 new-age and "lifestyle" channels, and 4 dance music channels (including disco!). Adding to the diversity, you'll also find 4 Latin music channels, 3 world music channels (including reggae), 3 classical music channels, and 2 kids' channels (including Radio Disney).

On the non-music front, you get 12 news channels (including CNN, Fox News, BBC, and XM Public Radio), 12 talk and entertainment channels (including progressive and conservative channels, plus a channel just for truckers), 6 sports channels (including two NASCAR channels), and 9 college sports channels (AAC, Big 10, and Pac 10). There are also 3 comedy channels and 20 local traffic/weather channels offering real-time info for selected metropolitan areas.

In terms of music programming, I find XM a little more adventurous than SIRIUS. I particularly like the deep cuts played on the Soul channel and the variety of classic and emerging artists on the Hear Music channel. (Yeah, it's sponsored by Starbucks.) There's also a lot to like in terms of sports; XM has deals to broadcast college football and basketball, major league baseball, and NASCAR and IRL races. Audio-wise, XM sounds just a tad better than SIRIUS to my ears; it doesn't suffer from the audio compression that the competing system sometimes exhibits.

ON THIS DAY: EDISON INVENTS PHONOGRAPH (1877)

On November 21, 1877, Thomas Edison announced the invention of his "talking machine," a tinfoil cylinder recorder. He appears to have envisioned it as a machine for business dictation.

GADGET OF THE WEEK: DELPHI XM MYFI

Delphi's MyFi is, simply put, a portable XM radio receiver. It includes a built-in audio recorder that can store up to five hours of XM programming and play it back when you're indoors or out of satellite range. Buy it for $299.99 from www.xmradio.com/myfi/.

November 22, 2006

WEDNESDAY

THIS WEEK'S FOCUS
SATELLITE RADIO

SIRIUS SATELLITE RADIO

SIRIUS (www.sirius.com) is currently number two in the U.S. satellite radio market, with about half the number of subscribers as the competing XM service. For $12.95 per month you get more than 130 total channels, including 65 channels of commercial-free music.

The programming divides into 13 pop music channels (including soft-rock and oldies channels), 17 rock music channels (including alternative and classic rock), and 6 country, folk, and bluegrass channels. There are also 4 hip-hop channels, 5 R&B/urban music channels, and 6 electronic/dance music channels, plus 6 jazz/standards channels, 3 classical music channels, and 5 Latin/world music channels.

Non-music programming includes 13 news channels (including CNN, Fox News, BBC, three NPR channels, and three regional Weather Channel channels), 8 sports channels, and 24 talk/entertainment channels (including two conservative channels, two liberal channels, two comedy channels, and Radio Disney). Like XM, SIRIUS also offers news/traffic info for 20 major metropolitan areas.

To my tastes, SIRIUS is a little bit more attuned to the news radio junkie than is XM Radio; SIRIUS is also adding shock jock Howard Stern to its schedule, which will be a big drawing card. Music-wise, the programming tends to be a little safer than what XM offers, with some notable exceptions (such as Little Steven's Underground Garage). On the sports front, SIRIUS broadcasts NFL, NBA, and NHL games, various college sports, and English soccer games. It's great variety!

ON THIS DAY: S.O.S. ADOPTED (1906)

A century ago today, the International Radio Telegraphic Convention in Berlin adopted the S.O.S. radio distress call. Seventy years later, ABBA turned it into a worldwide pop hit!

HARDWARE OF THE WEEK: SOUNDBLASTER AUDIGY 4

Expand your PC's sound capabilities with the Soundblaster Audigy 4. This is an outboard sound box that replaces your internal sound car, adding 24-bit audio, 7.1-channel surround sound, tons of audio inputs and outputs, and a whole lot more. Buy it for $299.99 from www.soundblaster.com.

November 23, 2006

THURSDAY

THIS WEEK'S FOCUS
SATELLITE RADIO

SATELLITE RADIO IN THE CAR

Chances are, your current car radio doesn't have XM or SIRIUS radio built in (though some car makers are including satellite radio as an add-on amenity), which means you need some sort of add-on unit if you want to go the satellite radio route. Fortunately, there are a lot of good models available, for both systems; here are some of my favorites.

For my money, the best solution for XM listeners is the XM Commander, an all-in-one satellite receiver package. You get the XM Commander itself, which is the controller and display for the system; a separate remote control, for operating the system from the rear seat; a hide-away tuner box; a low-profile satellite microantenna; and all the cables and mounting accessories you need to complete the installation. The XM Commander installs in your dash, the tuner box can be put under a seat or in the trunk, and the antenna goes on the outside of your vehicle. It all integrates seamlessly into your car's current sound system, via either a direct line-in connection or transmitted to your radio over an unused FM channel. The XM Commander system costs just $169.95; learn more at www.xmradio.com/commander/.

If you don't need a permanent satellite radio installation, check out Delphi's XM SKYFi2. This is a transportable receiver that you can carry between your car and your home or office. The SKYFi2 connects to your car radio via FM and includes a 30-minute recording buffer so you won't miss any of your favorite XM programming. Car and home adapter kits are available, as is the tabletop SKYFi Audio System. Buy it for $119.99 from www.xmradio.com/skyfi2/.

If you prefer SIRIUS to XM, consider the Sportster, a transportable receiver similar to the XM Roady2. It connects to your car radio via FM and includes its own wireless remote control. Car, home, and boombox kits are available; it sells for just $99.99 from www.sirius.com.

And both SIRIUS and XM now make dealer-installable boxes that add satellite radio to compatible car audio systems. See your aftermarket installer for more details.

ON THIS DAY: PENCIL SHARPENER PATENTED (1897)

On this date in 1897, black American inventor John Lee Love received a patent for his pencil sharpener. It's the same kind of portable, hand-operated pencil sharpener still in use today.

SOFTWARE OF THE WEEK: REPLAY RADIO

Recording your favorite Internet radio stations just got a whole lot easier. Replay Radio lets you capture Internet radio programming, broadcast radio, XM radio, and podcasts, and then save them to your hard disk or MP3 player. Buy the program for $49.95 from www.replay-video.com/replay-radio/.

November 24, 2006

FRIDAY

THIS WEEK'S FOCUS
SATELLITE RADIO

SATELLITE RADIO AT HOME

Listening to satellite radio at home requires installing some sort of satellite radio antenna. Most antennas can be positioned indoors, near a properly facing window. (If that doesn't work, you'll need to run wire outside and position the antenna there.)

My favorite home satellite receiver is Polk Audio's XRt12. This is an XM component tuner for your home theater system. It sits in your equipment rack like any other audio component and connects to your audio receiver; from then on it's like using any other audio component. It's XM radio in your living room, at the highest possible quality. Buy it for $299.95 from www.polkaudio.com.

If a table radio is more your style, check out the P7131 XM radio from Grundig/Porsche Design, which combines an AM/FM tuner, shortwave radio, and XM satellite radio tuner, all in a very stylish device. It also has two full-range speakers, a powered subwoofer, and a digital clock radio with "his and her" dual alarms. Find out more at www.grundigradio.com.

Another popular XM radio for home use is Polk's I-Sonic. This puppy receives XM satellite radio, standard analog AM/FM radio, and the new digital AM and FM HD Radio. It also has a built-in CD/DVD player and an auxiliary port you can use to connect your iPod or similar device. Connect it to your TV to watch DVDS, or just settle back to listen to practically any type of audio entertainment available. At $599 it's a bit pricy, however; find out more at www.polkaudio.com.

If you're a SIRIUS fan, you should check out the Tivoli Model Satellite. This is a desktop model, like the famed Tivoli Model One, that combines an AM/FM radio with a SIRIUS satellite tuner, which lets you listen to all that great SIRIUS programming in your bedroom or den. Tivoli's sound can't be beat; buy it for $299.99 from www.tivoliaudio.com.

ON THIS DAY: *ORIGIN OF SPECIES* PUBLISHED (1859)

On November 24, 1859, Charles Darwin's *The Origin of Species by Means of Natural Selection* was first published. Darwin argued that species are the result of a gradual biological evolution in which nature encourages, through natural selection, the propagation of those species best suited to their environments. (And he was right, you know.)

WEBSITE OF THE WEEK: XMFAN.COM

If you're an XM Radio fan, check out XMFan.com, a site run by and for XM aficionados. The site offers news, reviews, and a lively bunch of user forums. It's located at www.xmfan.com.

November 25/26, 2006

SATURDAY/SUNDAY

THIS WEEK'S FOCUS
SATELLITE RADIO

HD RADIO: DIGITAL RADIO WITHOUT THE SATELLITE

Satellite radio's great, but sometimes you want a local perspective—which is why terrestrial radio isn't going away. In fact, traditional AM and FM radio is getting better than ever, thanks to the digital development called HD Radio.

HD Radio adds static-free digital sound to existing AM and FM radio stations. Thanks to digital transmission, AM sounds more like traditional FM radio, and FM now has CD-quality sound. HD Radio also lets broadcasters include various data services in their signal, such as song titles, artist names, traffic updates, weather forecasts, sports scores, and other such information displayed on a scrolling text screen.

Unlike satellite radio, HD Radio is free for anyone to receive—it beams out over the airwaves just like traditional analog radio. You do need a special HD Radio receiver, however, but no additional antenna is required.

Where can you get an HD Radio receiver? Well, the list of manufacturers is still kind of small, but growing. Tabletop radios for the home are currently available from Boston Acoustics, Polk Audio, Radiosophy. Yahama is including HD Radio in some of their audio/video receivers. And aftermarket HD Radios for your car are available from Alpine, JVC, Kenwood, Panasonic, and Sanyo. Look for more variety in the future—including OEM automotive radios from Delphi and Visteon.

You can learn more about HD Radio at www.ibiquity.com/hdradio/. The site also includes a list of stations currently broadcasting digitally. (For listeners of my radio program, station KFI in Los Angeles is licensed to broadcast in HD, but isn't doing so as yet. Stay tuned!)

ON THIS DAY: FIRST LION IN AMERICA (1716)

On November 26, 1716, the first lion to be seen in America was exhibited by Captain Arthur Savage (that's really his name!) at his house on Brattle Street in Boston. The lion was supposedly tamed for exhibition.

BLOG OF THE WEEK: ORBITCAST

Here's a blog that's out of this world. (Sorry about that!) More precisely, Orbitcast is a blog about all things related to XM and SIRIUS satellite radio. Take a look at www.orbitcast.com.

November 27, 2006

MONDAY

THIS WEEK'S FOCUS
ONLINE SHOPPING

WHY ONLINE SHOPPING IS (GENERALLY) SAFE

Some Internet newbies have reasonable concerns about whether the information they provide online is secure. They fear that once their credit card numbers are in cyberspace, anyone can grab them. Some users fear that information will be hijacked between themselves and the retailer's website; other users fear that the website itself can be hacked and the numbers stolen.

There's little reason to fear online shopping, however. While it is possible for a dedicated hacker to intercept the transmission of private information over the Internet, this type of illicit activity is extremely rare—because it's extremely difficult. If someone wants to steal a few credit card numbers, it's easier to use a low-cost radio scanner to listen in on cordless phone calls, or to go dumpster-diving for carbons behind a local restaurant. Hacking into secure Internet transmissions is a lot more work.

As to stealing credit card numbers from website databases, it happens—although rarely. (Those rare instances receive a lot of publicity, however.) In almost all cases, security breaches of this magnitude are implicitly the cause of the site itself for not having adequate security measures on hand. Although even the biggest sites can be hacked, there is nothing you, as a user, can do to keep this from happening; it's the responsibility of the website to protect the information it stores on its own servers.

Even if the worst happens and your credit card information *is* stolen, you're still protected. That's because your credit card company bears the brunt of any fraudulent activity perpetrated on your account; you might be liable for a small amount (typically $50), but anything more than that amount is the credit card company's responsibility.

The bottom line is that providing your credit card information to a secure website is much safer than handing your credit card to a complete stranger dressed as a waiter in a restaurant or giving it over a cordless phone. In addition, all major credit card companies limit your liability if your card gets stolen, whether that's on the Web or in the so-called real world. So, go ahead and use your credit card online—there's little to worry about!

ON THIS DAY: ELECTRIC MOTOR INVENTED (1834)

On this date in 1834, Thomas Davenport invented the first commercially successful electric motor. Davenport used the motor to power a number of his other inventions.

FACT OF THE WEEK

Google has issued a list of the most popular product searches for 2004 on its Froogle site (froogle.google.com). The most searched-for consumer electronics product was the **iPod**; the top clothing query was for **bikini**; the most popular computer goods search was **tablet pc**; the top query in the sports and hobbies category was for **poker chips**; the most searched-for brand name was **louis vuitton**; and the most popular shoe search was **ugg boots**.

November 28, 2006

TUESDAY

THIS WEEK'S FOCUS
ONLINE SHOPPING

PRODUCT REVIEWS ONLINE

If you're like me, you find it extremely useful to read what other users have to say about an item before you decide to buy. Your fellow consumers will give you the unvarnished pros and cons based on actual use, and tell you whether or not they think the product was a good deal.

There are many sites on the Web that provide forums for customers' product reviews. Here's the best of the bunch:

- **Amazon.com** (www.amazon.com). Yeah, I know, Amazon.com is a retailer, but the fact remains that Amazon hosts one of the largest databases of customer product reviews on the Internet.
- **ConsumerREVIEW.com** (www.consumerreview.com). This site collects customer reviews from a bevy of specialty sites for audio equipment, automobiles, computers, sporting goods, PC and video games, digital cameras, and the like.
- **ConsumerReports.org** (www.consumerreports.org). This is the online arm of the venerable *Consumer Reports* magazine, which presents independent tests and reviews of all types of products.
- **ConsumerSearch** (www.consumersearch.com). This site aggregates a variety of professional reviews about each product listed; the reviews are written by pros, not by customers.

- **Epinions.com** (www.epinions.com). This is my favorite product review site. What's great about this site is that users can write reviews about virtually anything—not just products, but also services, retailers, and locales.
- **ReviewFinder** (www.reviewfinder.com). This site offers links to reviews of various types of electronic equipment. It's a good gateway to lots of other reviews across the Internet.

ON THIS DAY: MARINER 4 LAUNCHED (1964)

The Mariner 4 spacecraft was launched from Cape Kennedy on this date in 1964. The Mariner 4 was destined to be the first satellite to transmit close-up photographs of Mars, flying as close as 6,118 miles to the planet's surface.

GADGET OF THE WEEK: GODOGGO AUTOMATIC FETCH MACHINE

Here's a gizmo for your doggie: an automatic fetching machine. The GoDogGo looks like one of those automatic tennis ball shooting machines, with a big green bucket of balls on the top. All you have to do is train Fido to fetch the tennis balls—and, if you're really good, to put the balls back in the bucket! Buy it for $149.95 from www.buygodoggo.com.

November 29, 2006

WEDNESDAY

THIS WEEK'S FOCUS
ONLINE SHOPPING

COMPARISON SHOPPING ONLINE

The most important development in the history of online shopping is the creation of the price comparison site. This is a site that uses software called a *shopping bot* (short for *robot*) to search a large number of online retailers for current prices on available products. This product and pricing information is used to create a large database that you, the consumer, can access at will. In essence, you let the price comparison site do your shopping for you; all you have to do is evaluate the results.

When it comes to price comparison sites, there are five that consistently attract the most consumer traffic and get my personal approval:

- **BizRate** (www.bizrate.com). This is one of the better sites in terms of finding additional bargains and promotions. I also like the site's merchant reviews and ratings.
- **Froogle** (froogle.google.com). This is Google's shopping search engines, unique in that it's completely objective; Froogle doesn't take money for its listings, as the other sites do.
- **mySimon** (www.mysimon.com). Even though mySimon doesn't search nearly as many merchants as some of the other sites, it's quite easy to use—and quite popular.
- **Shopping.com site** (www.shopping.com). This site, recently acquired by eBay, not only offers price comparisons, it also incorporates customer reviews from Epinions.com. It's my personal favorite of all the price comparison sites.
- **Yahoo! Shopping** (shopping.yahoo.com). This is a full-featured shopping search engine, complete with some pretty neat product comparison features.

All these sites operate in pretty much the same fashion. You can either browse through product categories or search for specific products, brands, or item numbers. Results from participating merchants are then displayed; you can sort this list by merchant, price, and so on. It's a great way to comparison shop without visiting all the individual sites yourself!

ON THIS DAY: PONG ANNOUNCED (1971)

On November 29, 1971, Atari Corporation announced PONG, an electronic table tennis game designed for arcades. Atari eventually ported PONG to the home, where it became the first hugely successful home video game.

HARDWARE OF THE WEEK: MSI BLUETOOTH STAR USB HUB

Here's a USB hub that's more than a USB hub. Yes, the Bluetooth Star hub provides three extra USB connections, but it's also a Bluetooth receiver that supports up to seven separate Bluetooth wireless devices. Buy it for $50 from www.msicomputer.com.

November 30, 2006

THURSDAY

THIS WEEK'S FOCUS
ONLINE SHOPPING

USING ONLINE COUPONS

When you're shopping online, one way to save money is to use an online coupon. Or, to be more accurate, an online coupon *code* (sometimes called a promotional code), because there's really no way to use a printed coupon with an online merchant. What you get instead is a code you can enter when you check out at an online merchant; the coupon savings are deducted from your order when you check out.

There are a large number of sites that specialize in online coupons. How you apply an online coupon depends on which site you visit. Some online coupon sites totally automate the process; all you have to do is click a link on the coupon site and you're taken to the appropriate product page at a participating retailer. Other sites just give you the coupon codes for you to apply on your own—which typically means entering the promotional code sometime during the checkout process.

Here are some of the most popular sites that offer online coupons and promotions:

- Bargain Boardwalk (www.bargainboardwalk.com)
- Bargain Shopping.org (www.bargainshopping.org)
- CouponMountain (www.couponmountain.com)
- Daily eDeals (www.dailyedeals.com)
- DealofDay.com (www.dealofday.com)
- eCoupons (www.ecoupons.com)
- KovalchikFarms.com (www.kovalchikfarms.com)
- MyCoupons (www.mycoupons.com)
- Specialoffers.com (www.specialoffers.com)

And here's a bonus tip. Sites like www.couponpages.com and www.hotcoupons.com let you print out physical coupons you can use at local retail stores. It's not online shopping—but it will save you money!

ON THIS DAY: MASON JAR PATENTED (1858)

On this date in 1858, John Landis Mason received a patent for the jar that bears his name. The Mason jar was a shoulder-seal jar with a zinc screw cap; the cap screwed down onto the shoulder of the jar to form a tight seal.

SOFTWARE OF THE WEEK: BEST PRICE

Best Price is a shopping bot that runs from your own PC. Just enter your search criteria and Best Prices searches 16 of the most popular comparison shopping sites, delivering you the best prices out there. It's a free download, available at www.winshare.com/bpindex.htm.

December 2006

December 2006

SUNDAY	MONDAY	TUESDAY	WEDNESDAY	THURSDAY	FRIDAY	SATURDAY
					1 1913 First U.S. drive-in service station opens in Pittsburgh, PA	**2** 1942 First controlled nuclear chain reaction demonstrated
3 1967 First successful human heart transplant	**4** 1996 First electric vehicle to be mass-produced (the EV1, manufactured by General Motors) introduced	**5** 1879 Patent issued for first automatic telephone switching system	**6** 1877 First sound recording made by Thomas Edison	**7** 1999 Recording Industry Association of America (RIAA) sues Napster for alleged music piracy	**8** 1980 Beatles legend John Lennon shot and killed in New York	**9** 1987 Windows 2.0 released
10 1993 Hubble telescope repaired by astronauts on space shuttle Endeavor	**11** 1884 First dental anesthesia (nitrous oxide) used on a patient	**12** 1899 Golf tees are patented in Boston, MA	**13** 1977 Ethernet is patented by Xerox	**14** 1967 DNA created in a test tube	**15** 1854 The first street cleaning machine used in Philadelphia, PA	**16** 1917 Arthur C. Clarke, author of *2001: A Space Odyssey*, born
17 1903 Orville and Wilbur Wright makes first flight	**18** 1946 Steven Spielberg born	**19** 1975 The Altair 8800, considered by many to be the first personal computer, goes on sale	**20** 1996 Carl Sagan dies	**21** 1898 Radium discovered	**22** 1882 First string of Christmas lights made by Thomas Edison's assistant	**23** 1947 First transistor created
24 ChristmasEve 1936 First radioactive isotope medicine is administered	**25** Christmas Day 1642 Sir Isaac Newton born	**26** 1982 *Time* magazine names the computer "Man of the Year"	**27** 1845 Ether anesthetic first used in childbirth	**28** 1869 Chewing gum is patented in Mount Vernon, OH	**29** 1987 Russian cosmonaut ends record-setting 326-day stay in space	**30** 1924 Astronomers discover another galaxy in addition to the Milky Way
31 1938 Alcohol breath test device invented						

December 1, 2006

FRIDAY

THIS WEEK'S FOCUS
ONLINE SHOPPING

REDUCE YOUR SHIPPING COSTS

The one drawback to shopping online is that the item must be shipped to you—and shipping costs money. If you don't watch it, you'll end up paying a lot by buying online and having it shipped to you than you would by buying it—and picking it up yourself—locally.

Even worse, some less-reputable dealers lure you in with an unusually low price on a product, and then sock it to you with an unjustifiably high shipping/handling charge. I was recently shopping for a DVD player and found one merchant charging $10 less for the unit than other competing retailers. The catch? That retailer charged $24.95 for shipping/handling, when competitors were offering free shipping. So, I could "save" $10 on the product but end up spending a total of $15 more on the total order. Yikes!

What you want to look for are those merchants that offer some sort of deal on shipping. You might not be able to find out shipping charges until you enter a merchant's checkout, but that still lets you back out if the charges are too high. I like those merchants (such as Amazon.com) that offer free shipping if your order is more than a certain dollar amount. Free shipping is the best deal there is!

Another thing to look for are sites that offer more than one option for shipping your order. You can choose to ship each item in your order individually; this will get you each item faster if one of the items is back ordered or pre-ordered but cost you more in shipping costs. Or, you can choose to combine all the items into a single shipment, which might (or might not) be slower but will save you on shipping. Some sites even let you combine multiple orders into a single order for shipping purposes—which could earn you free shipping, depending.

Finally, some retailers offer yet another way to save money on shipping by choosing a slower shipping method. If you need to get the item right away, you can choose a faster and more expensive shipping option. But if you're in no hurry, choose the slowest method and pay less—or, depending on the retailer, nothing at all.

ON THIS DAY: FIRST WHITE HOUSE TELEPHONE (1878)

On the first day of December in 1878, Alexander Graham Bell himself installed the first telephone in the White House. President Rutherford B. Hayes made his first outgoing call to Bell, who had moved to another phone 13 miles away; Hayes's first words instructed Bell to speak more slowly.

WEBSITE OF THE WEEK: SALESHOUND

Aside from shopping online, you can also use the Web to find the best deals at traditional bricks and mortar stores. SalesHound (www.saleshound.com) lists advertised sales at hundreds of national and local retailers. All you have to do is enter your ZIP Code, and SalesHound displays all the sales and deals it knows about.

December 2/3, 2006

SATURDAY/SUNDAY

THIS WEEK'S FOCUS
ONLINE SHOPPING

TRACKING YOUR SHIPMENT

What happens after you place your order is the hard part—you wait. Although it's nice to get that immediate gratification from shopping *right now* online, you don't get the same immediate gratification in terms of actually receiving what you purchased. Depending on the speed of the retailer and the shipping method chosen, you might have to wait anywhere from a few days to a few weeks to receive your order. Try to be patient.

If an online retailer provides a tracking number for the order you placed, you can use this number to check the status of your delivery. The retailer might provide a direct link to the shipping service's tracking page, or you might have to manually enter the tracking number at the shipping service's site. In any case, being able to track your shipment enables you to know when you can expect to receive the package at your door.

If you need to enter a tracking number manually, here are the URLs for the major shipping services' tracking pages:

- United States Postal Service: www.usps.com/shipping/trackandconfirm.htm
- UPS: www.ups.com
- FedEx: www.fedex.com

Just go to the appropriate page, enter the tracking number, and view the progress of your shipment.

And here's a bonus tip: Google provides tracking information for just about every major shipping service right from their main search box. Just enter the tracking number and click the Google Search button. Google will figure out which shipping service the number applies to, look up the delivery status, and display it for you. Pretty neat!

ON THIS DAY: JOHN BACKUS BORN (1924)

John Backus was part of the IBM programming team that developed the FORTRAN programming language. FORTRAN was the first successful high-level programming language. Backus was born on December 3, 1924.

BLOG OF THE WEEK: SHOPPING BLOG

Well, you can probably tell from the name that Shopping Blog is all about shopping—in particular, hot shopping trends and new products. See it for yourself at www.shoppingblog.com.

December 4, 2006

MONDAY

THIS WEEK'S FOCUS
GEEK GIFT GUIDE

SHOPPING FOR YOUR FAVORITE GEEK

Shopping for a tech geek isn't as easy as it sounds. No doubt you're tempted to head over to the local Sharper Image store and buy something flashy. While that might work (nothing against Sharper Image—I personally like their stuff), you're probably not going to get off that easy. For one thing, Sharper Image merchandise, as neat as it might look to you, typically isn't cutting edge. In addition, Sharper Image stuff isn't particularly technical in nature. No, you'll probably have to shop elsewhere.

Okay, you're thinking. Maybe you'll head over to CompUSA or Best Buy. Good try, but not the best idea. That's because these stores carry merchandise targeted at the general consumer, and your tech geek is definitely not a general consumer. The types of items your geek friend drools over are unlikely to be found in the aisles of a big box store, or in a mall.

Geek stuff is highly specialized, and typically found at specialized retailers. In the bricks and mortar world, you're in luck if you have a Fry's Electronics in your town; Fry's is pretty much a tech-lover's superstore, with more oddball computer peripherals and geek gadgets than you can imagine. I can spend an entire afternoon just browsing the aisles of my local Fry's, considering just how big a power supply I should buy for that old PC I'm fixing up. It's that kind of place.

If you don't have a Fry's nearby, you'll probably end up doing your geek gift shopping from a catalog or online. If that scares you—well, your geek pal probably scares you, too. Get over it.

Shopping online has its benefits—chief of which is that you can usually find what you're looking for. For example, many geek gifts can be purchased directly from the manufacturer. I also like Gadget Universe (www.gadgetuniverse.com) and ThinkGeek (www.thinkgeek.com), both of which carry a lot of really cool stuff. Hey, it's the Internet—do a little searching and see what you find.

As for precisely what to buy the geek on your list, it's best to ask rather than guess. What looks cool and cutting edge to you might be so yesterday's news to the up-to-date tech geek. And if you need some gift ideas, check out my companion book, *Leo Laporte's 2006 Gadget Guide*.

ON THIS DAY: OMAR KHAYYAM DIES (1131)
Persian poet, mathematician, and astronomer Omar Khayyam passed away on this date in 1131. He studied Euclidian geometry, contributed to the theory of parallel lines, led work on compiling astronomical tables, and contributed to the reform of the Persian calendar.

FACT OF THE WEEK
MasterCard research recently found that nearly half of all U.S. consumers carry $20 or less in their wallets. Fully 86% of the people they surveyed said they would like to lessen the number of times they use cash.

December 5, 2006

TUESDAY

THIS WEEK'S FOCUS
GEEK GIFT GUIDE

SHOPPING FOR A COMPUTER GEEK

What distinguishes a true computer geek from the average computer user is the type of computer that he has—how it performs and how it's "tricked out" with various types of peripherals. While you might be happy with any $600 out-of-the-box PC you can pick up at Best Buy, the computer geek spends two to three times as much on the basic unit, and then upgrades key components for better performance or more specialized needs. For the computer geek, it's all about performance; speed rules.

If you're buying a complete computer system for a computer geek, be prepared to spend some big bucks. You'll want a system with the fastest Intel or AMD processor available (somewhere north of 3GHz); at least 1GB of RAM; a huge hard drive (200GB bare minimum); and a big LCD monitor (17" or larger). If the system you're looking at isn't quite fast enough or big enough out of the box, feel free to have the retailer or manufacturer customize it as necessary.

It also helps if the PC's system unit is cool looking. Plain beige or black cases are yesterday's news; computer geeks like cases with see-through sides and flashing lights and groovy paint jobs. For the uninitiated, just look at the units sold by Alienware, Falcon Northwest, and other specialty manufacturers. That kind of hip design impresses.

So, you want a fast system with lots of storage and a cool-looking case. Unfortunately, it won't come cheap. Be prepared to spend two grand or more on this type of system—and then have your geek recipient complain that you skimped on this, that, or the other.

It's a lot less expensive to buy the computer geek accessories for his existing system. I'd stick with external peripherals, of which there are a lot to choose from. A bigger monitor is always good, as is a wireless keyboard and/or mouse. You also can't go wrong with a USB expansion port or a USB flash memory device. Just remember, style is every bit as important as performance; make sure that whatever it is you buy looks fairly cool.

ON THIS DAY: FOLDING CHAIR PATENTED (1854)

On this date in 1854, Aaron H. Allen was issued a patent for a folding chair as an "Improvement in Self-Adjusting Opera-Seat" for theaters and other public building. The pivoted seat was constructed to assume and retain a vertical position when the occupant rises from it.

GADGET OF THE WEEK: SCOOBA ROBOTIC FLOOR WASHER

The folks behind the Roomba robotic vacuum cleaner have just come out with a similar product for hard floors, called the Scooba. The Scooba is a self-propelled cleaning robot that vacuums, washes, and dries wood and tile floors in a single pass. It uses a specially formatted cleaning solution to get your floors sparkling clean. Check it out at www.irobot.com.

December 6, 2006

WEDNESDAY

THIS WEEK'S FOCUS
GEEK GIFT GUIDE

GIFTS FOR THE ROAD WARRIOR GEEK

Geeks need lots of gadgets to keep them in touch when they're away from home. Even if he's just going down the street to the local video store (although why would he—he's probably a NetFlix customer!), there's still the need to check email messages, make a few cell calls, and maybe drop by a WiFi hotspot for some quick surfing. You can never be too in touch!

The essential road warrior gadget is the cell phone. Even if your pal already has one, you can always give him a newer, hipper one. Look for a model with multi-band capability, a multi-megapixel digital camera, capability to play back digital audio and video files, and lots of cool ringtones and games. It also helps if the phone is small and has a hip design, like the Motorola Razr. Functionality (and lots of it) comes first, but you can't ignore the cool.

And while we're on the topic of cell phones, don't forget the accessories. I'm talking car kits, carrying holsters, wireless headsets, and the like. There's a lot to choose from here, and they all make great stocking stuffers.

Higher-end geeks prefer smartphones to cell phones. These are phones with built-in PDA functions—or PDAs with cell phone functions. It doesn't matter. The point is, these uber-gadgets are uber-cool and uber-functional. They're also uber-expensive, so this type of gift is probably reserved for those you especially care for.

Then, of course, there's the PDA itself. If your geek pal already has a cell phone and doesn't want a smartphone (they are big and bulky), give him a PDA instead. Make sure you find out which operating system he prefers, Palm or PocketPC—they're not compatible.

Assuming your road warrior geek already has a notebook PC (and that you don't want to spend the bucks to buy him a new one), consider all the wonderful notebook accessories you can give. I'm talking about cases, desktop stands and docking stations, security devices, extra batteries, and the like. And, as with desktop computer geeks, road warriors just love those USB keychain storage devices. You can't have too many of them!

ON THIS DAY: FIRST SOUND RECORDING (1877)

On December 6, 1877, Thomas Edison recited "Mary Had a Little Lamb" into his new recording machine, thus creating the first sound recording of the human voice.

HARDWARE OF THE WEEK: BUFFALO TERASTATION STORAGE DRIVE

If you have a whole-house network, you can use a single network attached storage drive to back up all the PCs on your network. Buffalo's TeraStation NAS is easy to set up and practically invisible when in use, and it comes in three extra-large capacities—0.6 terabytes, 1.0 terabytes, or 1.6 terabytes. Prices range from $799.99 to $1,999.99.

December 7, 2006

THURSDAY

THIS WEEK'S FOCUS
GEEK GIFT GUIDE

GIFTS FOR THE HARD-DRIVING GEEK

Hot rodders like to customize their cars and fine-tune all the equipment under the hood. In contrast, hard-driving tech geeks like to customize the equipment in the car's passenger compartment—in particular, the car's audio system.

High-tech car owners don't settle for factory sound; they yank out the OEM system and install a high-end aftermarket audio system. That means a new in-dash receiver, as well as better (and more) speakers all around. Look for a unit with a really cool-looking faceplate, ideally one capable of displaying moving graphics. Something relatively theft-proof would be nice, as well.

Of course, one advantage to going with a custom car audio system is that your geek pal isn't limited to simple AM/FM/CD sources. You can help your pal expand his system by giving an aftermarket XM or SIRIUS satellite radio receiver, or an FM transmitter for your pal's iPod. If an entire system is in the budget, consider one with DVD playback capabilities. This type of system has a fold-out LCD display in the dash, for watching movies when parked.

Another good auto-related gift is an in-car GPS receiver. An automotive GPS unit comes with either a windshield or dashboard mount and is typically battery powered, although some can also tap into the car's DC power. Most car GPS devices come specially configured for road travel, with a variety of points of interest—gas stations, restaurants, hotels, ATMs, and so on—preprogrammed into memory. You also get built-in road maps, of course, as well as those turn-by-turn driving instructions. It's a tech geek's dream device.

Finally, consider giving some sort of automotive computers. These are devices that let you monitor your car's performance, by reading data from the car's control chip. Some of these devices provide diagnostic readouts; others offer monitoring functions that are perfect if your geek friend wants to track how his kids (or spouse) are driving. It's all very high-tech, and very appealing to any cutting edge tech geek.

ON THIS DAY: JET STREAM DISCOVERED (1934)

Pilot Wiley Post discovered the jet stream when he flew his plane *Winnie Mae* into the stratosphere over Bartlesville, Oklahoma. The flight took place in December 1934. In case you're wondering, jet streams are strong wind currents found in the earth's atmosphere at high altitudes. They're thousands of miles long and hundreds of miles wide, and help to move weather patterns around the earth.

SOFTWARE OF THE WEEK: PICASA

Picasa is a fairly robust, easy-to-use, and totally free image management and editing program. It's distributed by Google and looks and acts a lot like Apple's iPhoto. (If anything, it has a few more features.) It's great for managing all your digital photos; download it from picasa.google.com.

December 8, 2006

FRIDAY

THIS WEEK'S FOCUS
GEEK GIFT GUIDE

GIFTS FOR THE MUSIC-LOVIN' GEEK

I'm not sure why it is, but most tech geeks are also big-time music fans. Many techies are also amateur musicians, and even those that aren't still love to listen to music—so much so that they're never without their tunes.

Of course, a tech geek carries his tunes around with him on a portable music player. If your geek pal doesn't yet have one of these gizmos, there's your gift right there. Head down to your local electronics store (or let your mouse do the walking) and plop down the bucks for an iPod or something similar. While you can't go wrong with the iPod proper, your geek pal might prefer something a little less Apple-trendy, like those units offered by Creative Labs. If he wants to ride with the Apple crowd, you can choose between different capacity iPods, or the smaller iPod Nano or even smaller iPod Shuffle.

How big a unit you buy depends on how many tunes your friend wants to carry around with him. The smaller flash memory players, such as the iPod Shuffle, can hold 100–500 songs. The mid-size MicroDrive players, such as the iPod Mini, can hold 2,000–3,000 songs. The larger hard drive players, such as the iPod proper, can hold 10,000–20,000 songs. Make your choice accordingly—and remember, you can never have too much storage capacity.

If your pal already has a portable music player, there are lots and lots of accessories you can give. I did several days worth of articles on iPod accessories alone earlier in the book, so you know what I'm talking about—cases, holsters, battery packs, add-on cameras, voice recorders, FM transmitters and car kits, external speakers, you name it. Any self-respecting portable audio geek has a whole bag full of accessories for his portable music player, so the more the merrier.

Finally, you can't go wrong when you give a good set of earphones. Make sure you find out whether your pal prefers the smaller earbud or larger traditional headphone design, and give accordingly. (Or, if your pal does a lot of airline traveling, fork over the bucks for a good set of noise-canceling headphones; this is an indispensable accessory for the music-loving road warrior.)

ON THIS DAY: ECKERT-MAUCHLY COMPUTER CORP. INCORPORATED (1947)

On December 8, 1947, J. Presper Eckert and John Mauchly, the inventors of the ENIAC computer, incorporated their new company. The Eckert-Mauchly Computer Corporation was responsible for developing the BINAC and UNIVAC computers; the company was acquired by Remington Rand in 1950.

WEBSITE OF THE WEEK: CRAIG'S LIST

If you don't know about Craig's List yet, you're behind the curve. Craig's List is a kind of online classified ad site, complete with items and services for sale, personal ads, job listings, and discussion forums. Find the site for your city at www.craigslist.org.

December 9/10, 2006

SATURDAY/SUNDAY

THIS WEEK'S FOCUS
GEEK GIFT GUIDE

GIFTS FOR THE HOME THEATER GEEK

When your geek pal moves into the living room, the world of potential tech gifts gets a whole lot bigger. That's because there's always something fun to buy for that ever-expanding home theater system.

First, consider whatever new component your pal has been thinking of replacing. That might mean a new audio/video receiver, or DVD player, or—if you're really flush—a big screen TV. Heck, even a small-screen TV is good, if it's flat; an LCD TV for the bedroom would be a great gift.

Of course, every home theater system revolves around the audio/video receiver. If you have some spare cash, it's astounding how good a receiver $500 to $1,000 will buy. If your geek pal's system is more than 3-4 years old, it's a sure bet that you can buy a more powerful, more flexible receiver than what he's currently using. That would be one heck of a gift.

Now, if you really want to throw your pal for a a loop, spring for a Media Center PC, like those offered by HP and Niveus Media. A PC in the living room—you'll win big points for that one. Along the same lines, a digital media hub wouldn't be a bad idea; this is a gizmo that lets him beam music from his main PC to his home theater system, wirelessly. It's a cool idea that all techies will take well to.

Home theater gifts don't have to be big and splashy, however. Universal remotes are always good (look for one he can program from his PC), as are power regulators, surge suppressors, and the like. For that matter, audio/video cables make good gifts; make sure you go for the gold-plated models, of course.

Finally, new speakers will please the heck out of any self-respected techie. You don't have to buy a whole new system, either. Go for a good subwoofer, or even an upgraded center speaker, and you'll see a big smile come Christmas morning. Just don't go for anything cheap and cheesy; buy quality equipment if you want to impress your techie friends.

ON THIS DAY: FIRST NON-SOLAR PLANET DISCOVERED (1984)

On December 10, 1984, the National Science Foundation reported the discovery of the first planet outside our solar system. The planet was orbiting a star some 21 million light years from Earth.

BLOG OF THE WEEK: GIZMODO

This is my favorite blog in the whole world. I read it every morning, and then in the afternoon, too. Why? Because Gizmodo is all about tech gadgets—and I love tech gadgets! If *you* love tech gadgets, check out Gizmodo at www.gizmodo.com. It's a fun site!

December 11, 2006

MONDAY

THIS WEEK'S FOCUS
AMAZON.COM

NAVIGATING AMAZON

Amazon.com started out as an online bookstore, and still garners a large part of its revenues from book sales. But over the years, the folks at Amazon have added one new product category after another, to the point where the site now sells almost every type of product imaginable.

Because of everything that Amazon offers, navigating the site isn't always as easy as you'd like. So let's start with the home page and see what's where.

The left side of the home page is where you'll find Amazon's search box. Also on the left are links to Amazon's individual "stores" for specific product categories; you can browse through these stores to find the products you want.

A quick word on searching the Amazon site. If you just enter a query into the search box, you'll get results from every single category that Amazon carries, most of which are irrelevant to your search. It's a much better idea to pull down the category list and select a specific category for your search. This way you can limit your searches to just books, or CDs, or DVDS, or whatever it is that you're really looking for.

Site navigation remains constant from page to page, thanks to the navigation bar found at the top of every page. The navigation bar presents a series of tabs that take you directly to major categories and key sections of the site; click a tab, and subcategories are presented below the bar. To see all the available stores, hover your mouse over the See All Product Categories tab; this displays a nice little pop-up with all the available stores listed. This is the quickest way to navigate from one store to another.

And here's something interesting. After your first visit to the site, Amazon presents a customized version of its home page that features products it thinks you'll be interested in, based on your past browsing and purchasing. It's Amazon personalized just for you—which isn't a bad thing.

ON THIS DAY: LAST MOON MISSION (1972)

On December 11, 1972, the Apollo XVII mission landed on the lunar surface for what would be the final manned mission to the moon to date. For the record books, the last two men to walk on the moon were Gene Cernan and Harrison Schmitt.

FACT OF THE WEEK

Amazon's Inside This Book feature calculates a number of different statistics, one of the most interesting being "words per dollar." This is simply the number of words in a book divided by its selling price. Presumably, the more words per dollar, the better bargain you're getting. For example, Toni Morrison's *Beloved* delivers 5,388 words per dollar; Mark Haddon's *The Curious Incident of the Dog in the Night-Time* delivers 6,438 words per dollar; and William Faulkner's *Absalom, Absalom!* delivers a whopping 12,779 words per dollar.

December 12, 2006

TUESDAY

THIS WEEK'S FOCUS
AMAZON.COM

TODAY'S DEALS

It's easy enough to search or browse through the Amazon site to find specific products. And when you find a product, you'll often find that Amazon has already discounted the price. (That's one of the nice things about Amazon—lots of discounted prices.) But where do you go on the site to find even *bigger* savings?

The first place to find bargains on Amazon is the Today's Deals page. You get there by clicking the Welcome tab and then clicking Today's Deals under the navigation bar. Or, if you want to look at current deals in a particular category, go to that store page and click Today's Deals under that page's navigation bar. (Not available with all stores.)

The Today's Deals page lists all current promotions, discounts, sales, rebates, and the like across the entire Amazon.com site. Big deals are featured directly on this page; you can also browse through the deals offered in each of Amazon's main product categories.

What kinds of deals are we talking about? Well, most of these deals are manufacturer-sponsored promotions, and many apply across multiple items. For example, you might find factory-sponsored rebate promotions, extra discounts, free shipping, and that sort of thing.

The deals found on the Today's Deals page are the equivalent of the in-store promotions you might find at a traditional bricks and mortar retailer. Placement on the Today's Deals page is the online version of an in-store display or endcap. Featured placement on this page is almost always paid for by the manufacturer.

ON THIS DAY: GOLF TEE PATENTED (1899)

What did we do before the invention of the golf tee? The tee was patented on this date in 1899 by George F. Grant, an African-American dentist in Boston. (A *dentist*?)

GADGET OF THE WEEK: ROBOSAPIEN V2 TOY ROBOT

The Robosapien is a surprisingly advanced toy robot, capable of shuffling around aimlessly and making farting noises—just like a lot of programmers I know. The new Robosapian V2 is bigger than the original (almost 32" tall), with more advanced movements. There's even a new stereo sound detection system that allows him to respond and react to noises. Buy one for anyone on your Christmas list (or for yourself) for just $200, from www.robosapienonline.com.

December 13, 2006

WEDNESDAY

THIS WEEK'S FOCUS
AMAZON.COM

CLEARANCE MERCHANDISE AND THE FRIDAY SALE

The Amazon.com Outlet is where you can find all manner of clearance merchandise from Amazon and its retail partners. This section of the Amazon site is the equivalent of an outlet mall, with closeout products presented in all major product categories.

You access the Amazon.com Outlet by clicking the Outlet link when you expand the See All Product Categories tab. To view category-specific outlet stores, click a category link (on the left side of the page).

The products presented in the Amazon.com Outlet are typical clearance items—merchandise that the manufacturer has too many of and has priced to move. These are often prior-year or discontinued items, but can also be regular-line merchandise that the manufacturer is overstocked on. In any case, the Amazon.com Outlet is a prime destination for serious bargain hunters; the selection here is constantly changing.

In addition to the new items offered at clearance prices in the Amazon.com Outlet, Amazon also offers factory-reconditioned products direct from the manufacturers. These are items that have been returned to the manufacturer for whatever reason, and then cleaned and returned to like-new condition. These refurbished items are offered at substantial savings over equivalent new items, and come with full manufacturer warranties. You access the Factory-Reconditioned Products page from the Amazon.com Outlet page; just click Factory Reconditioned under the navigation bar.

And here's another tip, not solely limited to clearance merchandise. Savvy Amazon shoppers know that Friday is the day to find the site's best bargains. That's because every Friday Amazon conducts a one-day sale, called the Friday Sale. Selected merchandise is put on sale for 24 hours only; prices return to normal at midnight (Pacific time) and quantities are limited.

You find the Friday Sale by going to the Amazon.com Outlet page and clicking The Friday Sale under the navigation bar.

ON THIS DAY: WERNER VON SIEMENS BORN (1816)

German scientist and electrical engineer Werner von Siemens was born on this date in 1816. He played an important part in the development of the telegraph industry, and parlayed his success into the multinational business that bears his name.

HARDWARE OF THE WEEK: CREATIVE GIGAWORKS PC SPEAKER SYSTEM

The THX-certified GigaWorks 7.1 system from Creative gives you six two-way satellite speakers (front, surround, and rear lefts and rights), a center channel speaker, and a powerful subwoofer. Naturally, this system features a full complement of digital and analog inputs, and handles all the popular 7.1-channel surround-sound formats. Buy it for $499.99 from www.creative.com.

December 14, 2006

THURSDAY

THIS WEEK'S FOCUS
AMAZON.COM

BUY IT USED—AND SAVE

In addition to the factory-new merchandise that Amazon offers, you can also find a variety of used products from Amazon Marketplace sellers. These are individuals and small merchants who offer their goods through the Amazon site. You place your order with and pay Amazon, but the merchandise is shipped by the individual Marketplace seller.

Where do you find this used merchandise? There are two places, actually.

To view all of Amazon's used merchandise in one place, try the Used page. You get here by going to the Amazon.com Outlet page, and then clicking Used under the navigation bar. From here, used merchandise is available in several different product categories; click the individual category link to see the used merchandise available.

Another way to find a used item is to search for that item new. When you open a product listing page, if a used version is available, you'll see a "buy used" link in the More Buying Choices box. Click the link and you'll see a list of Amazon Marketplace sellers with used copies for sale; you can place your order from here.

All used items you order are added to your standard Amazon Shopping Cart, and you check out and pay for that item as you would with any other Amazon item. That means you don't have to pay the Amazon Marketplace merchant separately; you pay Amazon as you would normally, and Amazon pays the merchant for you. The item is shipped direct from the merchant, however, so you might not get the same shipping terms and deals as you would with Amazon's regular merchandise. Still, it's a great way to get some good deals—and to find merchandise that might not otherwise be available from Amazon proper.

The other place to find used merchandise on the Amazon site is at Amazon Auctions (auctions.amazon.com). Amazon Auctions is an online auction marketplace that works pretty much like eBay. Just remember that when you shop at Amazon Auctions, you're not actually buying an item, you're placing a bid on it. You get to buy the item only if you have the high bid at the end of the auction process.

ON THIS DAY: PLANCK'S QUANTUM PHYSICS (1900)

On this day in 1900, German physicist Max Planck made public his ideas on quantum physics at a meeting of the German Physics Society. Planck demonstrated that in certain situations energy exhibits characteristics of physical matter, something unthinkable at the time.

SOFTWARE OF THE WEEK: FEEDDEMON

FeedDemon is an RSS aggregator for Windows. You use FeedDemon to keep track of all your favorite RSS news, blog, and podcast feeds; whenever there's a new entry, you'll see it. Buy it for $29.95 from www.bradsoft.com/feeddemon/.

December 15, 2006

FRIDAY

THIS WEEK'S FOCUS
AMAZON.COM

SAVE MONEY ON SHIPPING

Okay, now you've learned where to find the best bargains on the Amazon.com site. But there's even more money to be saved, even if you purchase an item at little or no discount.

The easiest way to save money on your order is to cut out the shipping charge. Amazon offers free Super Saver Shipping if your order totals $25 or more, with some qualifications. The big qualification is that selected merchandise is excluded from the offer, specifically apparel, baby products, toys, video games, certain oversize items, products from third-party merchants, and products that don't include the Super Saver Shipping icon. Also know that Super Saver Shipping isn't automatically applied to your order; you have to select this option manually during the checkout process.

One more thing: Selecting Super Saver Shipping will add three to five days to your order. So, this isn't a good option if you need your merchandise quickly. On the other hand, if you don't mind waiting a few extra days, it can represent significant savings.

While we're on the topic of shipping, you don't have to choose Super Saver Shipping to save on shipping charges. When you order more than one item, you can choose to group all the items into a single shipment, which will reduce your shipping charges no matter which shipping method you choose. This is opposed to splitting your order into multiple shipments, which might get you your items faster, but will cost more.

You make your choice during the checkout process, on the Place Your Order page. Select the Group My Items Into as Few Shipments as Possible option to combine your order into a single shipment and save a few bucks—or select the I Want My Items Faster option to ungroup your order, with corresponding higher shipping charges.

And, whichever type of order grouping you select, you also have a choice of shipping speed. You can choose from Standard Shipping (3-7 business days), Two-Day Shipping (2 business days), or One-Day Shipping (1 business day). Obviously, Standard Shipping costs less than One- or Two-Day Shipping. Save money by choosing the Standard Shipping option.

ON THIS DAY: FIRST STREET CLEANING MACHINE (1854)

The world's first mechanized street cleaning machine was put into operation on this date in 1854. The machine utilized a series of brooms attached to a cylinder mounted on a cart; the cylinder was turned by a chain connected to the cart's wheels. The first street to be cleaned? In foul, fetid, fuming, foggy, filthy Philadelphia, of course.

WEBSITE OF THE WEEK: AMAZON WATCH

No, this isn't the Amazon.com store that sells wrist watches. It's a site devoted to defending the environment and indigenous peoples of the Amazon basin. Do the Earth a favor and check it out at www.amazonwatch.org.

December 16/17, 2006

SATURDAY/SUNDAY

THIS WEEK'S FOCUS
AMAZON.COM

BECOME AN AMAZON ASSOCIATE

Here's a tip you can use if you have your own website. You can sign up to be an Amazon Associate and direct users to purchase Amazon items from your site. For every purchased place, you earn a sales commission—between 4% to 10%, according to sales volume and other variables.

The Amazon Associates program is free to join. All you have to do is sign up for the program, put a few links to Amazon on your website, and let Amazon do the rest. When someone clicks through your Associate link to the Amazon site, a 24-hour shopping window is opened. Anything that user adds to his or her shopping cart for the next 24 hours is eligible for Associate referral fees—even if the actual purchase happens at a later time. The money you earn is paid quarterly, either via check, direct deposit to your checking account, or Amazon gift certificate.

Adding an Associates link requires some rudimentary knowledge of HTML, of course, although Amazon makes the process as easy as cutting and pasting a line or two of code. You can get more detailed instructions—as well as sign up for the Associates program—by clicking the Associates Program link on the See All Product Categories tab.

Now here's the real secret that Amazon doesn't want you to know. You can purchase from your own Associate links—and earn commissions on every item you buy from Amazon! That's right, you become your own referral, and earn a commission on every purchase you make. Those Amazon Associate referral fees are now like cash rebates on your own purchases. It's a cool deal, but only workable as long as you run your own website or web pages. And don't forget to ask all your friends and family to route their Amazon purchases through your site as well!

ON THIS DAY: END OF THE WORLD (1919)

According to seismographer and meteorologist Albert Porta, a conjunction of six planets on December 17, 1919, would cause a magnetic current that would pierce the sun and thereby engulf the Earth in flames. It appears that he was wrong.

BLOG OF THE WEEK: BLOG OF A BOOKSLUT

If you're a book lover, you'll love Bookslut's blog. It's all about books, with links to tons of great book reviews. Give it a read at www.bookslut.com/blog/.

December 18, 2006

MONDAY

THIS WEEK'S FOCUS
HIGH TECH FOR THE HOLIDAYS

GIFT IDEAS

It's holiday time in Leoville, and if you're like me you have at least one person on your list for whom you have no idea what to give. Thank heaven we have the Internet—which makes gift-choosing just a little bit easier.

Your first stop in gathering gift ideas should be the Christmas Gift Directory (www.christmasgifts.com). This site is a huge compendium of popular gifts, organized by type—art, auto accessories, body and bath, and so on. Just click a category to see some recommendations from participating online retailers.

A similar site is Holiday Gift Ideas (www.holidaygiftideas.com), which also goes the gift-by-category route. In addition, this one offers directories for types of people—babies, children, teenagers, mothers, fathers, and so on. It makes it a little easier to choose a more appropriate gift.

If you're an Amazon.com shopper, check out that site's Gifts section. (Click the Gifts link on the See All Product Categories tab.) You can browse suggested gifts by category, or view the most wished for, most gifted, and top rated gifts. Amazon also lets you view suggested gifts by price point, useful when you're on a budget.

Finally, I can't help but mention one of my favorite sites on the Web, period, as part of the whole gift-giving thing. Surprise.com offers gift ideas by categories, like all the other sites do, but also provides a bit of gift-targeting intelligence. You start by describing the person you're shopping for (age, gender, personality, lifestyle, interests), and then the site offers up some pretty good guesses as to what that person might like. You end up with suggested gifts for the avid reader, the fine arts lover, for someone who still watches cartoons, and other specific personality types. Check it out at www.surprise.com.

ON THIS DAY: FIRST CELESTIAL PHOTOGRAPH (1839)

On December 18, 1839, chemistry professor John William Draper took a daguerreotype of the moon, the first celestial photograph made in the United States. He exposed the plate for 20 minutes using a five-inch telescope, producing an image one inch in diameter.

FACT OF THE WEEK

A recent study by DoubleClick (www.doubleclick.com) found that roughly half of all online shoppers conduct product research well before making an online purchase. They typically use a normal search engine, such as Google or Yahoo!, and do their research several weeks in advance.

December 19, 2006

TUESDAY

THIS WEEK'S FOCUS
HIGH TECH FOR THE HOLIDAYS

CHRISTMAS CRAFTS

The holiday season is prime time for tons of fun arts and crafts projects—and the Internet is a prime place to find the information you need to create these projects. What kinds of things am I talking about? You're limited only by your imagination, but some of the most popular holiday crafts projects include

- Candle holders
- Candy Christmas trees
- Gift boxes
- Holiday cards
- Ornaments
- Decorations
- Tissue paper flowers
- Wreaths

And so on. You can find instructions for these and other projects at a number of different websites, including AllCrafts.net (www.allcrafts.net/xmas.htm), Craftown (www.craftown.com/xmas.htm), and Make-Stuff.com (www.make-stuff.com/hollidays/).

An upcoming holiday is also a good excuse to spend some quality time with your children. It's fun for the whole family to get together and make ornaments, decorations, and such. And, again, you can use the Internet to help you get started. You can find lots of holidaycrafts for kids at KidsDomain (www.kidsdomain.com/craft/), Kids Kreate (www.kidskreate.com/holiday_index.htm), EnchantedLearning.com (www.enchantedlearning.com/crafts/christmas/), and similar websites.

ON THIS DAY: ALTAIR 8800 INTRODUCED (1974)

On this date in 1974, the Altair 8800 microcomputer was first offered for sale as a do-it-yourself computer kit. The kit was sold by Ed Roberts' Micro Instrumentation and Telemetry System (MITS) company at a price of $395. The machine was featured on the cover of the January 1975 issue of *Popular Electronics*, and demand exceeded MITS's wildest expectations.

GADGET OF THE WEEK: SLEEPTRACKER WATCH

Wake up at the wrong time and you're grumpy all day. Better to ease yourself awake at the optimal time during your sleep cycle, which is what the Sleeptracker does. It looks like a normal wrist watch, but it actually monitors your sleep cycles for those almost-awake moments—and then gently wakes you when you're most alert. Buy it for $149 from www.sleeptracker.com.

December 20, 2006

WEDNESDAY

THIS WEEK'S FOCUS
HIGH TECH FOR THE HOLIDAYS

HOLIDAY GREETING CARDS

I swear, a good quarter of my friends send me holiday cards that they make themselves, using a digital photo of their kids and a graphics program of some sort. Some of these cards look appropriately cheesy (which is okay), but others are remarkably professional looking. It all depends on what program you use, and how much time you put into it.

Just about any graphics program can be used to create personalized holiday cards. Some of my friends use Adobe Photoshop CS or Photoshop Elements; others use Microsoft Publisher or Paint Shop Pro. They'll all do the job, of course, even if they're not specifically suited for it. In fact, using Photoshop to create holiday cards is a bit of overkill, if you ask me.

A better choice is a program designed specifically for creating greeting cards. (Duh!) There are several good greeting card programs on the market today, including ArcSoft Greeting Card Creator (www.broderbund.com), Greeting Card Factory (www.novadevelopment.com), Hallmark Card Studio (www.hallmarksoftware.com), PrintMaster Greeting Card Deluxe (www.broderbund.com). These programs all feature numerous holiday-themed templates that help you get started fast with your own greeting cards.

Of course, most crafts software features some sort of greeting card functionality. Since these programs also let you create scrapbooks and other crafts projects, they may be a better bet for year-round use. The best of these programs include Print Explosion Deluxe (www.novadevelopment.com), Print Perfect Deluxe (www.cosmi.com), Print Shop (www.broderbund.com), and PrintMaster (www.broderbund.com).

When you go to print your greeting cards, you'll need some sort of thick card stock paper (maybe even the glossy variety), as well as a quality color printer. And if you have a large recipient list, expect to give that printer a long and hard workout.

ON THIS DAY: BROADWAY LIGHTS UP (1880)

On the evening of December 20, 1880, New York's Broadway was first lighted by electric arc lamps, which preceded Thomas Edison's incandescent light bulb.

HARDWARE OF THE WEEK: USB CHRISTMAS TREE

The perfect gadget for the holidays, this festive little gizmo is just what it says—a USB-powered Christmas tree. It stands a little over 5" tall, is lit by a series of miniature LEDs, and cycles through six different colors. Buy it for $17 from www.everythingusb.com.

December 21, 2006

THURSDAY

THIS WEEK'S FOCUS
HIGH TECH FOR THE HOLIDAYS

ONLINE GREETING CARDS

Yesterday we discussed creating holiday cards on your own computer. But let's not forget greeting cards of the online variety—the kind you can send without printing a single sheet of paper.

There are a large number of online greeting card sites on the Web. Many are free, others charge some sort of subscription or per-card fee. They all work pretty much the same way: You choose the card you want to send, personalize it with your name and greeting, and supply the email address of the recipient. The recipient receives an email notifying them of the card; they click the link in the email, and they're taken to the online greeting card site to view their card in the web browser.

Most of these sites let you choose your cards in advance and schedule delivery several days in advance. This way you don't have to stay up late on Christmas Eve just to send all your online cards. Do your ordering beforehand, and let the site do the work for you.

So with the holidays approaching, it's time to visit one of these sites and schedule your online greeting cards for delivery. Here are some of the most popular sites:

- 123Greetings. com (www.123greetings.com)
- BlueMountain.com (www.bluemountain.com)
- Yahoo! Greetings (greetings.yahoo.com)
- Greeting Cards.com (www.greeting-cards.com)
- Hallmark (www.hallmark.com)
- Regards. com (www.regards.com)

And, for those on your list who've been both naughty and nice, check out Kinky Cards (www.kinkycards.com). Definitely not for children!

ON THIS DAY: *SNOW WHITE* PREMIERES (1937)

Walt Disney's first full-length animated film premiered in Los Angeles on December 21, 1937. *Snow White and the Seven Dwarfs* pioneered the use of the multi-plane camera to achieve an effected of depth, and introduced the concept of animated human characters modeled on live actors. The film took two years and $1.5 million to create.

SOFTWARE OF THE WEEK: CREATING KEEPSAKES SCRAPBOOK DESIGNER

A digital scrapbook is just like an old-fashioned scrapbook or photo album, except that it's created on your computer using digital photos. You can share your scrapbook pages electronically, or print them out on a color printer to pass around to your friends and family. One of the best scrapbook programs around is Creating Keepsakes Scrapbook Designer; buy it for $9.99 from www.broderbund.com.

December 22, 2006

FRIDAY

THIS WEEK'S FOCUS
HIGH TECH FOR THE HOLIDAYS

SANTA CLAUS ON THE WEB

Because the big guy only works one day a year, he has lots of time on his hands—and how better to spend that kind of free time than goofing around on the Web? That's right, kiddies, Santa Claus is online and ready to play.

Perhaps the best place to get your Kringle fix is at Claus.com (www.claus.com). This site has lots of fun games and activities for your kids, and also provides a way for them to send their wish lists to Old Saint Nick. Just click the Post Office graphic and your youngsters can email Santa about anything their little hearts desire—such as Barbie dolls and Hot Wheel cars.

Another good Santa-related site is Northpole.com (www.northpole.com). This site features lots of stories and games, as well as yet another opportunity to email Kris K. with their Christmas wishes.

Santa Claus Online (www.santaclausonline.com) features a nifty little clock that counts down the days (and hours and minutes and seconds) until Christmas arrives. Of course, there are lots of holiday games and such, and one more opportunity to email the big fellow. (Santa must have one heck of an inbox!)

One Christmas Eve finally arrives, you can track Santa's progress at www.noradsanta.org. This site is run by the North American Aerospace Defense Command (NORAD)—you know, the guys who normally protect against missile attacks—and features official radar tracking of Santa and his sleigh as he makes his way to good little boys and girls all around the globe. It's kind of a neat idea, and very well implemented. Bookmark it to visit on Christmas Eve!

ON THIS DAY: LINCOLN TUNNEL OPENS (1937)

The day after *Snow White* opened in 1937, New York's Lincoln Tunnel was first opened to traffic. The tunnel passes 1.5 miles under the Hudson River and connects Weehawken, New Jersey, with Manhattan. It was designed by Othmar H. Ammann, who also designed many of the 20th century's greatest bridges.

WEBSITE OF THE WEEK: CHRISTMAS.COM

Ho, ho, ho! Christmas.com is the site for everything Christmas-related, from cards and gifts to screen savers and sheet music. Check it out at www.christmas.com.

December 23/24, 2006

SATURDAY/SUNDAY

THIS WEEK'S FOCUS
HIGH TECH FOR THE HOLIDAYS

HOLIDAY MEALS ONLINE

You know what I like most about the holidays? Other than getting together with friends and family, that is—or receiving all sorts of cool gifts? Well, I like *everything* about the holidays, but I especially like the food. Nothing beats a glazed Christmas ham with all the trimmings, or a batch of sugary Christmas cookies, or... well, just about anything you can cook up over the holidays. For some reason, holiday food tastes better than food during the rest of the year.

Thanks to the Internet, lots of great holiday recipes can be shared by all. Whether you're talking cookies, yams, or turkeys, there's a recipe for it online. Here are some of my favorite Christmas recipe sites:

- All Recipes (holiday.allrecipes.com)
- Back of the Box Recipes (www.backofthebox.com/recipes/holiday.html)
- Fabulous Foods (www.fabulousfoods.com/holidays/xmas/xmas.html)
- Razzle Dazzle Recipes (www.razzledazzlerecipes.com/christmas/)
- Recipe Link (www.recipelink.com/holiday/merrylinks.html)
- Recipe*zaar (www.recipezaar.com/r/17/160/259/)
- Santa's Net Recipes (www.santas.net/recipes.htm)

And if you or someone in your family is diabetic, check out the low-sugar, low-carb recipes and cooking advice offered by the American Diebetes Association (www.diabetes.org/nutrition-and-recipes/holiday-meals.jsp) and Fabulous Foods (www.fabulousfoods.com/features/diabetic/diabrecipes.html). You don't need sugar to serve up great-tasting holiday meals!

ON THIS DAY: MAN ORBITS THE MOON (1968)

On December 23, 1968, the three Apollo 8 astronauts became the first men to orbit the moon. The crew included Frank Borman, James Lovell, Jr., and William Anders. Apollo 8 was also the first manned flight to escape the influence of Earth's gravity.

BLOG OF THE WEEK: MAKE BLOG

This is a blog from *Make* magazine, chock full of interesting do-it-yourself projects, all with a technology bent. It's all quite fun, and found at www.makezine.com/blog/.

December 25, 2006

MONDAY

THIS WEEK'S FOCUS
YOUR NEW PC

THERE'S NO NEED TO FEAR...

For those of us who work with computers day in and day out, a PC gets to be a little like a toaster—an appliance that you expect will do what you need it to do, without a lot of fuss and muss. But when I talk to new computer users, I find that they view PCs a bit differently. In particular, a lot of people are afraid of their computers. They think if they press the wrong key or click the wrong button that they'll break something or will have to call in an expensive repairperson to put things right.

You and I know that isn't true—but try convincing a newbie!

The important thing for any new computer user to know is that it's really difficult to break your computer system. Yes, it's possible to break something if you drop it (I've damaged more than a few wayward mice in my day), but in terms of breaking your system through normal use, it just doesn't happen that often.

It *is* possible to make mistakes, of course. You can click the wrong button and accidentally delete a file you didn't want to delete or turn off your system and lose a document you forgot to save. You can even take inadequate security precautions and find your system infected by a computer virus. But in terms of doing serious harm just by clicking your mouse, it's unlikely.

So if you're a new computer user, you shouldn't be afraid of the thing. Your computer is a tool, just like a hammer or a blender or a camera. After you learn how to use it, it can be a very useful tool. But it's *your* tool, which means *you* tell *it* what to do—not vice versa. Remember that you're in control and that you're not going to break anything, and you'll have a lot of fun—and maybe even get some real work done!

LEARN MORE ABOUT YOUR NEW PC

If you want to learn more about how your new computer works, check out the book *Easy Computer Basics*. It's a four-color, highly visual guide to using a new computer, written by the co-author of this book, Michael Miller. You should be able to find it at all major bookstores.

ON THIS DAY: CENTIGRADE SCALE DEVISED (1741)

On Christmas Day, 1741, astronomer Anders Celsius devised the Centigrade temperature scale. Celsius divided the fixed-point range of the Fahrenheit scale (the freezing and boiling temperatures of water) into 100 equal divisions. Curiously, he set the freezing point to 100 and the boiling point to 0, making one wonder if Celsius was a bit dyslectic. Fortunately, the scaling was reversed after his death.

FACT OF THE WEEK

The Computer Industry Almanac (www.c-i-a.com) reports that the number of PCs in use worldwide surpassed 820 million in 2004, and is projected to top 1 billion by 2007. The U.S. has the most PCs in use (223 million), nearly three times the number-two country (Japan, with 69 million).

December 26, 2006

TUESDAY

THIS WEEK'S FOCUS
YOUR NEW PC

TAKING A LOOK INSIDE

The most important piece of hardware in your computer system is the *system unit*. This is the big, ugly box that houses your disk drives and many other components. In other words, all the good stuff in your system unit is inside the case.

With most system units, you can remove the case to peek and poke around inside. Go ahead, I know you want to—and I won't let you hurt anything while you're peeking.

To remove your system unit's case, start by turning off your PC and then unplugging it from the wall outlet. Once it's unplugged, look for some big screws or thumbscrews on either the side or back of the case. (Even better—read your PC's instruction manual for instructions specific to your unit.) With the screws loosened or removed, you should then be able to either slide off the entire case, or pop open the top or back.

When you open the case on your system unit, you see all sorts of computer chips and circuit boards. The really big board located at the base or side of the computer (to which everything else is plugged into) is called the *motherboard*, because it's the "mother" for your microprocessor and memory chips, as well as for the other internal components that enable your system to function. This motherboard contains several slots, into which you can plug additional *boards* (also called *cards*) that perform specific functions.

Most PC motherboards contain six or more slots for add-on cards. For example, a video card enables your microprocessor to transmit video signals to your monitor. Other available cards enable you to add sound and modem/fax capabilities to your system. If you need to upgrade your system's audio or video capabilities, you do so by inserting a new card into one of these open slots.

Got all that? Then put the case back together, plug it back in, and turn your system on again. Good job!

ON THIS DAY: RADIUM DISCOVERED (1898)

On this date in 1898, scientist Marie Curie discovered the radioactive element radium. She was experimenting with pitchblende, a common uranium ore, and observed that the ore was more radioactive than refined uranium. This indicated the presence of another, even more radioactive element—radium.

GADGET OF THE WEEK: EGG & MUFFIN TOASTER

Now there's a single kitchen gadget that can make your entire Egg McMuffin. The Egg & Muffin Toaster has slots to toast your muffin, a tray to warm your Canadian bacon, and an attached egg cooker to poach your eggs. It's an all-in-one breakfast sandwich solution, available for $49.99 from www.eggandmuffintoaster.com.

December 27, 2006

WEDNESDAY

THIS WEEK'S FOCUS
YOUR NEW PC

YOUR PC'S MICROPROCESSOR: THE MAIN ENGINE

Remember the big motherboard inside your computer's system unit? Well, buried somewhere on that motherboard is a specific chip that controls your entire computer system. This chip is called a *microprocessor* or a *central processing unit (CPU)*.

The microprocessor is the brains inside your system. It processes all the instructions necessary for your computer to perform its duties. The more powerful the microprocessor chip, the faster and more efficiently your system runs.

Microprocessors carry out the various instructions that let your computer compute. Every input and output device hooked up to a computer—the keyboard, printer, monitor, and so on—either issues or receives instructions that the microprocessor then processes. Your software programs also issue instructions that must be implemented by the microprocessor. This chip truly is the workhorse of your system; it affects just about everything your computer does.

This is why a fast microprocessor is important. The faster it can process all those instructions, the faster your system runs. A slow microprocessor will make your system feel sluggish, especially if you're performing processing-intensive activities, such as working large files or doing digital photo or video editing.

Different computers have different types of microprocessor chips. Many IBM-compatible computers use chips manufactured by Intel. Some use Intel-compatible chips manufactured by AMD and other firms. But all IBM-compatible computers that run the Windows operating system use Intel-compatible chips.

In addition to having different chip manufacturers (and different chip families from the same manufacturer), you'll also run into microprocessor chips that run at different speeds. CPU speed today is measured in gigahertz (GHz). A CPU with a speed of 1GHz can run at one *billion* clock ticks per second! The bigger the gigahertz number, the faster the chip runs.

ON THIS DAY: JOHANNES KEPLER BORN (1571)

German astronomer Johannes Kepler was born on this date in 1571. Kepler formulated the three major laws of planetary motion which enabled Newton to devise the law of gravitation, based on the observation of elliptical planetary orbits.

HARDWARE OF THE WEEK: LOGITECH MX3100 CORDLESS DESKTOP

This is my new favorite keyboard/mouse combo. Both the keyboard and mouse are wireless; the keyboard has a ton of extra buttons for specific operations, as does the laser mouse. Even better, the keyboard has a nice firm feel, and the mouse is just about the smoothest you'll find. Buy the combo for $149.95 from www.logitech.com.

December 28, 2006

THURSDAY

THIS WEEK'S FOCUS
YOUR NEW PC

COMPUTER MEMORY: TEMPORARY STORAGE

Before your PC's microprocessor can process any instructions you give it, those instructions must be stored somewhere, in preparation for access by the microprocessor. These instructions—along with other data processed by your system—are temporarily held in the computer's *random access memory (RAM)*. All computers have some amount of memory, which is created by a number of memory chips. The more memory that's available in a machine, the more instructions and data that can be stored at one time.

Memory is measured in terms of *bytes*. One byte is equal to approximately one character in a word processing document. A unit equaling approximately one thousand bytes (1,024, to be exact) is called a *kilobyte (KB)*, and a unit of approximately one thousand (1,024) kilobytes is called a *megabyte (MB)*. A thousand megabytes is a *gigabyte (GB)*.

Most computers today come with at least 256MB of memory, and it's not uncommon to find machines with 1GB or more. To enable your computer to run as many programs as quickly as possible, you need as much memory installed in your system as it can accept—or that you can afford. Extra memory can be added to a computer by installing a new memory module, which is as easy as plugging a "stick" directly into a slot on your system's motherboard. Just make sure you install the right type of memory for your particular PC. Most retailers will help you look up the model number of your PC and match the right type of memory to that particular model.

If your computer doesn't possess enough memory, its microprocessor must constantly retrieve data from permanent storage on the hard disk. (This is called *virtual memory*.) This method of data retrieval is slower than retrieving instructions and data from electronic memory. In fact, if your machine doesn't have enough memory, some programs will run very slowly (or you might experience random system crashes), and other programs won't run at all, because of this need to constantly access the hard disk. It's better to add more RAM to your PC to avoid this problem, but it's also good to keep enough free space on your hard disk in case virtual memory has to be used.

ON THIS DAY: FIRST U.S. TEST TUBE BABY (1981)

December 28, 1981, is the birthday of the first American baby conceived through *in vitro* fertilization. Elizabeth Jordan Carr was born at Norfolk General Hospital, weighing 5 lbs., 12 oz. The world's first test tube baby, Louise Brown, was born three years before, in England.

SOFTWARE OF THE WEEK: WINZIP

Windows XP comes with file compression/extraction built-in, but it's not the most intuitive or flexible thing in the world to use. If you plan on compressing a lot of your files and folders into Zip files, I recommend using WinZip instead. WinZip is a fully-featured compression/extraction utility, and it's a lot easier for new users to use. Download it for $29 from www.winzip.com.

December 29, 2006

FRIDAY

THIS WEEK'S FOCUS
YOUR NEW PC

HARD DISK DRIVES: LONG-TERM STORAGE

Another important physical component inside your computer's system unit is the *hard disk drive*. The hard disk permanently stores all your important data. Some hard disks today can store up to 500 gigabytes of data—and even bigger hard disks are on the way. (Contrast this to your system's random access memory, which stores only a few hundred megabytes of data, temporarily.)

A hard disk consists of numerous metallic platters. These platters store data *magnetically*. Special read/write *heads* realign magnetic particles on the platters, much like a recording head records data onto magnetic recording tape.

Before data can be stored on any disk, including your system's hard disk, that disk must first be *formatted*. A disk that has not been formatted cannot accept any data. When you format a hard disk, your computer prepares each track and sector of the disk to accept and store data magnetically.

Of course, when you buy a new PC, your hard disk is already formatted for you. And, in most cases, your operating system and key programs also are preinstalled. (And here's a bit of a caution: If you try to reformat your hard disk, you'll erase all the programs and data that have been installed—so don't do it!)

By the way, one of the most common mistakes made by computer newbies is to confuse the computer's hard disk storage with the system's temporary. A hard disk stores data physically, while random access memory stores data electronically. This makes hard disk storage (relatively) permanent, and memory storage quite temporary. They're two completely different things, used for two different purposes!

ON THIS DAY: FIRST TRANSISTORIZED HEARING AID (1952)

On this date in 1952, Sonotone Corporation began selling the Model 1010, the world's first transistorized hearing aid. It weighed 3.5 ounces, measured 3" × 1.5" × 0.6", and cost $229.50. The Model 1010 was actually a hybrid device, consisting of two subminiature pre-amplifier tubes and a single transistor as the final audio amplifier.

WEBSITE OF THE WEEK: TECH SUPPORT GUY

Got questions about your new PC? Got a problem you need fixed? Then visit the Tech Support Guy forums, where other computer users (and a few experts among them) will answer all your questions, for free. It's a great resource for new computer users, located at www.techguy.org.

December 30/31, 2006

SATURDAY/SUNDAY

THIS WEEK'S FOCUS
YOUR NEW PC

TURNING ON YOUR PC—FOR THE FIRST TIME

The first time you turn on your PC is a unique experience. A brand-new, out-of-the-box system will have to perform some basic configuration operations, which include asking you to input some key information. While the first-time startup operation differs from manufacturer to manufacturer, it typically involves some or all of the following steps:

- **Windows Product Activation**—You may be asked to input the long and nonsensical product code found on the label attached to the rear of your PC (or on your Windows installation CD, if you received one). Your system then phones into the Microsoft mother ship (via the Internet), registers your system information, and unlocks Windows for you to use.

- **Windows Registration**—This is where you're asked to input your name and other personal information, perhaps along with the Windows product code. This information then is phoned into the Microsoft mother ship (via the Internet) to register your copy of Windows with the company, for warranty purposes.

- **Windows Configuration**—During this process Windows asks a series of questions about your location, the current time and date, and other essential information. You also might be asked to create a username and password.

- **System Configuration**—This is where Windows tries to figure out all the different components that are part of your system, such as your printer, scanner, and so on. Enter the appropriate information when prompted; if asked to insert a component's installation CD, do so.

Many computer manufacturers supplement these configuration operations with setup procedures of their own. It's impossible to describe all the different options that might be presented by all the different manufacturers, so watch the screen carefully and follow all the onscreen instructions.

After you have everything configured, Windows finally starts, and then you can start using your system. Don't worry—the next time you power up, the normal startup process is a lot less complicated!

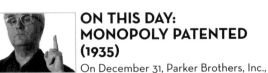

ON THIS DAY: MONOPOLY PATENTED (1935)

On December 31, Parker Brothers, Inc., received a patent for their Monopoly board game. The patent described it as "intended primarily to provide a game of barter, thus involving trading and bargaining" in which "much of the interest in the game lies in trading and in striking shrewd bargains." You betcha!

BLOG OF THE WEEK: BIGGESTSTARS BLOGS

When you need to get your daily celebrity fix, check out Biggeststars Blogs. This blog is chock full of celebrity news, gossip, and photos; find it at blog.biggeststars.com.

344

INDEX

Numbers

#1 CD Ripper, 261
2D charts, adding depth to presentations (PowerPoint), 304
4 to 3 aspect ratio (television), 63, 128
16 to 9 aspect ratio (television), 63
16-9 aspect ratio (HDTV), 128
123Greetings.com, 336

A

AAC audio compression (Advanced Audio Coding), 11
About.com, 241
Absolute Beginner's Guide to HDTV and Home Theater, 126
Absolute Beginner's Guide to Podcasting, 46
Absolute Beginner's Guide to Security, Spam, Spyware, and Viruses, 129, 276
access times to hard disks, PC upgrade considerations, 61
accessories for iPod, 112-113
Accoustica MP3 CD Burner, 261
Accoustica.com, 261
Act Labs PC USB Light Gun, 158
ActionTec Internet Phone Wizard, 92
Active@ Eraser utility, 171
ActMon spyware (legitimate), 172
Ad-Aware website, 171
add-ins (Windows Media Center)
    Comics for Media Center, 89
    MCE Customizer 2005, 89
    MCE mceWeather, 89
    MCE Outlook, 89
    mceAuction, 89
    My Netflix, 89
    Playlist Editor, 89
adding ports, 59
Addonics Pocket CDRW, 264
addresses (email)
    multiple use of, 131
    protecting from spammers, 131
    spamblocks, 131
Adobe GoLive, 189
Adobe Premiere Elements, 153
Adobe.com, Photoshop user statistics, 280

AeroCool AeroBase UFO Gaming Pad, 39
Age of Empires game, Easter egg, 98
The Agonist political blog, 41
airport information (Yahoo!), 16
ALAC lossless audio compression format, 12
Alanis Morisette's *Jagged Little Pill*, CD Easter egg, 97
album art
    displaying, 28
    searching, 28
Album Art Fixer, 13, 85
Aldrin, Edwin (Buzz), 20
Alien, Collector's Edition, 254
Aliens, UFO News Blog, 294
AllCrafts.net, 334
Allen, Paul, 4
Allmusic Database website, 14
AllPodcasts.com, 47
Allrecipes.com, 338
AllTheWeb website, 184
Aloha Bob PC Relocator, 60
Alta Vista website, 184
Altair 8800 computer, 334
Altec Lansing inMotion (iPod), 111
AltoMP3 CD Ripper, 261
Amateur-Sports.com, 245
Amazon.com
    Associates, sign ups, 332
    Auctions, 330
    clearance items, 329
    Friday Sale items, 329
    Inside This Book feature, 327
    Marketplace items (used), 330
    online shopping product reviews, 314
    Outlet, 329
    page appearance, 327
    page personalization, 327
    search guidelines, 327
    site navigation, 327
    Super Saver Shipping, 331
    Today's Deals, 328
AmazonWatch.org, 331
AMD microprocessors, 341
Amelie, 254
America Online. *See* AOL
American Diabetes Association, Christmas recipes, 338

American Management Association (AMA), workplace surveillance survey, 72
AmeriSurplus.com, merchandise liquidators, 106
AMF Software websites, 179
analog television versus digital television, 123
Andale.com, third party eBay tools, 106
AndrewSullivan.com, political blogs, 41
Angelfire website, 188
animals
   Dolly, the Cloned Sheep, 52
   Global Pet Finder, 23
Ankle Biting Pundits, political blogs, 41
anonymous email services, 54
Answers.com, 241
ANT 4 MailChecking, 133
anti-spam software, 133
antivirus software, 276
   signature scanning, 279
AOL (America Online)
   AOL Hometown home page community, 188
   chat rooms, 122
AOL Instant Messenger (AIM), 118
   Buddy Chat, 119
   Files, sending, 120
   group chats, initiating, 119
AOL Search website, 184
AOL@SCHOOL website, 244
APC ES 725 Broadband UPS device, 24
Apollo 13 (movie), 255
Apollo 13 flight, 97
Apollo 14 moon golf, 37
Apollo 17, 327
Apple AirPort Express, 110
Apple Computer
   1977-1983 history, 163
   1985-1996 history, 164
   1997-current history, 165
   history, 163
   introduction of Lisa model, 19
   Jobs, Steve, 53
   Lisa computer, 163
   ThinkSecret.com, insider information and rumors, 166
   Unofficial Apple Weblog, 167
Apple HD Cinema Display monitor, 164

Apple iBook, 166
Apple II computer, introduction of, 150
Apple IIc, 108
Apple iMac, development of, 165
Apple iPod, 108
   accessories
      Belkin Digital Camera Link, 113
      Belkin iPod Media Reader, 113
      Belkin iPod Voice Recorder, 112
      Belkin TunePower, 112
      Denision ice Link Plus, 113
      Griffin BlueTrip, 112
      Griffin iBeam, 113
      H20 Audio Underwater iPod Housing, 113
      naviPlay Remote, 112
      Solio Solar Charger, 112
   album art software, 111
   Bluetooth connections to home audio systems, 109
   celebrity playlists, 207
   connecting to PCs, 109
   development of, 165
   iPoditude blog, 113
   speakers
      Altec Lansing inMotion, 111
      Bose SoundDock, 111
      Cube Travel Speakers, 111
      JBL OnStage, 111
      Macally PodWave, 111
   usage statistics, 108
Apple iTunes
   development of, 165
   music stores, 13
Apple PowerBook, 166
Apple.com
   iPod Shuffle player, 204
Archie tool, pre-WWW archiving functions, 9
Archos Gmini XS200, 200
ArcSoft Greeting Card Creator, 335
area code information (Yahoo!), 16
Armstrong, Neil, 191
ARPANET, 284
arrows on desktop shortcuts, removing, 29
artificial snow, 301
Ask Jeeves! website, 184
aspect ratios
   HDTV, 128
   televisions, 63
AssetMetrix.com, Windows XP usage, 22

Association for Computing Machinery (ACM), 245
Atanasoff-Berry computer, 264
Atari 80 Classic Games CD-ROM, 229
Atari 2600, 227
Atari 5200, 228
Atari 7800 ProSystem, 229
Atari Flashback, 196
Atari Jaguar game system, 233
Atari PONG, 226, 315
Atatio.com, 271
ATI HDTV Wonder card, PC upgrades, 65
Auction Guild, eBay daily update resource, 102
Auction Mills websites, eBay Trading Assistant franchise, 107
AuctionBytes.com, eBay daily update resource, 102
auctions
   Auction Guild, 102
   AuctionBytes.com, 102
   Dumb Auctions websites, 92
   eBay
      add-on toolbar, 105
      birddogging practices, 104
      daily update access, 102
      feedback system, 103
      merchandise liquidators, 106
      PowerSeller program, 105
      third party management tools, 106
      Trading Assistants, 107
Audacity audio editor, 46
audio
   CDs
      burning, 264
      copying, 266
   Creative GigaWorks PC Speaker System, 329
   downloading
      file-trading networks, 14
      online music stores, 13
   Internet radio, 15
   presentations, adding to (PowerPoint), 303
   Soundblaster Audigy 4, 309
   surround sound
      Cubase SX recording studio program, 217
      Dolby, 214
      Dolby Digital discrete format, 215
      Dolby Digital EX discrete format, 215
      Dolby Pro Logic, 214

blogs

Dolby Pro Logic IIx, 214
DTS discrete format, 215
Fantasia, 217
format comparisons, 216
history of, 218-219
Sony Surround Sound Headphones, 215
Surround Sound Discography, 218
audio compression formats
   AAC (Advanced Audio Coding), 11
   lossless
      ALAC, 12
      FLAC, 12
      WMA Lossless, 12
   MP3 (MPEG-1 Level 3), 11
   WMA (Windows Media Audio), 11
audio editors for podcasts, 46
audio players, Consumer Electronics Association sales statistics, 200
Audio REVIEW website, 251
Audio Video Science (AVS) Forum websites, 67
audio/video receivers
   inputs and outputs, 247
   number of channels, 247
   power considerations, 247
AudioBlog.com, podcasting services, 46
Audiograbber, 261
AudioTronic iCool Scented MP3 Player, 203
Austin Powers – The Spy Who Shagged Me, DVD Easter egg, 96
AVG Anti-Virus, 276
Aviation, Boeing 747 Jet, 21
AVSoft websites, Album Art Fixer utility, 85

B

background colors
   documents, adding (Word), 178
   web pages, cycling, 295
background music, adding to web pages, 300
backgrounds
   desktop, setting with web pages, 25
   graphics, adding to Internet Explorer toolbar, 34
   images, web pages, scrolling effect, 297
   Wallpaperizer tool, 292
BackoftheBox.com, 338

Backpacks, SoundSak SonicBoom speakers, 157
backups
   Buffalo TeraStation Storage Drive, 323
   preventative maintenance, 80
Bad Pics Fixed Quick, 281
BadJocks.com sports blog, 42
ballpoint pen, introduction of, 286
Bally Professional Arcade, 227
Bandwidth Place website, connection speed tester, 3
Bargain Boardwalk website, 316
Bargain Shopping website, 316
bargains, eBay birddogging practices, 104
BASIC language, 143
batteries
   cell phones, FreePlay FreeCharge tool, 269
   notebooks, extending life of, 240
The Beatles, *Sgt. Pepper's Lonely Hearts Club Band*, CD Easter egg, 97
Beaucoup.com metasearcher, 185
beer, ancient discovery of, 294
Belkin Digital Camera Link, 113
Belkin iPod Media Reader, 113
Belkin iPod Voice Recorder, 112
Belkin TunePower, 112
Benz, Karl, 25
Berners-Lee, Tim, 138
Best Price, online shopping comparisons, 316
Beyond Bullets website, PowerPoint presentations, 306
bifocal eyeglasses, 136
BigBlog.com, news blog, 42
BiKiKii Anonymous Remailer websites, 54
BINAC computer, 325
birddogging practices (eBay), 104
Birdseye Frozen Foods, marketing of, 293
birth control pill, development of, 124
bit rates for digital music, 10
BitTorrent websites, 14
BizRate.com, online shopping comparisons, 315
BJ Pinchbeck Homework Helper website, 244
Blaze Media Pro conversion utility, 205, 261

BlazeMP.com, 261
blocking spam (Outlook 2003), 132
blog hosting services for podcasts
   AudioBlog.com, 46
   LibSyn.com, 46
   OurMedia.org, 46
   PodBus.com, 46
Blog Search Engine, 37
blog syndication services, 46
BlogChalking.com, 39
Blogger.com, host community, 39
blogosphere, 40
blogs (web logs), 37
   Biggeststars Blogs, 344
   blogosphere, 40
   cell phones, MobileBurn.com, 273
   Cooking for my Kids blog, 246
   Dave Barry's Blog, 95
   Die is Cast Blog, 161
   directories
      Blog Search Engine, 37
      Blogwise, 37
      Daypop.com, 37
      Feedster.com, 37
      Globe of Blogs.com, 37
      Weblogs.com, 37
   FeedDemon aggregator (RSS), 330
   Game*Blogs, 199
   Gizmodo.com, 326
   HDTV Blog, 128
   hosting communities
      Blogger.com, 39
      TypePad.com, 39
   hosting services for podcasts
      AudioBlog.com, 46
      LibSyn.com, 46
      OurMedia.org, 46
      PodBus.com, 46
   iPoditude.com, 113
   Jake Ludington's Media Blab, 122
   JavaScript Weblog, 300
   Jazz & Blues Music Reviews, 258
   Laptop Review, 240
   Leoville.com, 37
   Make Magazine, 338
   Matt Goyer's Media Center, 89
   moblogs/photoblogs
      BusyThumb.com, 42
      Photoblogs.org, 42
      TextAmerica.com, 42
   movie
      DVDVerdict.com, 42
      filmfodder.com, 42

blogs

   GreenCineDaily.com, 42
   Tagline.com, 42
   MP3PlayerBlog.com, 207
  music
   FreshTuneage.com, 42
   PeopleTalkTooLoud.com, 42
  New York media, Gawker.com, 54
  news
   BigBlog.com, 42
   Plastic.com, 42
   Slashdot.org, 3
   UnknownNews.net, 42
  The Office Weblog, 181
   Online Music Blog, 261
  Orbitcast.com, 312
  organization of, 38
  Outrageous eBay Auctions, 107
  Pew Internet and American Life Project usage statistics, 176
  photo, 42
  PhotographyBLOG, 149
  podcasts, Adam Curry, 48
  political
   The Agonist, 41
    AndrewSullivan.com, 41
   Ankle Biting Pundits, 41
   Captain's Quarters, 41
   Daily Kos, 41
   The Decemberist, 41
   Electablog, 41
   Eschaton, 41
   eTalkingheads, 41
   INDC Journal, 41
   Little Green Footballs, 41
   PoliBlog, 41
   Powerline, 41
   Stage Left, 41
   Talking Points Memo, 41
   Wonkette.com, 41
  political leanings, Pew Internet & American Life Project findings, 37
  portals
   BlogChalking.com, 39
   BlogUniverse.com, 39
   Blogwise.com, 39
   Eatonweb.com, 39
   Globe of Blogs.com, 39
   Weblogs.com, 39
  postings as privacy threat, 73
  PostSecret Blog, 267
  PVRblog.com, 155
  range of, 38
  reasons for, 38

  RescueComp Blog, 225
  Search Engine Blog, 187
  searching, 37
  ShoppingBlog.com, 320
  sports
   BadJocks.com, 42
   Fanblogs.com, 42
  Spyware Warrior blog, 175
  tech
   BoingBoing.net, 42
   Gizmodo.com, 42
   Slashdot.org, 42
   Techdirt.com, 42
  Tech Blog website, 62
  technology humor, Boing Boing, 9
  Thomas Hawk's Digital Connection, 193
  UFO News Blog, 294
  Unofficial Apple Weblog, 167
  Unofficial Google Weblog, 140
  Unofficial Photoshop Blog, 285
  Video Game Blog, 234
  Weblog Review, 41
  WiFi Networking News blog, 213
  WilWheaton.net, 219
  Windows OS future releases, LonghornBlogs.com, 27
  Windows XP tricks, TechWhack.com, 36
  WornBlog.com, 279
  Yahoo! Search Blog feature, 21
BlogUniverse.com portal, 39
Blogwise, 37-39
Blue Crush, 254
blue screen of death, 223
BlueMountain.com, 336
boating devices (GPS), 159
Boeing 747 Jet, introduction of, 21
BoingBoing.net
  tech blog, 42
  technology humor, 9
books
  Bookslut's Blog, 332
  Easter eggs, 100
  Google Print feature, 140
  Metacritic.com, 257
bookshelf speakers, 250
Bookslut's Blog, 332
Boolean logic, 292
Boolean search operators, 187
Borden's condensed milk, 219
Bose Noise Canceling Headphones, 281

Bose SoundDock (iPod), 111
Boston Computer Society, 42
branching presentations, creating (PowerPoint), 302
Briefcase (Yahoo!), 21
broadband cable Internet connections, 4
Broadband Reports website connection speed tester, 3
broadband satellite Internet connections, 4
Buffalo TeraStation Storage Drive, 323
burning CDs, 261, 264
  Addonics Pocket CDRW, 264
  process, 264
  Roxio Easy Media Creator, 265
  tools, 261
  Windows Media Player, 265
Burroughs Adding Machine, 220
business laptops, 235
BusyThumbs.com, moblogs/photoblogs host, 42
Buylink.com, merchandise wholesalers, 106

C

cable Internet connections, 4
CamcorderInfo.com, 154
camcorders
  brand sales statistics, 150
  CamcorderInfo.com, 154
  camera steadiness, 155
  Digital8 format, 150
  DVD burners, HP dc5000 Movie Writer, 152
  DVD format, 150
  HDV, 150
   cost range, 152
   features, 152
  MicroMV format, 150
  microphone types
   lavaliere, 154
   shotgun, 154
   stereo, 154
   surround-sound, 154
   wireless, 154
  MiniDV format, 150
  monopods, 155
  prosumer
   cost, 153
   features, 153
  PVRblog.com, 155

selection criteria, 151
tripods, 155
cameras, JB1 James Bond 007 Spy Camera, 169
camcorder lighting systems, 151
camping devices, 161
CanaryWireless.com, 209
capacities of hard disks, PC upgrade considerations, 61
Captain's Quarters, political blogs, 41
car
    satellite radio receivers
        SIRIUS Satellite Radio, 310
        XM Radio, 310
    surround sound output, Soundblaster Audigy 4, 309
carbon paper, 267
Careerbuilder.com, 220
Carnivore (FBI spyware), 175
cars
    audio systems, online shopping criteria, 324
    ScanGauge Automotive Computer, 275
cartoon characters, Skeleton Systems websites, 90
cash versus credit card usage, 321
CatsDomain.com, 38
CD to MP3 Gecko, 261
CD-R discs, 264
CDA (CD Audio) file format, 264
CDMA (cell division multiple access) cell phones, 268
CDs
    Addonics Pocket CDRW, 264
    bit rates, 10
    burning, 261, 264
        Windows Media Player, 265
    copying, 266
    data, writing to, 267
    Easter eggs
        Alanis Morisette's Jagged Little Pill, 97
        Dave Matthews Band's Before These Crowded Streets, 97
        Frank Zappa and the Mothers of Invention's We're Only In It for the Money, 97
        Jimi Hendrix's Are You Experienced?, 97
        Santana's Supernatural, 97
        Sgt. Pepper's Lonely Hearts Club Band, 97
        Todd Snider's Songs for the Daily Planet, 97

freedb.org music database, 266
home theater showcase, 257
    10,000 Hz Legend (Air), 258
    Brothers in Arms (Dire Straits), 258
    Casino Royale (Soundtrack), 258
    Gershwin Collection (Dave Grusin), 258
    Strange Little Girls (Tori Amos), 258
Media Center PCs, living room use, 81
peer-to-peer file swapping, legalities of, 263
playlists (Windows Media Center), 85
Recording Industry Association of America sales statistics, 262
ripping (Windows Media Center), 85, 261-263
Roxio Easy Media Creator, 265
sampling rates, 10
storing (Windows Media Center), 85
celebrities
    Biggeststars Blogs, 344
    iPod playlists, 207
cell phones, 273
    batteries, FreePlay FreeCharge tool, 269
    CDMA (cell division multiple access), 268
    dual-band CDMA, 269
    dual-band TDMA/GSM, 269
    dual-mode, 269
    FDMA (frequency division multiple access), 268
    GSM (global system for mobile communications), 268
    headsets, choosing, 272
    ICE address entries, 294
    Mobile Phone Directory, 272
    MobileBurn.com blog, 273
    Motorola Razr V3, 273
    Nokia N90, 273
    online shopping criteria, 323
    selection criteria, 270
    Siemens S66, 273
    Sony Ericsson S710z, 273
    TDMA (time division multiple access), 268
    tri-band GSM, 269
    versus smartphones, 271
Centigrade temperature scale, 339
central processing units (CPUs), 341
CERN, 138

Channel F Video Entertainment game system, 226
chats
    Internet TeleCafe, 122
    MSN Chat, 122
    postings as privacy threat, 73
    Talk City, 122
    tips, 121
    Yahoo! Chat, 122
Chernobyl nuclear disaster (Russia), 110
children. See kids computing
China, software piracy issues, 100
Chipsets, PowerOpen Association, 66
Christmas card programs
    ArcSoft Greeting Card Creator, 335
    Greeting Card Factory, 335
    Hallmark Card Studio, 335
    Print Explosion Deluxe, 335
    Print Perfect Deluxe, 335
    Print Shop, 335
    PrintMaster Greeting Card Deluxe, 335
Christmas craft sites
    AllCrafts.net, 334
    Craftown, 334
    Enchanted Learning, 334
    Kids Kreate, 334
    KidsDomain, 334
    Make-Stuff.com, 334
Christmas Gift Directory website, 333
Christmas online cards
    123Greetings.com, 336
    BlueMountain.com, 336
    GreetingCards.com, 336
    Hallmark.com, 336
    KinkyCards.com, 336
    Regards.com, 336
    YahooGreetings.com, 336
Christmas recipes
    Allrecipes.com, 338
    American Diabetes Association, 338
    BackoftheBox.com, 338
    FabulousFoods.com, 338
    RazzleDazzleRecipes.com, 338
    Recipe*zaar.com, 338
    RecipeLink.com, 338
Christmas Santa Claus sites
    Claus.com, 337
    NORAD Santa, 337
    Northpole.com, 337
    Santa Claus Online.com, 337

Christmas shopping sites

Christmas shopping sites
   Christmas Gift Directory, 333
   Holiday Gift Ideas, 333
   Surprise.com, 333
Christmas.com, 337
cigarette smoking, 12
Cinemar MainLobby, home theater software, 66
Citizen Kane, 254
Citizen TASER X26c Pistol, 5
citywide WiFi, 208
Classicade Upright Game System, 228
Claus.com, 337
cleaning
   mouse, 78
   printers, 79
ClearType technology, enabling (Windows XP), 24
clip art, Clipart Connection website, 305
clock speeds for CPUs, 341
clocks, USB Alarm Clock, 216
closed-air earphones, 206
closing stuck programs (Windows XP), 33
clothing
   removable shirt collars, 36
   SCOTTeVEST, 29
   UglyDress.com, 93
CNET website, connection speed tester, 3
CNETSearch.com metasearcher, 185
Coca Cola, development of, 83
Coffee Break Worm game, 99
cold fusion, 78
Coleco Colecovision, 228
Coleco Telstar game system, 226
Colt, Samuel, 190
Comics for Media Center add-in, 89
Commodore Computer, 120
Commtouch.com, spam sources, 49
Compaq Presario, 225
comparison sites for online shopping
   Best Price, 316
   BizRate.com, 315
   Froggle.com, 315
   mySimon.com, 315
   Shopping.com, 315
   Yahoo! Shopping.com, 315
Computer Industry Almanac website
   notebook usage, 235
   number of PCs worldwide, 339
   smartphone sales, 268
computer timesharing, advent of, 117

ComputerHope.com, 224
computers
   online shopping criteria, 322
   TRADIC, 70
Concorde SST, 261
configuring PCs, initial setup process, 344
connections
   Internet
     broadband cable, 4
     broadband satellite, 4
   speeds, testing, 3
consumer camcorders, 151
Consumer Electronics Association, audio player sales statistics, 200
ConsumerReports.org, online shopping product reviews, 314
ConsumerREVIEW.com, online shopping product reviews, 314
ConsumerSearch.com, online shopping product reviews, 314
Control Panel (Windows XP), 23
Cookies, cleaning (NSClean tool), 72
copying CDs, 266
core memory, invention of, 56
Corel.com websites, 147
cosmological constant, 10
Couch Potato Tormentor TV remote, 64
Counterexploitation (CEXX).org, 172
CouponMountain website, 316
CouponPages.com website, 316
CPUs (central processing units), 341
Craftown, 334
crafts
   Christmas sites, 334
     AllCrafts.net, 334
     Craftown, 334
     Enchanted Learning, 334
     Kids Kreate, 334
     KidsDomain, 334
     Make-Stuff.com, 334
     Creating Keepsakes Scrapbook Designer, 336
Craig's List, online classifieds, 325
Creating Keepsakes Scrapbook Designer, 336
Creative GigaWorks PC Speaker System, 329
Creative Zen Micro player, 203
Creative Zen Nano Plus player, 204
Creative Zen Touch, 200
Creative.com
   Zen Micro player, 203
   Zen Nano Plus player, 204

credit cards
   identity theft, 69
   online shopping safety, 313
   versus cash usage, 321
Crouching Tiger, Hidden Dragon, 254
CRT projectors, rear-projection televisions (RPTVs), 64
Crystal Graphics PowerPlugs, 304
Ctrl+Alt+Del combination, stuck programs, closing, 33
Cubase SX recording studio program, 217
Cube Travel Speakers (iPod), 111
current events, Google News feature, 140
Curry.com, podcast resources, 48
Cute CD DVD Burner, 261
Cybermoon Studios, animated humor site, 121
cycling background colors in web pages, 295

D

Daily eDeals website, 316
Daily Kos, political blogs, 41
Daily Rotten website, 95
Das Boot, 255
data backups, preventative maintenance, 80
data CDs, writing to, 267
Dave Barry's Blog, 95
Dave Matthews Band's *Before These Crowded Streets*, CD Easter egg, 97
Daylight Savings Time (DST), 85
Daypop.com, 37
Days of Heaven, 255
dead computers, troubleshooting, 221
DealofDay.com website, 316
Death Clock website, 91
The Decemberist political blog, 41
Deck Keyboard, 184
defragmenting hard disks, 75
Dell Axim X50v Pocket PC PDA, 302
Dell UltraSharp 24 LCD Display monitor, 297
Delphi XM MyFi receiver, 308
Delphi XM SKIFi2 receiver, 310
Denali Edition Media Center PC, 83
Denision Ice Link Plus, 113
Denning, Peter, virtual memory concept, 8
desktop
   backgrounds, setting with web pages, 25
   screen savers (Marine Aquarium), 93

Theme Manager 2005, 25
Windows XP Taskbar
  hiding, 27
  moving, 27
desktop icons, shortcut arrows, removing, 29
desktop replacement notebooks, 235
Desktop Search feature (Google), 138
device upgrades, pre-planning checklist, 55
diagnostic tools
  PC Certify utility, 78
  PC Pitstop websites, 79
dictionary definitions (Yahoo!), 16
Dictionary.com, 241
Die Another Day, DVD Easter egg, 96
Die is Cast Blog, 161
digital audio
  downloading
    file-trading networks, 14
    online music stores, 13
  Internet radio, 15
digital cameras
  digital SLRs, 143
  Epson Perfection 4180 Photo Scanner, 143
  JB1 James Bond 007 Spy Camera, 169
  Lyra Research, men versus women and preferences, 141
  product reviews, 148
  prosumer models, 142
  SanDisk Photo Album, 142
  selection criteria, 141
Digital Hotspotter (WiFi), 209
digital light projection (DLP), rear-projection televisions, 65
digital music
  album art
    displaying, 28
    searching, 28
  Album Art Fixer, 13
  Allmusic Database, 14
  audio compression
    lossless, 10-12
    lossy, 10-11
  Creative I-Trigue Speaker System, 12
  Fluxblog website, 15
  Jupiter Research survey, 10
  sampling rates, 10
  StickAx Music Mixer, 11
digital music players, hard drive types
  Archos Gmini XS200, 200
  Creative Zen Touch, 200
  iRiver H340, 200
  Olympus m-robe 500i, 200
digital photography
  Bad Pics Fixed Quick, 281
  camera reviews, 148
  images, vault storage, 149
  lighting guidelines, 147
  moblogs
    BusyThumbs.com, 42
    Photoblogs.org, 42
    TextAmerica.com, 42
  Paint Shop Pro, editing functions, 147
  photoblogs
    BusyThumbs.com, 42
    Photoblogs.org, 42
    TextAmerica.com, 42
  PhotographyBLOG, 149
  red eye effect, removing (Photoshop), 282
  shooting tips, 148
Digital Photography Review website, 148
digital photos
  Picasa tool, image management functions, 324
  Wallflower 2 Multimedia Picture Frame, 136
digital refrigerator voice recorders, 50
digital SLR cameras, 143
digital television
  confusion with HDTV terminology, 123
  formats
    EDTV (enhanced definition digital television), 124
    HDTV 720p (high definition digital television), 124
    HDTV 1080i (high definition digital television), 124
    HDTV 1080p (high definition digital television), 124
    SDTV (standard definition digital television), 124
  HDTV
    aspect ratios, 128
    HDTV Blog, 128
    pixels, number of, 126
    progressive scanning, 127
    smart expand mode, 128
    zoom mode, 128

HDTV advantages
  low flicker, 125
  picture sharpness, 125
  sound quality, 125
  widescreen display, 125
  sale statistics, 63
  TViX Digital Movie Jukebox, 125
  versus analog television, 123
digital video
  camcorders
    CamcorderInfo.com, 154
    PVRblog.com, 155
  camera selection criteria, 151
  camera steadiness, 155
  Digital8 format, 150
  DVD burners, HP dc5000 Movie Writer, 152
  DVD format, 150
  editing tools (Adobe Premiere Elements), 153
  HDV camcorders
    cost range, 152
    features, 152
  MicroMV format, 150
  microphone types
    lavaliere, 154
    shotgun, 154
    stereo, 154
    surround-sound, 154
    wireless, 154
  MiniDV format, 150
  momopods, 155
  NPD TechGroup.com, brand sales rankings, 150
  PC upgrades
    hard disk sizes, 60
    memory, 60
    microprocessors, 60
    monitor sizes, 60
  prosumer camcorders, 153
  Sunpak Readylight 20, 151
  tripods, 155
digital video recorders
  electronic programming guides (EPGs), 248
  size/storage considerations, 248
  TiVos, 248
Digital8 format, 150
DigitalNetworks.com
  Rio Carbon Player, 203
  Rio Forge Sport Player, 204
DigitalPodcasts.com, 47
DIMM memory module, 56
Direct Connect Magazine websites, 81

direct-dial transatlantic phone service, 59
directories (podcasts)
   AllPodcasts.com, 47
   DigitalPodcasts.com, 47
   iPodder.org, 47
   iPodderX.com, 47
   iTunes.com, 47-48
   Podcast.net, 47
   PodcastAlley.com, 47
   PodcastBunker.com, 47
   PodcastDirectory.com, 47
   PodcastingStation.com, 47
   PodcastPickle.com, 47
   Syndic8.com, 47
DIRECWAY Satellite Internet, 4
discrete surround sound
   Dolby Digital, 215
   Dolby Digital EX, 215
   DTS, 215
Disk Defragmenter utility, 75
disk drives, formatting, 343
Display Properties utility
   special effects, activating (Windows XP), 35
Display Properties utility (Windows XP), 26
documents (Word)
   background colors, adding, 178
   footers, adding, 180
   headers, adding, 180
   multiple-column layouts, creating, 181
   numbered lists, creating, 176
   section breaks
      deleting, 179
      inserting, 179
   styles
      applying, 177
      elements of, 177
      modifying, 177
      predesigned templates, 177
   watermarks, adding, 178
Dogpile.com metasearcher, 185
Dolby Digital discrete surround sound, 215
Dolby Digital EX discrete surround sound, 215
Dolby Surround Sound, 214
DolbyPro Logic IIx Surround Sound, 214
DolbyPro Logic Surround Sound, 214
Dolly, the Cloned Sheep, 52
Double-Data-Rate SDRAM (DDR SDRAM), 56
DoubleClick.com, online shopping research survey, 333
downloading
   digital music
      file-trading networks, 14
      online music stores, 13
   programs from FTP sites (WS_FTP utility), 7
   Tweak UI tool, 34
drawing tablets, Wacom Cintiq 21UX Touchscreen Pen Tablet, 303
Dreamweaver, 189
drive interface types, hard disks, PC upgrade considerations, 61
DTS discrete surround sound, 215
dual-band CDMA cell phones, 269
dual-band TDMA/GSM cell phones, 269
dual-mode cell phones, 269
Dumb Auctions website, 92
Dumb Bumpers website, 92
Dumb Criminals website, 92
Dumb Laws website, 92
Dumb Warnings website, 92
DuPont nylon stockings, 281
DVD Information website, 253
DVDs
   annual sales statistics, 253
   Easter eggs, 96
      Austin Powers – The Spy Who Shagged Me, 96
      Die Another Day, 96
      The Incredibles, 96
      Star Wars Episode 1 – The Phantom Menace, 96
   format, 150
   home theater showcase
      Alien, Collector's Edition, 254
      Amelie, 254
      Apollo 13, 255
      Blue Crush, 254
      Citizen Kane, 254
      Crouching Tiger, Hidden Dragon, 254
      Das Boot, 255
      Days of Heaven, 255
      Finding Nemo, 254
      The Hulk, 254
         Kill Bill Vol. 1, 256
      Lord of the Rings, 256
      Master and Commander, 255
      The Matrix, 256
         Saving Private Ryan, 255
      Seabiscuit, 254
      Sky Captain and the World of Tomorrow, 256
      Spider-Man 2, 256
      Star Wars Episode III, 256
      The Hulk, 254
      The Matrix, 256
   Media Center PCs, living room use, 81
   Movie Label 2006, 126
   players, shopping considerations, 249
   recorders, HP dc5000 Movie Writer, 152
DVDVerdict.com movie blog, 42
Dvorak keyboard versus QWERTY keyboard, 127
DVRs, electronic program guide (EPG), 84
dynamite, 186

E

earbuds, 206
Earhart, Amelia, 161
earphones, 206
   Bose Noise Canceling Headphones, 281
   closed-air, 206
   open-air, 206
   Sony Surround Headphones, 215
Easter Egg Archive website, 100
Easter eggs
   CDs
      Alanis Morisette's Jagged Little Pill, 97
      Dave Matthews Band's Before These Crowded Streets, 97
      Frank Zappa and the Mothers of Invention's We're Only In It for the Money, 97
      Jimi Hendrix's Are You Experienced?, 97
      Santana's Supernatural, 97
      Sgt. Pepper's Lonely Hearts Club Band, 97
      Todd Snider's Songs for the Daily Planet, 97
   DVDs, 96
      Austin Powers – The Spy Who Shagged Me, 96
      Die Another Day, 96
      The Incredibles, 96
      Star Wars Episode 1 – The

Flash memory players

Phantom Menace, 96
  games
    Age of Empires, 98
    Halo 2, 98
    King's Quest VIII, 98
    Roller Coaster Tycoon 2, 98
  literary, 100
  software
    Final Cut Pro, 99
    Grand Theft Auto - San Andreas, 101
    Microsoft PowerPoint, 99
    mIRC, 99
    Mozilla Firefox, 99
    Paint Shop Pro, 99
***Easy Computer Basics*, 339**
Easy Music CD Burner, 261
Eatonweb.com blog portal, 39
eBay
  add-on toolbar, 105
  bargains, birddogging, 104
  The Chatter Newsletter, 102
  daily updates, accessing, 102
  feedback system, 103
  merchandise sources, 106
  My Summary page, 102
  Outrageous eBay Auctions blog, 107
  PEZ dispense pages, 102
  PowerSeller program, 105
  third party management tools, 106
  Trading Assistants
    franchises, 107
    qualifications, 107
eCoupons website, 316
Edirol R-1 Portable Music Recorder, 44
Edison, Thomas, 273
eDonkey website, 14
EDTV (enhanced definition digital television) format, 124
EFI PrintMe website, 239
Egg & Muffin Toaster, 340
eHomeUpgrade.com, 252
Einstein's Theory of Relativity, 75
Einstein, Albert, 10
Electablog.com political blog, 41
Electronic Frontier Foundation (EFF) website, 73
elevator, invention of, 49
email
  addresses, multiple use of, 131
  anonymous services
    BiKiKii Anonymous Remailer, 54
    Offshore Mixmaster Anonymous Remailer, 54
    Riot Anonymous Remailer, 54
    W3 Anonymous Remailer, 54
  common problems, troubleshooting, 53
  Email Sentinel Pro utility, 132
  GMail (Google), 53, 139
  HTML, potential problems, 51
  mailing groups, creating (Outlook), 52
  mass mailings
    managing (Outlook), 52
    sending (Outlook), 52
  phishing schemes, 74
  POP accounts, 49
  protocols
    IMAP, 50
    POP3, 50
    SMTP, 50
  remailing services, 54
  spam, 129
    Absolute Beginners Guide to Security, Spam, Spyware, and Viruses, 129
    address gathering techniques, 130
    address protection, 131
    anti-spam software, 133
    blockers/filters, 129
    blocking in Outlook 2003, 132
    deleting, 129
    filtering services, 133
    Spam Kings Blog, 134
    Spamhaus.org, 133
    statistical surveys, 129
    tracing and reporting spammer abuse, 134
  spam sources, 49
  Thunderbird program (Mozilla Firefox browser), 52
  web-based accounts, 49
Email Sentinel Pro utility, 132
EmailAddressManager.com, 132
Emailias.com, 133
emergency startup disks, troubleshooting tips, 224
employees and workplace privacy, 72
encryption
  enabling (Windows XP), 34
Enchanted Learning websites, 334
ENIAC computer, 44, 262
ENIGMA machine, 187
Entertainment Software Association
  online gaming statistics, 295
  player statistics, 226
  sales statistics, 194
environment, AmazonWatch.org, 331
Epinions.com, online shopping product reviews, 314
Epson Perfection 4180 Photo Scanner, 143
Eschaton.com political blog, 41
ESSA-1 weather satellite, 35
eTalkingheads, political blog, 41
Eva Solo Magnetimer, 97
EvaSolo.com, 97
EverythingUSB.com, 335
exercise and fitness, running performance GPS monitors, 157
Extended Data Out EDO DRAM (EDO DRAM), 56

F

FabulousFoods.com, 338
Fairchild Semiconductor, 22
Fanblogs.com sports blog, 42
Fantasia, surround sound innovation, 217
Fast-Page-Mode DRAM (FPM DRAM), 56
FBI, Carnivore spyware, 175
FDMA (frequency division multiple access) cell phones, 268
feathering image edges (Photoshop), 281
FedEx, product shipment tracking, 320
FeedBurner.com, podcasting syndication services, 46
FeedDemon aggregator (RSS), 330
Feedster.com, 37
Fender guitars, 211
Ferris wheel, development of, 164
files
  encryption
    enabling Windows XP, 34
  file-trading networks, 14
  migration, Aloha Bob PC Relocator, 60
  proper organization, preventative maintenance, 80
  sending via MSN Messenger, 120
  sending via Yahoo! Messenger, 120
filmfodder.com movie blog, 42
Final Cut Pro, Easter egg, 99
Finding Nemo, 254
Fish-N-Log Professional Suite, 159
fishing-finding devices, 158
FLAC lossless audio compression format, 12
Flash memory players
  Creative Zen Nano Plus, 204
  iPod Shuffle, 204
  iRiver iFP-900 Series, 204
  Rio Forge Sport Player, 204
  Verbatim Store 'n' Go, 270

flat screen televisions

flat screen televisions
   LCD, 68
   plasma, 67
floor-standing speakers, 250
fluoridated drinking water, introduction of, 24
flying cars, 50
FM radio, advent of, 147
foam rubber, 176
fonts on web pages
   changing, 193
   resizing, 193
food
   Christmas recipes
      Allrecipes.com, 338
      American Diabetes Association, 338
      BackoftheBox.com, 338
      FabulousFoods.com, 338
      RazzleDazzleRecipes.com, 338
      Recipe*zaar.com, 338
      RecipeLink.com, 338
   Egg & Muffin Toaster, 340
   Eva Solo Magnetimer, 97
footers in documents, adding (Word), 180
formatting hard disks, 343
FORTRAN language, 104, 249
Frank Zappa and the Mothers of Invention's *We're Only In It for the Money*, CD Easter egg, 97
fraud and identity theft, 69
   handling, 71
freedb.org music CD database, 266
FreePlay FreeCharge tool, 269
FreshTuneage.com music blog, 42
Froggle.com
   online shopping comparisons, 315
   online shopping searches, most popular products, 313
FrontBridge.com, spam statistics, 129
FrontPage (Microsoft), 189
FTC (Federal Trade Commission) website, identity theft resources, 71
FTP sites, program downloads, 7
Fujita tornado scale, 280

G

Gadget Universe, geek merchandise, 321
GAIM, 118
Galileo, Vatican apology, 287
Game Jamboree website, 99
Game*Blogs, 199

gamepad controllers, 197
games, 195
   Act Labs PC USB Light Gun, 158
   Atari Flashback, 196
   building, minimum requirements, 196
   Coffee Break Worm, 99
   controllers
      choosing, 197
      Gravis Xterminator Force Feedback Gamepad, 198
      Logitech Cordless Rumblepad 2, 198
      Logitech MOMO Racing Wheel, 199
      MonsterGecko PistolMouse FPS, 199
      Saitek Cyborg evo Wireless, 198
      Saitek X52 Flight Control System, 198
      ThrustMaster Enzo Ferrari Force Feedback Wheel, 199
      X Arcade Solo, 199
      Zboard Modular Keyset, 199
   Easter eggs
      Age of Empires, 98
      Grand Theft Auto - San Andreas, 101
      Halo 2, 98
      King's Quest VIII, 98
      Roller Coaster Tycoon 2, 98
   Entertainment Software Association, online gaming statistics, 295
   IGN.com, 198
   Microsoft Xbox 360, 195
   Nintendo Revolution, 195
   PlayStation 3, 195
   RollerCoaster Tycoon 3, 197
   Spherex RX2 Game Chair, 195
   systems, choosing, 194
GarageBand Guitar Cable, 163
gaslight, introduction of, 27
Gates, Bill, 4
Gawker.com media blog, 54
GB (gigabytes), 342
geek products
   car audio systems, 324
   cell phones, 323
   computers, 322
   Gadget Universe, 321
   home theater components, 326
   music players, 325
   PDAs, 323
   smartphones, 323
   ThinkGeek.com, 321

gift sites
   Christmas Gift Directory, 333
   Holiday Gift Ideas, 333
   Surprise.com, 333
gigabytes (GB), 342
Gizmodo.com, 42, 326
glamour glows for skin, creating (Photoshop), 285
Global Pet Finder, 23
Globe of Blogs.com, 37-39
Gmail (Google), 139
Goddard, Robert, 72
GoDogGo Automatic Fetch Machine, 314
golf, ProAim Golfing Goggles, 189
Google, 183
   Answers feature, 140
   Catalogs feature, 140
   Desktop Search, 138
   Gmail service, 139
   Groups feature, 140
   home pages, personalizing, 138
   Image Search, album art searches, 28
   Images, search filters, 136
   Language Tools, 139
   Local feature, 140
   Maps feature
      address searches, 137
      satellite photos, 137
   News feature, 140
   Print feature, 140
   Ridefinder feature, 140
   search statistics, 135
   search tricks and tips, 135
   top 10 search list, 90
   The Unofficial Google Weblog, 140
   Usenet archive, 73
   Video feature, 140
Gopher tool, pre-WWW archiving functions, 9
GORP website, outdoor travel and recreation resources, 160
Gossip, Biggeststars Blogs, 344
government resources, THOMAS website, 8
GPS (global positioning system)
   boating devices, 159
   camping devices, 161
   devices
      cost range, 156
      primary user market, 156
      running performance monitors, 157
   Global Pet Finder, 23
   military development, 156

high-tech outdoors

satellite signal transmission process, 156
scuba diving devices, 160
TomTom Navigator, 76
TrimTrac GPS Security Locator, 70
Grand Theft Auto - San Andreas, hidden sex scenes Easter egg, 101
Graphire Bluetooth drawing tablet, 282
Gravis Xterminator Force Feedback Gamepad, 198
GreenCineDaily.com movie blog, 42
Greenwich Mean Time (GMT), 204
Greeting Card Factory, 335
greeting cards
  online, 336
    123Greetings.com, 336
    BlueMountain.com, 336
    GreetingCards.com, 336
    Hallmark.com, 336
    KinkyCards.com, 336
    Regards.com, 336
    YahooGreetings.com, 336
  papers, 335
  programs, 335
    ArcSoft Greeting Card Creator, 335
    Greeting Card Factory, 335
    Hallmark Card Studio, 335
    Print Explosion Deluxe, 335
    Print Perfect Deluxe, 335
    Print Shop, 335
    PrintMaster Greeting Card Deluxe, 335
GreetingCards.com, 336
Griffin
  AirClick USB Remote, 190
  BlueTrip, 109, 112
  iBeam Light, 113
  PowerMate, 131
  Radio SHARK AM/FM Radio/Receiver, 45
GriffinTechnology.com, 131, 163
Grokster websites, 14
Grundig P7131 receiver, 311
GSM (global system for mobile communications) cell phones, 268

H

H2O Audio Underwater Housing, 113
Hall & Partners.com, USB flash drive usage survey, 301
Hallmark Card Studio, 335
Hallmark.com, 336
Halo 2 game, Easter egg, 98
Ham, the Monkey, space flight exploits, 29
handouts for presentations, printing (PowerPoint), 306
hard disks
  Active@ Eraser utility, 171
  defragmenting, 75
  error scans, 75
  formatting, 343
    Macintosh, SuperDuper cloner, 165
  PC upgrades
    access time specifications, 61
    capacity specifications, 61
    digital video editing purposes, 60
    drive interface type specifications, 61
    spin rates specifications, 61
hard drives
  digital music players, 200
    Archos Gmini XS200, 200
    Creative Zen Touch, 200
    iRiver H340, 200
    Olympus m-robe 500i, 200
  LaCie Bogger Disk Extreme, 291
  palmOne LifeDrive, 296
  Seagate, 51
  WD Passport Pocket Hard Drive, 276
hardware
  adding external hardware, 62
  APC ES 725 Broadband UPS device, 24
  digital video editing, required PC upgrades, 60
  hard disk drives, 343
  Linksys WRT54GS Wireless Router, 6
  Logitech diNova Cordless Desktop, 104
  Logitech MediaPlay Cordless Mouse, 18
  Music, Creative I-Trigue Speaker System, 12
  PC upgrades, ATI HDTV Wonder card, 65
  Philips RC9800i WiFi Remote Control, 82
  preventative maintenance, 76
    keyboards, 77
    mouse, 78
    printers, 79
  Seagate External Hard Drives, 51
  Sunbeam 20-in-1 Superior Panel, 59
  Tom's Hardware Guide websites, 61
  upgrades, pre-planning checklist, 55
  USB Café Pad, 98
Harvard Mark I computer, 18
Hasbro VideoNow Color video player, 242
HD Radio
  Ibiquity.com, 312
  receivers, 312
HDTV (high definition digital television)
  720p format, 124
  1080i format, 124
  1080p format, 124
  Absolute Beginner's Guide to HDTV and Home Theater, 126
  advantages
    low flicker, 125
    sharpness of picture, 125
    sound quality, 125
    widescreen display, 125
  aspect ratios, 128
  confusion with digital television terminology, 123
  Dell UltraSharp 24 LCD Display monitor, 297
  HDTV Blog, 128
  HDTVoice.com, 127
  How HDTV and Home Theater Work, 126
  numbers in U.S. household, 123
  pixels, number of, 126
  programming, locating, 253
  progressive scanning, 127
  smart expand mode, 128
  zoom mode, 128
HDTVoice.com, 127
HDV (high-definition) camcorders, 150-152
HDV format, 150
headers in documents, adding (Word), 180
headphones, Bose Noise Canceling Headphones, 281
headsets for cell phones, choosing, 272
high-tech outdoors
  Act Labs PC USB Light Gun, 158
  camping devices, 161
  fishing-finding devices, 158
  GPS devices
    boating devices, 159
    cost range, 156
    military development, 156

high-tech outdoors

primary user market, 156
satellite signal transmission process, 156-157
scubu diving devices, 160
SoundSak SonicBoom speakers, 157
Hindenburg disaster, 122
Hitchhiker's Guide to the Galaxy, 271
Hitwise.com, reference site popularity, 241
hoaxes
  urban legends
    common tales, 292-293
    Snopes.org, 293
  viruses, 287
  War of the Worlds, 286
Holiday Gift Ideas website, 333
Hollerith Tabulating Machine Company, 9
home pages
  communities, 188
  Google personalization options, 138
home satellite radio receivers
  SIRIUS Satellite Radio, 311
  XM Radio, 311
home theater
  Audio REVIEW websites, 251
  Audio Video Science (AVS) Forum websites, 67
  audio/video receivers
    inputs and outputs, 247
    number of channels, 247
    power considerations, 247
  Cinemar MainLobby software, 66
  components, online shopping criteria, 326
  digital video recorders
    electronic programming guides (EPGs), 248
    size/storage considerations, 248
    TiVos, 248
  DVD players, shopping considerations, 249
  DVDs, annual sales statistics, 253
  eHomeUpgrade.com, 252
  HDTV programming, locating, 253
  Home Theater Blog website, 68
  household components statistics, 247
  Logitech Harmony 880, 248
  Media Center PCs, living room use, 81
  physical layout considerations, 251
  showcase CDs, 257
    10,000 Hz Legend (Air), 258
    Brothers in Arms (Dire Straits), 258
    Casino Royale (Soundtrack), 258
    Gershwin Collection (Dave Grusin), 258
    Strange Little Girls (Tori Amos), 258
  showcase DVDs
    Alien, Collector's Edition, 254
    Amelie, 254
    Apollo 13, 255
    Blue Crush, 254
    Citizen Kane, 254
    Crouching Tiger, Hidden Dragon, 254
    Das Boot, 255
    Days of Heaven, 255
    Finding Nemo, 254
    The Hulk, 254
      Kill Bill Vol. 1, 256
    Lord of the Rings, 256
    Master and Commander, 255
    The Matrix, 256
      Saving Private Ryan, 255
    Seabiscuit, 254
    Sky Captain and the World of Tomorrow, 256
    Spider-Man 2, 256
    Star Wars Episode III, 256
  Sonos Digital Music System, 249
  speaker systems
    bookshelf, 250
    choosing, 250
    floor-standing, 250
    satellite, 250
  universal remote controls, 252
Home Theater Blog website, 68
home WiFi networks, configuring, 211
Homebrew Computer Club, 61
Homework Center website, 244
homework help sites
  AOL@SCHOOL, 244
  BJ Pinchbeck's Homework Helper, 244
  Homework Center, 244
  Jishka, 244
  KidInfo.com, 244
  MadSci Network, 244
  Open Door, 244
  WebMath, 244
  Word Central, 244
HostIndex.com website, 189
hosting communities (blogs)
  Blogger.com, 39
  TypePad.com, 39

hosting services for websites
  cost, 189
  locating, 189
HostSearch website, 189
Hot Coffee module, *Grand Theft Auto San Andreas* **Easter egg, 101**
hot swappable devices, 62
HotBot website, 184
HotDog PageWiz Editor, 191
HotJobscom, 220
hotspots
  citywide WiFi, 208
  WiFi, locating, 209
***How HDTV and Home Theater Work,* 126**
HP dc5000 Movie Writer, 152
HP-35 Calculator, introduction of, 6
Hron & Hardart Automat Restaurant, 154
HTML (Hypertext Markup Language), 190
  email, potential problems, 51
  fonts
    changing, 193
    resizing, 193
  HotDog PageWiz Editor, 191
  hyperlinks, inserting, 192
  tags, 190-191
  web page editors, 189
hubs
  Apple AirPort Express, 110
  Keyspan USB Mini Hub, 71
  MSI Bluetooth Star USB Star, 315
Hughes Network Systems, broadband satellite Internet, 4
***The Hulk,* 254**
hydrogen bomb, U.S. testing, 291
hyperlinks in web pages, inserting, 192

I

iArt software, iTunes album artwork, 111
Ibiquity.com, HD Radio resources, 312
IBM
  Harvard Mark I computer, 18
  incorporation date, 160
  origins of, 9
  Selective Sequence Electronic Calculator (SSEC), 23
IBM 5150 PC, 213
IBM 650 Mainframe, 175
IBM Model 704, 221
IBM System/360 Mainframe, 94

ICE cell phone entry (urban legend), 294
ICQ.com, 118
IDC.com, computer manufacturer sales rankings, 162
identity theft, 69
   FTC (Federal Trade Commission) resources, 71
   handling, 71
   phishing schemes, 74
   preventing, 70
IEEE 802.11 standard (WiFi), 213
IGN.com, 198
iLounge, 112
images
   Bad Pics Fixed Quick, 281
   digital photo vaults, 149
   edges, feathering (Photoshop), 281
   glamour glows, adding (Photoshop), 285
   Google search filters, 136
   Picasa tool, management functions, 324
   red eye, removing (Photoshop), 282
   repairing (Photoshop), 280
   skin blemishes, removing (Photoshop), 284
   wrinkles, softening (Photoshop), 283
IMAP (Interactive Mail Access Protocol), 50
iMesh websites, 14
The Incredibles, DVD Easter egg, 96
INDC Journal political blog, 41
indoor photography, lighting guidelines, 147
infected computers, repairing, 277
InfoSports.net, 245
instant messaging, 118
   AOL Instant Messenger (AIM)
      Buddy Chat, invitations, 119
      group chats, initiating, 119
   client software
      AOL Instant Messenger, 118
      GAIM, 118
      ICQ, 118
      Miranda-IM, 118
      MSN Messenger, 118
      Trillian, 118
      Yahoo! Messenger, 118
   Files, sending
      via AOL Instant Messenger, 120
      via MSN Messenger, 120
      via Yahoo! Messenger, 120
   Internet Relay Chat (IRC), 120

peer-to-peer (P2P) connections, 117
Pew Internet and American Life Project usage statistics, 117
tips, 121
integrated circuit, invention of, 109
Intel Corporation, formation of, 189
Intel microprocessors, 341
interlaced scanning versus progressive scanning, 127
internal combustion engines, development of, 25
Internet
   connections
      broadband cable, 4
      broadband satellite, 4
      speeds, testing, 3
   file-trading networks, 14
   history, cleaning (Active@ Eraser), 171
   hours per day usage statistics, 28
   pre-WWW tools
      Archie, 9
      Gopher, 9
      Veronica, 9
      WAIS, 9
   security
      email spam, 133
      spyware, 171
      viruses, 274-276
Internet Explorer
   keyboard shortcuts, 5
   toolbar, background graphics, adding, 34
   versus Mozilla Firefox, 6
Internet Public Library, 243
Internet radio websites, 15
Internet Relay Chat (IRC), 120
Internet TeleCafe website, 122
Internet Video Game Museum, 233
IOGEAR MiniView Micro KVM Switch, 77
iPod (Apple), 108
   accessories
      Belkin Digital Camera Link, 113
      Belkin iPod Media Reader, 113
      Belkin iPod Voice Recorder, 112
      Belkin TunePower, 112
      Denision ice Link Plus, 113
      Griffin BlueTrip, 112
      Griffin iBeam, 113
      H2O Audio Underwater Housing, 113
      naviPlay Remote, 112
      Solio Solar Charger, 112

album art software, 111
Bluetooth connections to home audio systems, 109
celebrity playlists, 207
connecting to PCs, 109
iPoditude blog, 113
Shuffle player, 204
speakers
   Altec Lansing inMotion, 111
   Bose SoundDock, 111
   Cube Travel Speakers, 111
   JBL OnStage, 111
   Macally PodWave, 111
usage statistics, 108
iPodder.org, 47
iPodderX.com, 47
IPSwitch.com, FTP download utilities, 7
iRiver H10 player, 203
iRiver H340, 200
iRiver iFP-900 Series player, 204
iRiverAmerica.com
   H10 player, 203
   iRiver iFP-900 Series player, 204
iSoldIt website, eBay Trading Assistant franchise, 107
ISPs (Internet Service Providers)
   broadband cable connections, 4
   broadband satellite connections, 4
   connection speeds, testing, 3
iSuppli.com, digital television sales, 63
IT Garage websites, 80
ITtoolbox.com, PC upgrades survey, 55
iTunes
   playlists, creating, 110
   podcast subscriptions, 43
iTunes.com, 47-48

J

Jake Ludington's Media Blab website, 122
Jakks TV Games.com, 227
Java, birth of, 33
JavaScript Menu Master, 298
JavaScript Weblog, 300
JavaScript.com, web page tricks, 299
JB1 James Bond 007 Spy Camera, 169
JBL OnStage (iPod), 111
Jbox.com, 216
Jello-O, invention of, 140
Jemison, Mae (NASA), 275

Jimi Hendrix's *Are You Experienced?*, CD Easter egg, 97
Jishka websites, 244
JiWire.com, WiFi directory, 209
jobs
   Careerbuilder.com, 220
   HotJobs.com, 220
   Monster.com, 220
Jobs, Steve, 53, 163, 271
joystick game controllers, 197
Jupiter Research website
   broadband household survey, 75
   digital music surveys, 10
   home theater household statistics, 247
   online travel revenue, 156

K

Kaspersky Anti-Virus Personal, 276
Kazaa website, 14
Keeler polygraph machine, 34
Kellogg's Cornflakes, 64
keyboard shortcuts (Internet Explorer), 5
keyboards
   Deck Keyboard, 184
   Dvorak, 127
   Logitech diNova Cordless Desktop, 104
   Logitech MX3100 Cordless Desktop, 341
   maintenance, 77
   Ultra Antistatic Wrist Straps, 56
Keyspan USB Mini Hub, 71
Keyspan.com, 71
Kid Info website, 244
kids computing
   Cooking for my Kids blog, 246
   homework help
     AOL@SCHOOL, 244
     BJ Pinchbeck's Homework Helper, 244
     Homework Center, 244
     Jishka, 244
     KidInfo.com, 244
     MadSci Network, 244
     Open Door, 244
     WebMath, 244
     Word Central, 244
   Internet Public Library, 243
   Kids' Space website, 245
   LibDex, 243
   Library of Congress, 243
   Literature.org, 243

Microsoft Student 2006, 244
New York Public Digital Library Collections, 243
Questia.com, 243
refdesk.com, 243
safe search sites, 242
safety tips, 241
   Santa Claus
     Claus.com, 337
     NORAD Santa websites, 337
     Northpole.com, 337
     Santa Claus Online.com, 337
school work research study, 246
Smithsonian Institution Libraries, 243
sports
   Amateur-Sports, 245
   InfoSports, 245
   Kids Sports Bulletin Board, 245
   Kids Sports Network, 245
   MomsTeam, 245
   Play Football, 245
   Sports Illustrated for Kids, 245
   Youth Sports Instruction, 245
Kids Kreate websites, 334
Kids Sports Bulletin Board, 245
Kids Sports Network, 245
Kids' Space website, 245
KidsDomain website, 334
***Kill Bill Vol. 1*, 256**
King's Quest VIII game, Easter egg, 98
KinkyCards.com, 336
kitchens
   Eva Solo Magnetimer, 97
   VoiSec digital refrigerator voice recorder, 50
Kovalchik Farms website, 316

L

labels for DVDs (Movie Label 2006 software), 126
LaCie Bigger Disk Extreme hard disk, 291
Language Tools (Google), 139
LapCool 2 Notebook Cooler, 237
Laptop Review Blog, 240
laser slingshot, 38
Laserpod Light, 254
LAUNCH website, 15
lavaliere microphones, 154
LCD displays, ClearType technology, 24
LCD televisions, 66-68
Leap Second Day, 172

Leoville.com author blog, 37
Levi Strauss jeans, invention of, 134
LibDex, 243
Library of Congress, 243
LibSyn.com podcasting services, 46
lie detector, introduction of, 34
light guns, 197
lighting
   camcorders (Sunpak Readylight 20), 151
   digital photography, 147
LimeWire website, 14
Linksys Wireless Adapter, 210
Linksys WRT54GS Wireless Router, 6
Linksys.com, router technologies, 6
liquid-fuel rocketry, 72
Liquidation.com, merchandise liquidators, 106
Lisa computer (Apple), 19, 163
LISP programming language, 101
Listerine mouthwash, 129
Lists, numbered, creating (Word), 176
literary works, Easter eggs, 100
Literature.org, 243
Little Green Footballs political blog, 41
Live365 website, 15
Logitech
   Cordless Rumblepad 2, 198
   Football Mouse, 243
   Harmony 880 home theater remote, 248
   MediaPlay Cordless Mouse, 18
   MOMO Racing Wheel, 199
   MX1000 laser mouse, 178
   MX3100 Cordless Desktop, 341
   QuickCam Orbit, 119
   Z-5500 Speakers, 255
Logitech.com, 18, 104
LonghornBlogs.com, future Windows releases, 27
LookSmart website, 183-184
***Lord of the Rings*, 256**
lossless compression, digital music, 10-12
   ALAC format, 12
   FLAC format, 12
   WMA Lossless format, 12
lossy compression, digital music, 10-11
   AAC format (Advanced Audio Coding), 11
   MP3 format (MPEG -1 Level 3), 11
   WMA format (Windows Media Audio), 11

LucidLink Wireless Client, 211
Luddite Riots, 68
Lyra Research, men versus women brand preferences for digital cameras, 141

M

Mac OS X, troubleshooting guidelines and tools, 167
Macally PodWave (iPod), 111
Macintosh
    Apple HD Cinema Display monitor, 164
    music accessories, GarageBand Guitar Cable, 163
    SuperDuper cloner, 165
Mackinac Suspension Bridge, 170
MadSci Network websites, 244
Magnavox Odyssey game system, 226-227
mailing groups, creating (Outlook), 52
Mailshell.com, 133
MailWasher Web site, 133
maintenance
    data backups, 80
    file organization, 80
    keyboards, 77
    mouse, 78
    printers, 79
    system units, 76
    updates, 80
    virus and spyware updates, 80
Make Magazine, 338
Make-Stuff.com, 334
Mamma.com metasearcher, 185
man-made diamonds, development of, 46
manned maneuvering units (MMUs), untethered spacewalks, 38
maps. *See also* GPS
    Yahoo! Maps, 18
Marathon Computer RePorter, 222
Marine Aquarium Screen Saver, 93
Mariner 4 (NASA), 314
Mark I computer, 208
Maskbit.com, 261
Mason jars, 316
mass mailings (email)
    managing (Outlook), 52
    sending (Outlook), 52
Master and Commander, 255
The Matrix, 256
Matt Goyer's Media Center Blog, 89
Mattel Intellivision, 227

MB (megabytes), 342
McAfee AntiSpyware, 171
McAfee SpamKiller, 133
McAfee VirusScan website, 276
MCE Customizer 2005 add-in, 89
MCE Outlook add-in, 89
MCE Playlist Editor add-in, 89
mceAuction add-in, 89
mceWeather add-in, 89
MCSSoft website, 89
measurement conversions (Yahoo!), 16
media blogs, Gawker.com, 54
Media Center PCs
    building on your own, 83
    buying considerations, 82
    Denali Edition, 83
    living room use, 81
    Philips RC9800i WiFi Remote Control, 82
    required components, 82-83
    screen size, 82
    usage statistics, 81
Media Jukebox, 261
media players, portable (MicroDrive), 203
Meedio Essentials, 250
megabytes (MB), 342
memory
    notebooks, upgrading, 238
    storage terminology, 342
    upgrades, digital video editing purposes, 60
menus in web pages, building (JavaScript Menu Master), 298
merchandise liquidators
    AmeriSurplus.com, 106
    eBay sellers, 106
    Liquidation.com, 106
    My Web Wholesaler, 106
    Overstock.com, 106
merchandise wholesalers
    Buylink.com, 106
    Wholesale Central, 106
    Wholesale411, 106
MetaCrawler.com metasearcher, 185
Metacritic.com, 257
metasearches (Web)
    Beaucoup.com, 185
    CNETSearch.com, 185
    Dogpile.com, 185
    Mamma.com, 185
    MetaCrawler.com, 185
    OneSeek.com, 185

SearchSpaniel.com, 185
WebFerret.com, 185
WebTaxi.com, 185
metric system, creation of, 123
Michael Paulus website, cartoon character studies, 90
Michelangelo virus, 63
MicroDrive media players
    Creative Zen Micro, 203
    iRiver H10, 203
    Rio Carbon Player, 203
MicroMV format, 150
microphones for camcorders
    lavaliere, 154
    shotgun, 154
    stereo, 154
    surround-sound, 154
    wireless, 154
microprocessors
    AMD, 341
    function of, 341
    Intel, 341
    speeds, 341
    upgrades, digital video editing purposes, 60
Microsoft Corporation, birth of, 4
Microsoft FrontPage, 189
Microsoft Knowledge Base, 6, 224
Microsoft Office
    The Office Weblog, 181
    Woody's Office Portal website, 180
Microsoft PowerPoint, Easter egg, 99
Microsoft Student 2006, 244
Microsoft Word. *See* Word
Microsoft Xbox, 194
Microsoft Xbox 360, 195
microwave ovens, introduction of, 282
MIDI files in web pages, playing, 300
migrating files (Aloha Bob PC Relocator), 60
MiniDV format, 150
Miranda-IM, 118
mIRC software, 120
    Easter egg, 99
MIT Artificial Intelligence Project, 210
MIT Whirlwind computer, 105
    direct keyboard input, 177
Mitnick, Kevin, 255
Mobile Phone Directory website, 272
mobile phones. *See* cell phones
MobileBurn.com blog, 273

moblogs

moblogs
   BusyThumbs.com, 42
   Photoblogs.org, 42
   TextAmerica.com, 42
Model K Complex Number Calculator, 113
MomsTeam.com, 245
monitors
   ClearType technology, enabling (Windows XP), 24
   Macintosh, Apple HD Cinema Display, 164
   Upgrades, digital video editing purposes, 60
monopods for camcorders, 155
Monopoly game, patent issue, 344
Monster PowerCenter, 124
Monster.com, 220
MonsterCable.com, 124
MonsterGecko PistolMouse FPS, 199
Morpheus website, 14
Mosiac web browser, development of, 15
motherboards, 340-341
motorcycles, SportVue MC1 Heads-Up Display, 103
Motorola Ojo Personal Videophone, 91
Motorola Razr V3 cell phone, 273
mouse
   AeroCool AeroBase UFO Gaming Pad, 39
   Logitech diNova Cordless Desktop, 104
   Logitech Football Mouse, 243
   Logitech MediaPlay Cordless Mouse, 18
   Logitech MX1000 laser mouse, 178
   Logitech MX3100 Cordless Desktop, 341
   maintenance, 78
   patent issue, 305
   Razer Diamondback Mouse, 204
   two-button innovation, 118
   Ultra Antistatic Wrist Straps, 56
mouse pads, USB Café Pad, 98
mouseover alerts on web pages, displaying, 298
movie blogs
   DVDVerdict.com, 42
   filmfodder.com, 42
   GreenCineDaily.com, 42
   Metacritic.com, 257
   Tagline.com, 42

Movie Label 2006 software, 126
Mozilla Firefox
   downloading, 6
   Easter egg, 99
   Thunderbird email program, 52
   tips and tricks, 7
   versus Internet Explorer, 6
MP3 audio compression (MPEG-1 Level 3), 11
   Apple iPod, 108
      playlist creation, 110
   AudioTronic iCool Scented MP3 Player, 203
   Blaze Media Pro conversion utility, 205
   Fluxblog website, 15
   MP3.com, 206
   Oakley THUMP MP3 sunglasses, 183
   PlayerBlog.com, 207
   Replay Radio tool, 310
MP3.com, 206
MS-DOS, development of, 295
MSI Bluetooth Star USB Star, 315
MSN Chat website, 122
MSN Encarta.com, 241
MSN Messenger, 118
   files, sending, 120
MSN Music website, 13
MSN Search website, 184
multiple-layout columns in documents, creating (Word), 181
Museum of Bad Album Covers website, 94
music
   45 RPM records, 11
   album art
      displaying, 28
      searching, 28
   Album Art Fixer utility, 85
   Apple iPod, development of, 165
   Apple iTunes, development of, 165
   GarageBand Guitar Cable (Macintosh), 163
   Jazz & Blues Music Reviews blog, 258
   Metacritic.com, 257
   Museum of Bad Album Covers, 94
   playlists, creating (iPod), 110
   presentations, adding to (PowerPoint), 303
   Recording Industry Association of America sales statistics, 262
   StikAx Music Mixer, 11
   web pages, adding to, 300

music blogs
   FreshTuneage.com, 42
   PeopleTalkTooLoud.com, 42
music players
   flash memory types, 325
   High Capacity types, 325
   MicroDrive types, 325
   online shopping criteria, 325
music recorders for podcasts, Edirol R-1 model, 44
My Coupons websites, 316
My Netflix add-in (MCE), 89
My Web Wholesaler, merchandise liquidators, 106
My Yahoo! Website, RSS feeds, 20
mySimon.com, online shopping comparisons, 315

N

NASA Lunar Rover (Apollo 15), 229
National Center for Supercomputer Applications (NCSA), 15
naviPlay Remote, 112
NEC TurboGrafx-16, 233
Nestle's Instant Coffee, 194
Netscape Communications, birth of, 26
networks, home WiFi, configuring, 211
new viruses
   research analysis, 278
   user reporting, 278
   web searching, 278
New York Public Library Digital Library Collections, 243
news
   Daily Rotten website, 95
   Slashdot.org, 3
news blogs
   BigBlog.com, 42
   Plastic.com, 42
   UnknownNews.net, 42
Newseum website, 94
NewsGator RSS News Feeder, 40
newsgroups
   Google Groups, 140
   postings as privacy threat, 73
newspapers, Newseum.org, 94
NeXT Computer, 164, 271
Nielsen-Netratings.com, search engine popularity survey, 182
Nintendo
   64 game system, 234
   Entertainment System (NES), 229

GameBoy, 229
GameCube, 194
Revolution, 195
Super NES game system, 233
NiveusMedia.com, 83
Nobel, Alfred, 186
NOD32 Antivirus utility, 277
Nokia N90 cell phone, 273
NORAD Santa websites, 337
Northpole.com, 337
Norton AntiSpam Web site, 133
Norton AntiVirus Web site, 276
NotebookReview.com, 239
notebooks
   batteries, extending life of, 240
   business laptops, 235
   college student market, 237
   desktop replacement, 235
   LapCool 2 Notebook Cooler, 237
   Laptop Review Blog, 240
   memory upgrades, 238
   NotebookReview.com, 239
   printer options on the road, 239
   RMClock utility, 238
   shopping criteria, 236
   tablet PC, 235
   Targus DEFCOM MDP Motion Sensor, 236
   ultracompact, 235
   usage statistics, 235
Noyce, Robert, 22
NPDTech Group website, camcorder sales, 150
NSClean tool, browser cookie cleaner, 72
numbered lists, creating (Word), 176

O

Oakley THUMP MP3 sunglasses, 183
off-axis viewing (television), 63
The Office Weblog, 181
Offshore Mixmaster Anonymous Remailer website, 54
Olympus m-robe 500i, 200
OneSeek.com metasearcher, 185
online coupons
   Bargain Boardwalk, 316
   Bargain Shopping, 316
   CouponMountain, 316
   CouponPages.com, 316
   Daily eDeals, 316
   DealofDay.com, 316
   eCoupons, 316

Kovalchik Farms, 316
My Coupons, 316
Specialoffers.com, 316
online dating, usage statistics, 214
online greeting cards
   123Greetings.com, 336
   BlueMountain.com, 336
   GreetingCards.com, 336
   Hallmark.com, 336
   KinkyCards.com, 336
   Regards.com, 336
   YahooGreetings.com, 336
Online Music Blog, 261
online music services, individual song purchases, 263
online shopping, 314
   Amazon.com
      associates, signing up, 332
      Auctions (used), 330
      clearance items, 329
      Friday Sale items, 329
      Inside This Book feature, 327
      Marketplace items (used), 330
      Outlet, 329
      page appearance, 327
      page personalization, 327
      search guidelines, 327
      site navigation, 327
      Super Saver Shipping, 331
      Today's Deal, 328
   Christmas crafts
      AllCrafts.net, 334
      Craftown, 334
      Enchanted Learning, 334
      Kids Kreate, 334
      KidsDomain, 334
      Make-Stuff.com, 334
   Christmas gifts
      Christmas Gift Directory, 333
      Holiday Gift Ideas, 333
      Surprise.com, 333
   Christmas greeting card programs, 335
   comparison sites
      Best Price, 316
      BizRate.com, 315
      Froggle.com, 315
      mySimon.com, 315
      Shopping.com, 315
      Yahoo! Shopping.com, 315
   Craig's List, online classifieds, 325
   credit card safety, 313
   DoubleClick.com, research survey, 333

outdoor travel and adventure

   Froggle.com searches, most popular products, 313
   geek products, 321
      car audio systems, 324
      cell phones, 323
      computers, 322
      Gadget Universe, 321
      home theater components, 326
      music players, 325
      PDAs, 323
      smartphones, 323
      ThinkGeek.com, 321
   online coupons, 316
      Bargain Boardwalk, 316
      Bargain Shopping, 316
      CouponMountain, 316
      CouponPages.com, 316
      Daily eDeals, 316
      DealofDay.com, 316
      eCoupons, 316
      Kovalchik Farms, 316
      My Coupons, 316
      Specialoffers.com, 316
   product reviews
      Amazon.com, 314
      ConsumerReports.org, 314
      ConsumerREVIEW.com, 314
      ConsumerSearch.com, 314
      Epinions.com, 314
      ReviewFinder.com, 314
   SalesHound.com, brick and mortar store sales resources, 319
   shipments, tracking, 320
   shipping costs, reduction measures, 319
   shipping services
      FedEx, 320
      U.S. Postal Service, 320
      UPS, 320
   ShoppingBlog.com, 320
Open Directory website, 183-184
Open Door website, 244
open-air earphones, 206
Oppenheimer, Robert, 92
Oregon Scientific AWS888 Weather Forecaster, 130
OurMedia.org, podcasting services, 46
outdoor photography, lighting guidelines, 147
outdoor travel and adventure
   camping devices, 161
   GORP websites, 160

Outlook

Outlook
  mailing groups, creating, 52
  mass mailings
    managing, 52
    sending, 52
  spam, blocking, 132
Outrageous eBay Auctions blog, 107
Overstock.com, merchandise liquidators, 106

P

page fault errors, troubleshooting, 223
Paint Shop Pro
  Easter egg, 99
  photo-editing tool, 147
palmOne LifeDrive, 296
Panasonic BL-C30A Network Camera, 170
Panda Software website, 276
Panda TruPrevent, 276
paragraph styles (Word)
  applying, 177
  elements of, 177
  modifying, 177
  predesigned templates, 177
Pascal programming, 162
Pascal, Blaise, 162
Pauling, Linus, 306
PC Certify Lite tool, 223
PC Certify utility, 78
PC Pitstop website, 79
PC-cillin website, 276
PC-Diagnostics.com, 78
PCs
  Computer Industry Almanac, number in use worldwide, 339
  computer manufacturer sales rankings, 162
  Easy Computer Basics, 339
  Hardware, system units, 340
  iPod connections, 109
  maintenance and care checklist, 225
  Media Center PCs, buying considerations, 82
  Memory, storage terminology, 342
  microprocessors, function of, 341
  setup
    first-time startup operations, 344
    turning on components, 344
  speed tips
    hard disk defragmentation, 75
    hard disk error scans, 75
    unnecessary file/program deletion, 75
  system units
    motherboards, 340
    preventative maintenance, 76
  TechGuy.com, 343
  upgrades
    ATI HDTV Wonder card, 65
    digital video editing, 60
    hard disks, 61
      port additions, 59
    USB device sharing, IOGEAR MiniView Micro KVM Switch, 77
PDAs (personal digital assistants). See also smartphones
  Dell Axim X50v Pocket PC PDA, 302
  online shopping criteria, 323
  palmOne LifeDrive, 296
  TomTom GPS Navigator, 76
Pearl Echo spyware (legitimate), 172
peer-to-peer (P2P) connections, instant messaging, 117
peer-to-peer CD swapping, legality of, 263
Pentium chips, history, 77
PeopleTalkTooLoud.com music blog, 42
performance tips for PCs
  hard disk defragmentation, 75
  hard disk error scans, 75
  unnecessary file deletion, 75
  unnecessary program deletion, 75
personal security, Citizen TASER X26c Pistol, 5
pets, Global Pet Finder, 23
Pew Internet & American Life Project
  blogging statistics, 37, 176
  instant messaging usage, 117
  iPod usage statistics, 108
  podcast listening statistics, 43
  terrorism survey, 69
Philips RC9800i WiFi Remote Control, 82
Phillips head screwdriver, 180
phishing schemes, 74
photoblogs
  BusyThumbs.com, 42
  Photoblogs.org, 42
  TextAmerica.com, 42
Photoblogs.org, moblogs/photoblogs host, 42
photography, digital cameras
  digital SLRs, 143
  Epson Perfection 4180 Photo Scanner, 143
  men versus women brand preferences, 141
  prosumer models, 142
  SanDisk Photo Album, 142
  selection criteria, 141
PhotographyBLOG, 149
Photoshop
  images
    edges, feathering, 281
    red eye removal, 282
    repairing, 280
  Planet Photoshop website, 284
  Unofficial Photoshop Blog, 285
  user statistics, 280
Photoshop Elements, 283
Picasa tool, image management functions, 324
pixels (HDTV), 126
Planet Photoshop website, 284
plasma televisions, 67
Plastic.com news blog, 42
Play Football.com, 245
playlists
  CDs (Windows Media Center), 85
  iPod, creating, 110
PodBus.com, podcasting services, 46
Podcast.net, 47
PodcastAlley.com, 47
PodcastBunker.com, 47
PodcastDirectory.com, 47
PodcastingStation.com, 47
PodcastPickle.com, 47
podcasts
  Absolute Beginner's Guide to Podcasting, 46
  Audacity audio editor, 46
  blog hosting services
    AudioBlog.com, 46
    LibSyn.com, 46
    Ourmedia.org, 46
    PodBus.com, 46
  blog syndication services, FeedBurner.com, 46
  blogs, Adam Curry, 48
  creation requirements, 44
  directories
    AllPodcasts.com, 47
    DigitalPodcasts.com, 47
    iPodder.org, 47
    iPodderX.com, 47
    iTunes.com, 47-48
    Podcast.net, 47
    PodcastAlley.com, 47

PodcastBunker.com, 47
PodcastDirectory.com, 47
PodcastingStation.com, 47
PodcastPickle.com, 47
Syndic8.com, 47
hardware
 Edirol R-1 Portable Music Recorder, 44
 Griffin Radio SHARK AM/FM Radio/Receiver, 45
 requirements, 44
listening statistics, 43
publishing, 46
range of, 43
recording, 45
RSS feeds, subscribing, 43
save requirements, 44
subscriptions, 43

PoliBlog.com, 41
political blogs
 The Agonist, 41
 AndrewSullivan.com, 41
 Ankle Biting Pundits, 41
 Captain's Quarters, 41
 Daily Kos, 41
 The Decemberist, 41
 Electablog.com, 41
 Eschaton, 41
 eTalkingheads, 41
 INDC Journal, 41
 Little Green Footballs, 41
 PoliBlog, 41
 Powerline, 41
 Stage Left, 41
 Talking Points Memo, 41
 Wonkette.com, 41

Polk Audio I-Sonic receiver, 311
Polk Audio XRt12 receiver, 311
POP3 (Post Office Protocol 3), 50
 email accounts, 49
portable media players
 AudioTronic iCool Scented MP3 Player, 203
 earbuds, 206
 earphones
  closed-air, 206
  open-air, 206
 Flash memory players
  Creative Zen Nano Plus, 204
  iPod Shuffle, 204
  iRiver iFP-900 Series, 204
  Rio Forge Sport Player, 204

 MicroDrive
  Creative Zen Micro, 203
  iRiver H10, 203
  Rio Carbon Player, 203
 video, 205
portable music players, Apple iPod, 108
 connecting to PCs, 109
 playlist creation, 110
portals (blogs)
 BlogChalking.com, 39
 BlogUniverse.com, 39
 Blogwise.com, 39
 Eatonweb.com, 39
 Globe of Blogs.com, 39
 Weblogs.com, 39
ports
 adding, 59
 Marathon Computer RePorter, 222
 USB, 62
postings as privacy threat, 73
PostSecret Blog, 267
Powerline.com political blog, 41
PowerOpen Association, 66
PowerPoint
 presentations
  2D charts, depth additions, 304
  Beyond Bullets websites, 306
  branching types, creating, 302
  creating, 301
  music, adding to, 303
  publishing to Web, 305
  professional handouts, printing, 306
PowerSellers program (eBay), 105
pre-World Wide Web tools
 Archie, 9
 Gopher, 9
 Veronica, 9
 WAIS, 9
presentations
 2D charts, depth additions, 304
 Beyond Bullets websites, 306
 branching types, creating (PowerPoint), 302
 clip art, Clipart Connection website, 305
 creating (PowerPoint), 301
 music, adding to (PowerPoint), 303
 professional handouts, printing (PowerPoint), 306
 publishing to Web (PowerPoint), 305
preventing identity theft, 70

PVRblog.com

Print 2 Kinko's website, 239
Print Explosion Deluxe, 335
Print Perfect Deluxe, 335
print services
 EFI PrintMe, 239
 Print 2 Kinko's, 239
Print Shop, 335
printers
 maintenance tips, 79
 notebook options on the road, 239
PrintMaster Greeting Card Deluxe, 335
privacy
 Electronic Frontier Foundation (EFF) website, 73
 Internet postings, 73
 phishing schemes, 74
 Privacy.org, 74
 workplace issues, 72
Privacy.org website, 74
ProAim Golfing Goggles, 189
products, online shopping reviews
 Amazon.com, 314
 ConsumerReports.org, 314
 ConsumerREVIEW.com, 314
 ConsumerSearch.com, 314
 Epinions.com, 314
 ReviewFinder.com, 314
program freezes, closing (Windows XP), 33
programs
 anti-spam, 133
 antivirus, 276
 Start menu, adding to (Windows XP), 36
progressive scanning versus interlaced scanning, 127
prosumer camcorders, 151
 cost, 153
 features, 153
prosumer digital cameras, 142
ProtechDiagnostics.com, 223
protocols (email)
 IMAP, 50
 POP3, 50
 SMTP, 50
publishing
 podcasts, 46
 presentations to Web (PowerPoint), 305
push-button phones, 60
PVRblog.com, digital video blog, 155

363

quadraphonic sound

Q - R

quadraphonic sound, 218
Questia.com, 243
QuikDrop website, eBay Trading Assistant franchise, 107
QWERTY keyboard versus Dvorak keyboard, 127

racing wheels, 197
radio
 FM broadcasting, 147
 Griffin Radio SHARK AM/FM Radio/Receiver, 45
 HD
  iBiquity.com, 312
  receivers, 312
 Replay Radio tool, 310
 satellite
  Orbitcast.com, 312
  SIRIUS Satellite Radio, 307
  SIRIUS Satellite Radio, cost, 309
  SIRIUS Satellite Radio, home receivers, 311
  SIRIUS Satellite Radio, programming, 309
  SIRIUS Satellite Radio, Sportstyer receiver, 310
  XM Radio, 307
  XM Radio, car receivers, 310
  XM Radio, cost, 308
  XM Radio, home receivers, 311
  XM Radio, programming, 308
  XM Radio, XMFan.com, 311
 Tivoli Model One Radio, 118
 War of the Worlds, 286
radio (Internet), 15
Radio Shack TRS-80 Model 1 computer, 205
Radio-Locator website, 15
RadioMOI website, 15
RAM (random access memory), 342
 chips
  Double-Data Rate SDRAM (DDR SDRAM), 56
  Extended Data Out DRAM (EDO DRAM), 56
  Fast-Page-Mode DRAM (FPM-DRAM), 56
  Rambus DRAM, 56
  Synchronous Dynamic RAM (SDRAM), 56
Rambus DRAM, 56
Ranger 7 moon probe, 200
Razer Diamondback Mouse, 204

RazzleDazzleRecipes.com, 338
RCA Corporation, introduction of 45 RPM records, 11
read/write heads, 343
ReaderWare, 256
rear-projection televisions (RPTVs), 64
 digital light projection (DLP) technology, 65
 LCD projection technology, 66
Recipe*zaar.com, 338
RecipeLink.com, 338
recipes for Christmas
 Allrecipes.com, 338
 American Diabetes Association, 338
 BackoftheBox.com, 338
 FabulousFoods.com, 338
 RazzleDazzleRecipes.com, 338
 Recipe*zaar.com, 338
 RecipeLink.com, 338
Recording Industry Association of America, CD sales statistics, 262
recording podcasts, 45
red eye effect, removing from images (Photoshop), 282
refdesk.com, 243
reference sites
 About.com, 241
 Answers.com, 241
 Dictionary.com, 241
 MSN Encarta.com, 241
 Wikipedia.com, 241
Regards.com, 336
Registry
 desktop icons, removal of shortcut arrows, 29
 Internet Explorer toolbar, adding background graphics to, 34
 owner information, changing, 36
remailers (email), 54
remote computation, 241
Remote Spy spyware (legitimate), 172
removable shirt collars, 36
removing skin blemishes (Photoshop), 284
repairing images (Photoshop), 280
Replay Radio tool, 310
reporting new viruses, 278
ReviewFinder.com, online shopping product reviews, 314
RIMM memory module, 56
Rio Carbon Player, 203
Rio Forge Sport player, 204
Riot Anonymous Remailer website, 54

ripping CDs, 261
 tools, 261
 Windows Media Player, 85, 262-263
RMClock utility, 238
RoadBlock website, 133
Robosapien V2 Toy Robot, 328
Roentgen, Wilhelm, 41
Roller Coaster Tycoon 2, Easter egg, 98
RollerCoaster Tycoon 3, 197
root beer, invention of, 130
routers, Linksys WRT54GS Wireless Router, 6
Roxio Easy Media Creator, CD burning function, 265
RSS feeds
 incorporating in web pages (My Yahoo!), 20
 NewsGator, 40
 podcasts
  FeedBurner.com, 46
 subscribing, 43
Rubick's Cube, 185
running performance GPS monitors, 157

S

Safe mode, troubleshooting, 222
Saitek Cyborg evo Wireless, 198
Saitek X52 Flight Control System, 198
SalesHound.com, brick and mortar store sales resources, 319
sampling rates, digital music reproduction, 10
SanDisk Photo Album, 142
Santa Claus Online.com, 337
Santa Claus sites
 Claus.com, 337
 NORAD Santa, 337
 Northpole.com, 337
 Santa Claus Online.com, 337
Santana's *Supernatural*, CD Easter egg, 97
Sarnoff, David, 55
satellite Internet connections, 4
satellite radio
 Orbitcast.com, 312
 Replay Radio tool, 310
 SIRIUS Satellite Radio, 307
  cost, 309
  home receivers, 311
  programming, 309
  Sportster receiver, 310
  subscribers, 307

shopping

XM Radio, 307
   car receivers, 310
   cost, 308
   Delphi XML MyFi receiver, 308
   home receivers, 311
   programming, 308
   subscribers, 307
   XMFan.com, 311
satellite speakers, 250
***Saving Private Ryan*, 255**
saxophone, 166
ScanDisk utility, 75
ScanGauge Automotive Computer, 275
scanners, Epson Perfection 4180 Photo Scanner, 143
Schrapnel, Henry, 69
Scooba Robotic Floor Washer, 322
Scope's Monkey Trial, 121
SCOTTeVEST tech clothing, 29
scrapbooks, Creating Keepsakes Scrapbook Designer, 336
screen savers, Marine Aquarium, 93
screen sizes for televisions, 63
scrolling background images in web pages, 297
scuba diving devices, 160
Sculley, John, 163
SDTV (standard definition digital television) format, 124
***Seabiscuit*, 254**
Seagate.com, external hard drives, 51
search directories
   LookSmart, 183
   Open Directory, 183
   versus search engines, 183
Search Engine Blog, 187
search engines, 183
   AllTheWeb.com, 184
   AltaVista.com, 184
   AOLSearch.com, 184
   AskJeeves.com, 184
   Google, 183
      Answers feature, 140
      Catalogs feature, 140
      Desktop Search, 138
      Gmail Service, 139
      Groups feature, 140
      home page personalization, 138
      images, search filters, 136
      Language Tools, 139
      Local feature, 140
      Maps feature, address searches, 137

      News feature, 140
      Print feature, 140
      Ridefinder feature, 140
      search statistics, 135
      search tricks and tips, 135
      top 10 search list, 90
      The Unofficial Google Weblog, 140
      Video feature, 140
   guidelines, 186
   HotBot.com, 184
   metasearches
      Beaucoup.com, 185
      CNETSearch.com, 185
      Dogpile.com, 185
      Mamma.com, 185
      MetaCrawler.com, 185
      OneSeek.com, 185
      SearchSpaniel.com, 185
      WebFerret, 185
      WebTaxi.com, 185
   MSNSearch.com, 184
   Nielsen-Netratings survey, 182
   Search Engine Watch website, 186
   Teoma.com, 184
   versus search directories, 183
   Yahoo!
      company history, 16
      meta word searches, 19
      My Web feature, 20
      navigational tips, 17
      search shortcuts, 16
      toolbar downloads, 19
   Yahoo! Maps, 18
searching
   blog directories, 37
   podcasts, 47
   Web via metasearches, 185
SearchSpaniel.com metasearcher, 185
section breaks in documents
   deleting (Word), 179
   inserting (Word), 179
security
   email spam, 133
      address protection, 131
   spyware
      Counterexploitation (CEXX).org, 172
      dangers of, 168
      legitimate programs, 172
      potential damage, 169
      removing, 168, 170
      Spyware Guide.com, 170
      utilities, 171

   TrimTrac GPS Security Locator, 70
   viruses
      Absolute Beginner's Guide to Security, Spam, Spyware, and Viruses, 276
      antivirus software, 276
      defined, 274
      NOD32 Antivirus utility, 277
      prevention measures, 275
      reducing risks, 275
      repairing infections, 277
      research analysis, 278
      signature scanning, 279
      signs of infection, 274
      transmission methods, 274-275
      user reporting, 278
      Virus Bulletin, 278
      web searching, 278
      WornBlog.com, 279
   WiFi networks, 212
Sega Genesis game system, 233
Sega Master System (SMS), 229
Sega Saturn game system, 234
Selective Sequence Electronic Calculator (SSEC), 23
self-defense, Citizen TASER X26c Pistol, 5
sending files
   via AOL Instant Messenger, 120
   via MSN Messenger, 120
   via Yahoo! Messenger, 120
Serene Screen website, 93
Service Pack 2 (Windows XP)
   downloading, 22
   features of, 22
   installation of, 22
***Sgt. Pepper's Lonely Hearts Club Band*, CD Easter egg, 97**
shipments, online shopping, tracking, 320
shipping costs, online shopping
   reduction measures, 319
   money saving tips on Amazon.com, 331
shipping services
   FedEx, 320
   U.S. Postal Service, 320
   UPS, 320
Shirt-Pocket.com, 165
shooting tips in digital photography, 148
shopping. *See also* **online shopping**
   DVD players, desired features, 249
   Google Catalogs feature, 140

Shopping.com

Shopping.com, online shopping comparisons, 315
ShoppingBlog.com, 320
shortcut arrows in desktop icons, removing, 29
shotgun microphones, 154
SHOUTcast website, 15
Siemens S66 cell phone, 273
signature scanning (viruses), 279
Silicon Graphics, 26
SIMM memory module, 56
SIRIUS Satellite Radio, 307
    cost, 309
    home receivers, 311
    programming, 309
    Sportster receiver, 310
    subscribers, 307
Skeletal Systems website, cartoon character studies, 90
skin blemishes, removing (Photoshop), 284
Sky Captain and the World of Tomorrow, 256
Slashdot.org tech blog, 3, 42
Sleeptracker Watch, 334
slide shows
    presentations
        branching types (PowerPoint), 302
        creating (PowerPoint), 301
    SanDisk Photo Album, 142
    Wallflower 2 Multimedia Picture Frame, 136
Smartphones. *See also* PDAs
    Computer Industry Almanac website, sales statistics, 268
    online shopping criteria, 323
    typical features, 271
    versus cell phones, 271
Smithsonian Institution Libraries, 243
SMTP (Simple Mail Transfer Protocol), 50
Snopes.org, urban legend debunking, 293
softening wrinkles, age effects (Photoshop), 283
software
    anti-spam, 133
    antivirus, 276
    Cinemar MainLobby, 66
    Easter eggs
        Final Cut Pro, 99
        Grand Theft Auto - San Andreas, 101
        Microsoft PowerPoint, 99
        mIRC, 99

        Mozilla Firefox, 99
        Paint Shop Pro, 99
    FTP, WS_FTP program, 7
    piracy in China, 100
    Tweak UI tool, downloading, 34
solar power plant, 152
Solio Solar Charger, 112
Sonos Digital Music System, 249
Sony Ericsson S710z cell phone, 273
Sony
    Playstation, 234
    PlayStation 2, 194
    PlayStation 3, 195
    Surround Sound Headphones, 215
Sophos.com, Zafi-D virus reports, 274
Soundblaster Audigy 4, 309
SoundSak SonicBoom speakers, 157
space
    Aldrin, Edwin (Buzz), 20
    Apollo 13 flight, 97
    Apollo 14 moon golf, 37
    ESSA-1 weather satellite, 35
    Ham, the Monkey exploits, 29
    planetary discoveries, 17
    untethered spacewalks, 38
Space Needle (Seattle), 106
Spalding Adding Machine, 14
spam (email), 129
    Absolute Beginner's Guide to Security, Spam, Spyware, and Viruses, 129
    address gathering techniques, 130
    address protection, 131
    anti-spam software, 133
    blockers/filters, 129
    blocking in Outlook 2003, 132
    deleting, 129
    Email Sentinel Pro utility, 132
    filtering services
        Emailias, 133
        Mailshell, 133
        SpamCop, 133
        SpamMotel, 133
    inbound statistics, 129
    Spam Kings Blog, 134
    Spamhaus.org, 133
    tracing and reporting spammer abuse, 134
    worldwide sources, 49
Spam Kings Blog, 134
spamblocks, 131
SpamCop.net, 133
Spamhaus.org, 133
SpamKiller website, 133
SpamMotel.com, 133

speaker systems
    home theater
        bookshelf, 250
        choosing, 250
        floor-standing, 250
        satellite, 250
    Logitech Z-5500 Speakers, 255
    music
        Creative GigaWorks PC Speaker System, 329
        Creative I-Trigue Speaker System, 12
        iPod types, 111
special effects
    display properties, setting (Windows XP), 26
    Display Properties utility, activating (Windows XP), 35
    System Properties utility, activating (Windows XP), 35
Specialoffers.com website, 316
Spherex RX2 Game Chair, 195
Spider-Man 2, 256
spin rates for hard disks, PC upgrade considerations, 61
sports
    Fish-N-Log Professional Suite, 159
    fishing-finding devices, 158
    GPS boating devices, 159
    scuba diving devices, 160
sports blogs
    BadJocks.com, 42
    Fanblogs.com, 42
Sports Illustrated for Kids, 245
sports sites for kids
    Amateur-Sports.com, 245
    InfoSports, 245
    Kid Sports Network, 245
    Kids Sports Bulletin Board, 245
    MomsTeam, 245
    Play Football, 245
    Sports Illustrated for Kids, 245
    Youth Sports Instruction, 245
SportVue MC1 Heads-Up Helmet Display, 103
SportVue.net, 103
Spy Buddy uspyware (legitimate), 172
Spy Sweeper website, 171
SpyAgent spyware (legitimate), 172
Spybot Search & Destory website, 171
spyware
    Counterexploitation (CEXX).org, 172
    dangers of, 168
    definitions, preventative maintenance, 80

legitimate programs, 172
potential damage, 169
removing, 168-170
Spyware Guide.com, 170
Spyware Warrior Blog, 175
U.S. government, Carnivore, 175
utilities, 171
Webroot.com, infection statistics, 168
Spyware Warrior Blog, 175
Stage Left.info political blog, 41
Stanford Institute for Quantitative Study of Society, Internet usage per day, 28
Star Wars Episode 1 - The Phantom Menace, DVD Easter egg, 96
Star Wars Episode III, 256
Stardock.com, desktop customization products, 25
Start menu programs, adding (Windows XP), 36
Stealth Activity Reporter spyware (legitimate), 172
stereo microphones, 154
stereophonic sound, 218
StikAx Music Mixer, 11
Strategy Analytics website, HDTV usage statistics, 123
streaming audio, 15
stuck programs, closing (Windows XP), 33
styles
    applying (Word), 177
    elements of (Word), 177
    modifying (Word), 177
    predesigned templates (Word), 177
Sunbeam 20-in-1 Superior Panel, 59
Sunbeamtech.com, 59
Sunpak Readylight 20 video lighting system, 151
SuperDuper cloner (Mac OS X), 165
surgical zippers, 183
Surprise.com website, 333
surround sound
    Cubase SX recording studio program, 217
    Dolby, 214
    Dolby Digital discrete format, 215
    Dolby Digital EX discrete format, 215
    Dolby Pro Logic, 214
    Dolby Pro Logic IIx, 214
    DTS discrete format, 215
    Fantasia, 217
    format comparisons, 216
    history of, 218-219
    Sony Surround Sound Headphones, 215
    Soundblaster Audigy 4, 309
    Surround Sound Discography, 218
    Surround Sound Discography website, 218
    surround-sound microphones, 154
    surveillance in the workplace, 72
    survival kits, 17
    Suunto N3i SmartWatch, 287
    Symantec AntiVirus Research Center (SARC), 278
    Synchronous Dynamic RAM (SDRAM), 56
Syndic8.com, 47
system memory upgrades
    DIMMs, 56
    RAM chips, 56
    RIMMs, 56
    SIMMs, 56
System Properties utility (Windows XP), special effects, 26, 35
system units, 340
    maintenance tips, 76

T

tablet PC notebooks, 235
Tagline.com movie blog, 42
tags (HTML), 190-191
Talk City website, 122
Talking Points Memo political blog, 41
Targus DEFCOM MDP Motion Sensor, 236
Taser.com, self-protection devices, 5
Task Manager, end processes method, 33
Taskbar (Windows XP)
    hiding, 27
    moving, 27
TaySays.com, fishing software, 159
TDMA (time division multiple access) cell phones, 268
tech blogs
    BoingBoing.net, 42
    Gizmodo.com, 42
    Slashdot.org, 42
    Techdirt.com, 42
tech clothing, SCOTTeVEST, 29
Techdirt.com tech blog, 42
TechGuy.com, 343

Technology
    BoingBoing.net, 9
    Engadget.com blog, 101
    Luddite Riots, 68
    Tech Blog website, 62
TechWhack.com, Windows XP tweaks and tips, 36
telegraph, invention of, 137
telephones
    Action lec Internet Phone Wizard, 92
    direct-dial transatlantic phone service, 59
    first mobile phone call, 90
    Motorola Ojo Personal Videophone, 91
    push-button phones, 60
television
    ATI HDTV Wonder card, 65
    Couch Potato Tormentor TV remote, 64
    Dell UltraSharp 24 LCD Display monitor, 297
    digital formats
        digital light projection (DLP) technology, 65
        EDTV format, 124
        HDTV 720p format, 124
        HDTV 1080i format, 124
        HDTV 1080p format, 124
        HDTV advantages, 125
        sale statistics, 63
        SDTV format, 124
        versus analog, 123
    DVRs
        Windows Media Center features, 84
    flat-screen displays
        LCD, 68
        plasma, 67
    HDTV
        aspect ratios, 128
        HDTV Blog, 128
        number of pixels, 126
        progressive scanning, 127
        smart expand mode, 128
        zoom mode, 128
    household statistics thought the decades, 286
    rear-projection (RPTVs), 64
        LCD projection technology, 66
    Sarnoff, David, 55

television

selection criteria, 63
aspect ratios, 63
off-axis viewing, 63
screen sizes, 63
TViX Digital Movie Jukebox, 125
Telstar 1 satellite, 182
Teoma website, 184
Texas Instruments' Speak & Spell, 155
text translators, Language Tools (Google), 139
TextAmerica.com, moblogs/photoblogs host, 42
TheGreenButton.com, Windows Media Center resources, 85
Theme Manager 2005, 25
ThinkGeek.com, geek merchandise, 321
ThinkSecret.com, Apple insider information, 166
Thomas Hawk's Digital Connection blog, 193
THOMAS website, government and legislative resources, 8
Three Mile Island nuclear accident, 82
ThrustMaster Enzo Ferrari Force Feedback Wheel, 199
Thunderbird email program, 52
TIME Magazine, Machine of the Year declaration, 5
TiVo, 248
Tivoli Model One Radio, 118, 311
TOAST.net websites, connection speed tester, 3
Todd Snider's *Songs for the Daily Planet*, CD Easter egg, 97
Tom's Hardware Guide website, 61
TomTom GPS Navigator, 76
toothbrushes, VIOlight Toothbrush Sanitizer, 177
toothpaste tube, invention of, 135
Top of the Needle restaurant (Seattle), 106
TopHosts.com website, 189
tracking online shopping orders, 320
TRADIC Computer, 70
Trading Assistants (eBay)
franchises, 107
qualifications, 107
traffic information (Yahoo!), 16
Trans-Alaska Oil Pipeline, 163
Transistor, patent issue, 263
Treo 650 smartphone, 271
tri-band GSM cell phones, 269

Trillian, 118
Trimble.com, 70
Trimline Telephone, 279
TrimTrac GPS Security Locator, 70
Tripod website, 188
Tron, 181
troubleshooting tips, 220
bad WiFi connections, 210
dead computers, 221
email, common problems, 53
emergency startup disks, creating, 224
Mac OS X, tools and guidelines, 167
page fault errors, 223
PC maintenance and care checklist, 225
RescueComp Blog, 225
Safe mode, 222
TViX Digital Movie Jukebox, 125
Tweak MCE utility, 84
Tweak UI utility, downloading, 34
TweakXP.com, 35
two-button mouse, introduction of, 118
TypePad.com, blog host community, 39
typewriter, invention of, 193

U

U.S. Postal Service, product shipment tracking, 320
UFO News Blog, 294
UglyDress.com, 93
Ultra Antistatic Wrist Straps, 56
ultracompact notebooks, 235
UltraProducts.com, antistatic merchandise, 56
UNIVAC computer, 158, 325
Universal Product Code (UPC), 168
universal remote controls (TV components), 252
UnknownNews.net news blog, 42
Unofficial Apple Weblog, 167
Unofficial Google Weblog, 140
updates, preventative maintenance, 80
upgrades
ITtoolbox.com industry survey, 55
pre-planning checklist, 55
system memory
DIMMs, 56
RAM chips, 56
RIMMs, 56
SIMMs, 56

upgrading
notebook memory, 238
PCs
digital video editing, 60
hard disks, 61
port additions, 59
UPS (uninterruptible power supplies), APC ES 725 Broadband device, 24
UPS (United Parcel Service), product shipment tracking, 320
urban legends, 291
common tales, 292-293
ICE cell phone entry, 294
Snopes.org, 293
USB
Alarm Clock, 216
Café Pad, 98
Christmas Tree, 335
devices, sharing via IOGEAR MiniView Micro KVM Switch, 77
flash drives, Hall & Partners.com, usage survey, 301
Geek website, 33, 98
Mini-Aquarium, 33
ports, peripheral connections, 62
UseIt.com, top 10 web page design mistakes, 188
Usenet
Google archive, 73
Google Groups, 140
USS Nautilus submarine, 258

V

V1 rockets, 157
vacuum cleaner, invention of, 48
VariZoom VZ-LSP, camcorder steady support system, 155
vaults, digital photo storage, 149
Velvetta cheese, 148
Venexx Perfume Watch, 221
Verbatim Store 'n' Go, 270
Veronica tool, pre-WWW archiving functions, 9
Video Game Blog, 234
video games, 195
Atari
80 Classic Games CD-ROM, 229
5200, 228
7800 ProSystem, 229
Flashback, 196
Jaguar, 233
PONG, 226
Bally Professional Arcade, 227
building, minimum requirements, 196

Web-Radio website

Channel F Video Entertainment System, 226
Classicade Upright Game System, 228
Coleco Colecovision, 228
Coleco Telstar, 226
controllers
  choosing, 197
  Gravis Xterminator Force Feedback Gamepad, 198
  Logitech Cordless Rumblepad 2, 198
  Logitech MOMO Racing Wheel, 199
  MonsterGecko PistolMouse FPS, 199
  Saitek Cyborg evo Wireless, 198
  Saitek X52 Flight Control System, 198
  ThrustMaster Enzo Ferrari Force Feedback Wheel, 199
  X Arcade Solo, 199
  Zboard Modular Keyset, 199
Entertainment Software Association sales statistics, 194
history
  fifth generation, 233
  first generation, 226
  fourth generation, 229
  second generation, 227
  sixth generation, 234
  third generation, 228
IGN.com, 198
Internet Video Game Museum, 233
Jakks TV Games, 227
Magnavox Odyssey, 226-227
Mattel Intellivision, 227
Microsoft Xbox 360, 195
NEC TurboGrafx-16, 233
Nintendo
  64, 234
  Entertainment System (NES), 229
  GameBoy, 229
  Revolution, 195
  Super NES, 233
PlayStation 3, 195
RollerCoaster Tycoon 3, 197
Sega Genesis, 233
Sega Master System (SMS), 229
Sega Saturn, 234
Sony Playstation, 234
Spherex RX2 Game Chair, 195
systems, choosing, 194

video players, 205
  Hasbro VideoNow Color, 242
VIOlight Toothbrush Sanitizer, 177
Virtual Library Project, 233
virtual memory, 8
Virus Bulletin website, 278
viruses, 274
  Absolute Beginner's Guide to Security, Spam, Spyware, and Viruses, 276
  antivirus software, 276
  definitions, preventative maintenance, 80
  hoaxes, 287
  infection computers, repairing, 277
  Michelangelo virus, 63
  new
    research analysis, 278
    user reporting, 278
    web searching, 278
  NOD32 Antivirus utility, 277
  prevention measures, 275
  reducing risks, 275
  signature scanning, 279
  signs of infection, 274
  Symantec AntiVirus Research Center (SARC), 278
  transmission methods, 274-275
  Virus Bulletin website, 278
  WornBlog.com, 279
  Zafi-D, 274
VisiCalc spreadsheet program, 126
VoiSec digital refrigerator voice recorder, 50

W

W3 Anonymous Remailer website, 54
wacky websites
  Daily Rotten, 95
  Death Clock, 91
  Dumb Auctions, 92
  Dumb Bumpers, 92
  Dumb Criminals, 92
  Dumb Laws, 92
  Dumb Warnings, 92
  Museum of Bad Album Covers, 94
  Skeleton Systems, 90
  UglyDress.com, 93
Wacom Cintiq 21UX Touchscreen Pen Tablet, 303
Wacom.com, 282
WAIS tool, pre-WWW archiving functions, 9

Wal-Mart Music Downloads website, 13
Wallflower 2 Multimedia Picture Frame, 136
Wallpaperizer tool, 292
War of the Worlds radio broadcast, 286
watches
  Sleeptracker Watch, 334
  Suunto N3i SmartWatch, 287
  Venexx Perfume Watch, 221
watermarks
  documents, adding (Word), 178
  web pages, adding, 296
Wayback Machine website, history resources, 8
WD Passport Pocket Hard Drive, 276
Weather, Oregon Scientific AWS888 Weather Forecaster, 130
web browsers
  cookies (NSClean tool), 72
  Internet Explorer, adding background graphics to toolbar, 34
  Mosiac development, 15
  Mozilla Firefox
    Thunderbird program, 52
    tips and tricks, 7
    versus Internet Explorer, 6
web pages
  background colors, cycling, 295
  background scrolling effect, 297
  creating, 188
  fonts
    changing, 193
    resizing, 193
  HTML coding, 190
  hyperlinks, inserting, 192
  JavaScript Weblog, 300
  JavaScript.com tricks, 299
  menus, JavaScript Menu Master, 298
  mouseover alerts, displaying, 298
  music, adding to, 300
  opening link in new window, 299
  RSS feeds, incorporating (My Yahoo!), 20
  setting as desktop background, 25
  top 10 design mistakes, 188
  watermarks, adding, 296
Web Pages That Suck website, 192
web searches
  Boolean searching, operators, 187
  guidelines, 186
web-based email accounts, 49
Web-Radio website, 15

webcams

webcams
    Logitech QuickCam Orbit, 119
    Panasonic BL-C30A Network Camera, 170
WebFerret.com metasearcher, 185
WeblogReview.com, 41
Weblogs.com, 37-39
WebMath website, 244
Webroot.com, spyware statistics, 168
websites
    Accoustica.com, 261
    ActionTec.net, 92
    Active-Eraser.com, 171
    ActMon.com, 172
    Ad-Aware, 171
    Addonics.com, 264
    AllTheWeb, 184
    AlohaBob.com, 60
    Alta Vista, 184
    Alteclansing.com, 111
    AmazonWatch.org, 331
    AMF Software, 179
    Andale.com, third party eBay tools, 106
    Angelfire, 188
    anonymous remailers, 54
    ANT 4 MailChecking, 133
    AOL Hometown, 188
    AOL Instant Messenger, 118
    AOL Search, 184
    APC.com, UPS devices, 24
    Ask Jeeves!, 184
    AssetMetrix.com, Windows XP usage, 22
    Atari.com, 196
    Atatio.com, 271
    Auction Guild, 102
    Auction Mills, eBay Trading Assistant franchise, 107
    AuctionGuild.com, 102
    Audacity.SourceForge.net, 46
    Audio Video Science (AVS) Forum, 67
    Audiograbber, 261
    AVG Anti-Virus, 276
    AVSoft, 85
    Belkin.com, 112
    Beyond Bullets, 306
    BlazeMP, 261
    blogs
        Blog Search Engine, 37
        blogosphere, 40
        Blogwise, 37
        Daypop.com, 37
        Feedster.com, 37

        Globe of Blogs.com, 37
        Jake Ludington's Media Blab, 122
        Leoville.com, 37
        Matt Goyer's Media Center, 89
        organization of, 38
        range of, 38
        reasons for, 38
        Weblogs.com, 37
    Bookslut's Blog, 332
    Bose.com, 111
    CamcorderInfo.com, 154
    CatsDomain.com, 38
    Cbuenger.com, 89
    CdtoMP3Gecko.com, 261
    Clipart Connection, 305
    Commtouch.com, spam statistics, 49
    Computer Industry Almanac, 235
        smartphone sales, 268
    connection speed testers
        Bandwidth Place, 3
        Broadband Reports, 3
        CNET, 3
        TOAST.net, 3
    Consumer Electronics Association, audio player sales statistics, 200
    Corel.com, 147
    Counterexploitation (CEXX).org, 172
    Creative.com, 12
    Cute-CD-DVD-Burner.com, 261
    Cybermoon Studios, animated humor, 121
    Daily Rotten, 95
    Death Clock, 91
    DeckKeyboards.com, 184
    Die is Cast Blog, 161
    digital music, Album Art Fixer, 13
    Digital Photography Review, 148
    Direct Connect Magazine, 81
    DoubleClick.com, online shopping research survey, 333
    Dumb Auctions, 92
    Dumb Bumpers, 92
    Dumb Criminals, 92
    Dumb Laws, 92
    Dumb Warnings, 92
    DVD Information, 253
    Easter Egg Archive, 100
    EasyDVDCDBurner.com, 261
    eBay, daily update access, 102
    eHomeUpgrade.com, 252
    Electronic Frontier Foundation (EFF), 73
    EmailAddressManager.com, 132
    Emailias.com, 133

    Entertainment Software Association
        online gaming statistics, 295
        player statistics, 226
    Epson.com, 143
    EvaSolo.com, 97
    ExploreAnywhere.com, 172
    Flash memory players
        Apple.com, 204
        Creative.com, 204
        DigitalNetworks.com, 204
        iRiverAmerica.com, 204
    freedb.org music CD database, 266
    FreePlayEnergy.com, 269
    Froggle.com, online shopping product searches, 313
    FrontBridge.com, 129
    FTC (Federal Trade Commission), identity theft resources, 71
    GAIM, 118
    Game*Blogs, 199
    Global Pet Finder, 23
    Google.com, top 10 search list, 90
    GORP, outdoor travel and recreation resources, 160
    government and legislation, THOMAS, 8
    GriffinTechnology.com, 45, 109, 131, 163
    Grundig Radio, 311
    H2oAudio.com, 113
    HDTV Blog, 128
    HDTVoice.com, 127
    Home Theater Blog, 68
    HostIndex.com, 189
    hosting services
        cost, 189
        locating, 189
    HostSearch, 189
    HotBot, 184
    HTML editors, 189
    Ibiquity.com, HD Radio resources, 312
    ICQ, 118
    IDC.com, computer manufacturer sales rankings, 162
    IGN.com, 198
    Internet radio
        LAUNCH, 15
        Live365, 15
        Radio-Locator, 15
        RadioMOI, 15
        SHOUTcast, 15
        Web-Radio, 15
    Internet TeleCafe, 122
    Internet Video Game Museum, 233

websites

iPoditude.com, 113
IPSwitch.com, 7
iSold It, eBay Trading Assistant franchise, 107
iSuppli.com, digital television sales, 63
IT Garage, 80
ITtoolbox.com, 55
JavaScript.com, web page tricks, 299
Jazz & Blues Music Reviews blog, 258
JBCamera.com, 169
JBL.com, 111
Jbox.com, 216
jobs
  Careerbuilder.com, 220
  HotJobs.com, 220
  Monster.com, 220
Jupiter Research, 10
  broadband household survey, 75
  online travel revenue, 156
Kaspersky Anti-Virus Personal, 276
Keyspan.com, 71
kid reference libraries
  Internet Public Library, 243
  LibDex, 243
  Library of Congress, 243
  Literature.org, 243
  New York Public Library Digital Library Collections, 243
  Questia.com, 243
  refdesk.com, 243
  Smithsonian Institution Libraries, 243
Linksys.com, 6
Logitech.com, 18, 104, 119
LonghornBlogs.com, 27
LookSmart, 183-184
  Macally.com, 111
Mailshell.com, 133
MailWasher, 133
  Maskbit, 261
McAfee AntiSpyware, 171
McAfee VirusScan, 276
  MCE Customizer, 89
MCE-Software, 89
MCESoft, 89
mceWeather, 89
MediaJukebox.com, 261
merchandise liquidators, 106
merchandise wholesalers, 106
MicroDrive players
  Apple.com, 203
  Creative.com, 203
  DigitalNetworks.com, 203
  iRiverAmerica.com, 203

Microsoft Knowledge Base, 26, 224
Microsoft.com, XP Service Pack 2 downloads, 22
Miranda-IM, 118
Mobile Phone Directory, 272
moblogs/photoblogs
  BusyThumbs.com, 42
  Photoblogs.org, 42
  TextAmerica.com, 42
MonsterCable.com, 124
Motorola.com, cell phone lines, 273
movie blogs
  DVDVerdict.com, 42
  filmfodder.com, 42
  GreenCineDaily.com, 42
  Tagline.com, 42
MSN Chat, 122
MSN Messenger, 118
MSN Search, 184
  Museum of Bad Album Covers, 94
music, Allmusic Database, 14
music blogs
  Fluxblog, 15
  FreshTuneage.com, 42
  PeopleTalkTooLoud.com, 42
music trading
  BitTorrent, 14
  eDonkey, 14
  Grokster, 14
  iMesh, 14
  Kazaa, 14
  LimeWire, 14
  Morpheus, 14
  WinMX, 14
My Yahoo!, 20
news blogs
  BigBlog.com, 42
  Plastic.com, 42
  UnknownNews.net, 42
Newseum.org, 94
NewsGator.com, 40
Nielsen-Netratings.com, search engine popularity survey, 182
NiveusMedia.com, 83
Nokia.com, cell phone lines, 273
Norton AntiSpam, 133
Norton AntiVirus, 276
  NPDTech Group, camcorder sales, 150
NSClean.com, 72
online music
  Apple iTunes, 13
  MSN Music, 13
  Wal-Mart Music Downloads, 13

Online Music Blog, 261
  Open Directory, 183-184
  Oregon Scientific, 130
Pacrim Technologies, 111
Panasonic.com, 170
Panda Software, 276
PC Pitstop, 79
PC-cillin, 276
  PC-Diagnostics.com, 78
PearlEcho.com, 172
Pew Internet and American Life Project, 43
  blogging statistics, 176
  instant messaging usage, 117
  terrorism survey, 69
photoblogs, 42
PhotographyBLOG, 149
Planet Photoshop, 284
podcasts
  Adam Curry, 48
  AllPodcasts.com, 47
  AudioBlog.com, 46
    DigitalPodcasts.com, 47
  Edirol.com, 44
  FeedBurner.com, 46
    iPodder.org, 47
  iPodderX.com, 47
  iTunes.com, 47-48
  LibSyn.com, 46
  OurMedia.org, 46
  PodBus.com, 46
  Podcast.net, 47
  PodcastAlley.com, 47
  PodcastBunker.com, 47
  PodcastDirectory.com, 47
  PodcastingStation.com, 47
  PodcastPickle.com, 47
  Syndic8.com, 47
political blogs
  AndrewSullivan.com, 41
  Ankle Biting Pundits, 41
  Captain's Quarters, 41
  Daily Kos, 41
  Electablog, 41
  Eschaton, 41
  eTalkingheads, 41
  INDC Journal, 41
  Little Green Footballs, 41
  PoliBlog, 41
  Powerline, 41
  Stage Left, 41
  Talking Points Memo, 41
  Wonkette.com, 41
Polk Audio, 311

...et Blog, 267
...y.org, 74
...otechDiagnostics.com, 223
PVRblog.com, 155
QuikDrop, eBay Trading Assistant franchise, 107
RazerZone, 204
Recording Industry Association of America, CD sales statistics, 262
RemoteSpy.com, 172
RescueComp Blog, 225
RoadBlock, 133
SalesHound.com, brick and mortar store sales resources, 319
SanDisk.com, 142
Sausage.com, 191
ScanGauge.com, 275
Seagate.com, 51
Search Engine Blog, 187
Search Engine Watch, 186
Serene Screen, 93
Shirt-Pocket.com, 165
Siemens-Mobile.com, cell phone lines, 273
Skeletal Systems, cartoon character studies, 90
SleepTracker.com, 334
Snopes.org, urban legend debunking, 293
Solio.com, 112
SonyEricsson.com, cell phone lines, 273
Sophos.com, Zafi-D virus reports, 274
Spam Kings Blog, 134
SpamCop.net, 133
Spamhaus.org, 133
SpamKiller, 133
SpamMotel.com, 133
sports blogs
  BadJocks.com, 42
  Fanblogs.com, 42
  SportVue.net, 103
Spy Sweeper, 171
Spybot Search & Destroy, 171
Spytech.com, 172
Spyware Guide.com, 170
Spyware Warrior Blog, 175
Stardock.com, desktop customization, 25
Stikax.com, 11
Strategy Analytics, HDTV usage statistics, 123
Sunbeamtech.com, 59

SuuntoWatches.com, 287
Talk City, 122
Taser.com, 5
TaySays.com, 159
Tech Blog, 62
tech blogs
  BoingBoing.net, 42
  Gizmodo.com, 42
  Slashdot.org, 42
  Techdirt.com, 42
TechWhack.com, 36
TenTechnology.com, 112
TheGreenButton.com, Media Center resources, 85
ThinkSecret.com, Apple insider information, 166
Thomas Hawk's Digital Connection, 193
Tivoli Audio, 118
Tom's Hardware Guide, 61
Trillian, 118
Trimble.com, 70
TViX.com, 125
TweakXP.com, 35
UglyDress.com, 93
UltraProducts.com, 56
UnimitigatedRisk.com, 89
Unofficial Apple Weblog, 167
USB Geek, 33, 98
Vantecusa.com, 237
VIOlight.com, 177
VoiSec, 50
Wacom.com, 282
Wallflower Systems, 136
Web Pages That Suck, 192
Weblog Review, 41
Webroot.com, spyware infection instatistics, 168
WhistleCreek.com, 17
Wi-Fi Alliance, 212
Win-Spy.com, 172
Winamp.com, 261
Woody's Office Portal, 180
Yahoo!.com, business interest survey, 96
Yahoo! Chat, 122
Yahoo! Messenger, 118
Yahoo! Search Blog, 21
Yuansoft, 261
WebTaxi.com metasearcher, 185
Western Digital Media Center, 137
Wheeler Jump programming technique, 40
WhistleCreek.co, survival kits, 17

Wholesale Central, merchandise wholesalers, 106
Wholesale411, merchandise wholesalers, 106
Wi-Fi Alliance, 212
WiFi (IEEE 802.11 standard), 213
  bad connections, troubleshooting, 210
  citywide, 208
  Digital Hotspotter, 209
  directories
    JiWire, 209
    Wi-Fi-FreeSpot Directory, 209
    WiFi411, 209
    WiFinder, 209
  home networks, configuring, 211
  hotspots, locating, 209
  Linksys Wireless Adapter, 210
  LucidLink Wireless Client, 211
  security, 212
  Top 10 Cities List, 208
  Wi-Fi Alliance, 212
WiFi Networking News blog, 213
WiFi411.com, WiFi directory, 209
WiFiFreespot.com, WiFi directory, 209
WiFinder.com, WiFi directory, 209
Wikipedia.com, 241
WilWheaton.net blog, 219
Win-Spy spyware (legitimate), 172
Winamp Pro, 261
Windows 3.1, introduction of, 93
Windows 95, introduction of, 223
Windows Media Center
  add-ins
    Comics for Media Center, 89
    MCE Customizer 2005, 89
    MCE Outlook, 89
    mceAuction, 89
    mceWeather, 89
    My Netflix, 89
    Playlist Editor, 89
  DVR usage, electronic program guide (EPG), 84
  music
    CD playlists, 85
    CD ripping, 85
    WMA Lossless format, 85
  TheGreenButton website, 85
  Tweak MCE utility, 84
Windows Media Player
  CDs
    burning, 265
    ripping, 262-263

Windows Registry. See Registry
Windows Vista OS, 27
Windows XP
  AssetMetrix.com survey, 22
  ClearType technology, enabling, 24
  Control Panel features, 23
  file encryption
    enabling, 34
  owner information, changing, 36
  programs, adding to Start menu, 36
  Service Pack 2
    downloading, 22
    features of, 22
    installation of, 22
  special effects, activating, 26, 35
  stuck programs, closing, 33
  Task Manager, end processes, 33
  Taskbar
    hiding, 27
    moving, 27
  Tweak UI tool, 34
  tweaks and tips
    TechWhack.com, 36
    TweakXP.com, 35
WinMX website, 14
WinZip utility, file compression/extraction, 342
wireless devices, MSI Bluetooth Star USB Star, 315
wireless microphones, 154
WMA audio compression (Windows Media Audio), 11-12
  Blaze Media Pro conversion utility, 205
  CD storage, 85
Wonkette.com political blog, 41
Woody's Office Portal website, 180
Word (Microsoft), 176
  background colors, adding, 178
  footers, adding, 180
  headers, adding, 180
  multiple-column layouts, creating, 181
  numbered lists, creating, 176
  paragraph styles
    applying, 177
    elements of, 177
    predesigned templates, 177
  section breaks
    deleting, 179
    inserting, 179
  watermarks, adding, 178
  Wordware suite, 179
Word Central website, 244
Wordware suite, 179
workspace privacy, 72
Wozniak, Steve, 163, 212
Wrinkles, softening (Photoshop), 283
writing to data CDs, 267

X – Y – Z

X Arcade Solo, 199
x-ray photography, 13
XM Commander receiver, 310
XM Radio, 307
  car receivers, 310
  cost, 308
  Delphi XM MyFi receiver, 308
  home receivers, 311
  programming, 308
  subscribers, 307
  XMFan.com, 311
XMFan.com, 311

Y2K bug, 3
Yahoo!
  airport information, 16
  area code information, 16
  Briefcase, 21
  business interest survey, 96
  company history, 16
  dictionary definitions, 16
  measurement conversions, 16
  meta word searches, 19
  My Web feature, 20
  navigational tips, 17
  package tracking, 16
  Search Blog feature, 21
  search shortcuts, 16
  toolbar downloads, 19
  traffic information, 16
Yahoo! Chat website, 122
Yahoo! Geocities website, 188
Yahoo! Maps, 18
Yahoo! Messenger, 118
  files, sending, 120
Yahoo! Shopping.com, online shopping comparisons, 315
YahooGreetings.com, 336
Youth Sports Instruction, 245
Yuansoft.com, 261

Zafi-D virus, 274
Zboard Modular Keyset, 199
Zuse, Konrad, 165

LEOVILLE PRESS

NEW FROM LEO LAPORTE AND QUE PUBLISHING

LEO LAPORTE'S 2006 GADGET GUIDE

By Leo Laporte and Michael Miller

$19.99, 240 pp
ISBN 0-7897-3395-1

LEO LAPORTE'S PC HELP DESK

By Leo Laporte and Mark Edward Soper

$29.99, 790 pp
ISBN 0-7897-3394-3

LEO LAPORTE'S GUIDE TO MAC OS X TIGER

By Leo Laporte and Todd Stauffer

$24.99, 400 pp
ISBN 0-7897-3393-5

LEO LAPORTE'S GUIDE TO TIVO

By Leo Laporte and Gareth Branwyn

$29.99, 432 pp
ISBN 0-7897-3195-9